Biotechnological Innovations in Chemical Synthesis

BOOKS IN THE BIOTOL SERIES

The Molecular Fabric of Cells
Infrastructure and Activities of Cells

Techniques used in Bioproduct Analysis
Analysis of Amino Acids, Proteins and Nucleic Acids
Analysis of Carbohydrates and Lipids

Principles of Cell Energetics
Energy Sources for Cells
Biosynthesis and the Integration of Cell Metabolism

Genome Management in Prokaryotes
Genome Management in Eukaryotes

Crop Physiology
Crop Productivity

Functional Physiology
Cellular Interactions and Immunobiology
Defence Mechanisms

Bioprocess Technology: Modelling and Transport Phenomena
Operational Modes of Bioreactors

In vitro Cultivation of Micro-organisms
In vitro Cultivation of Plant Cells
In vitro Cultivation of Animal Cells

Bioreactor Design and Product Yield
Product Recovery in Bioprocess Technology

Techniques for Engineering Genes
Strategies for Engineering Organisms

Principles of Enzymology for Technological Applications
Technological Applications of Biocatalysts
Technological Applications of Immunochemicals

Biotechnological Innovations in Health Care

Biotechnological Innovations in Crop Improvement
Biotechnological Innovations in Animal Productivity

Biotechnological Innovations in Energy and Environmental Management

Biotechnological Innovations in Chemical Synthesis

Biotechnological Innovations in Food Processing

Biotechnology Source Book: Safety, Good Practice and Regulatory Affairs

BIOTECHNOLOGY BY OPEN LEARNING

Biotechnological Innovations in Chemical Synthesis

PUBLISHED ON BEHALF OF:

Open universiteit and **University of Greenwich (formerly Thames Polytechnic)**

Valkenburgerweg 167
6401 DL Heerlen
Nederland

Avery Hill Road
Eltham, London SE9 2HB
United Kingdom

BUTTERWORTH HEINEMANN

Butterworth-Heinemann
Linacre House, Jordan Hill, Oxford OX2 8DP

A division of Reed Educational and
Professional Publishing Ltd.

A member of the Reed Elsevier plc group

OXFORD BOSTON JOHANNESBURG
MELBOURNE NEW DELHI SINGAPORE

First published 1997

British Library Cataloguing in Publication Data
A catalogue record for this book is
available from the British Library

Library of Congress Cataloguing in Publication Data
A catalogue record for this book is
available from the Library of Congress.

ISBN 0 7506 0561 8

Composition by University of Greenwich
(formerly Thames Polytechnic)
Printed and Bound in Great Britain by
Martins the Printer, Berwick-upon-Tweed.

The Biotol Project

The BIOTOL team

OPEN UNIVERSITEIT, THE NETHERLANDS
Prof M. C. E. van Dam-Mieras
Prof W. H. de Jeu
Prof J. de Vries

UNIVERSITY OF GREENWICH (FORMERLY THAMES POLYTECHNIC), UK
Prof B. R. Currell
Dr J. W. James
Dr C. K. Leach
Mr R. A. Patmore

This series of books has been developed through a collaboration between the Open universiteit of the Netherlands and University of Greenwich (formerly Thames Polytechnic) to provide a whole library of advanced level flexible learning materials including books, computer and video programmes. The series will be of particular value to those working in the chemical, pharmaceutical, health care, food and drinks, agriculture, and environmental, manufacturing and service industries. These industries will be increasingly faced with training problems as the use of biologically based techniques replaces or enhances chemical ones or indeed allows the development of products previously impossible.

The BIOTOL books may be studied privately, but specifically they provide a cost-effective major resource for in-house company training and are the basis for a wider range of courses (open, distance or traditional) from universities which, with practical and tutorial support, lead to recognised qualifications. There is a developing network of institutions throughout Europe to offer tutorial and practical support and courses based on BIOTOL both for those newly entering the field of biotechnology and for graduates looking for more advanced training. BIOTOL is for any one wishing to know about and use the principles and techniques of modern biotechnology whether they are technicians needing further education, new graduates wishing to extend their knowledge, mature staff faced with changing work or a new career, managers unfamiliar with the new technology or those returning to work after a career break.

Our learning texts, written in an informal and friendly style, embody the best characteristics of both open and distance learning to provide a flexible resource for individuals, training organisations, polytechnics and universities, and professional bodies. The content of each book has been carefully worked out between teachers and industry to lead students through a programme of work so that they may achieve clearly stated learning objectives. There are activities and exercises throughout the books, and self assessment questions that allow students to check their own progress and receive any necessary remedial help.

The books, within the series, are modular allowing students to select their own entry point depending on their knowledge and previous experience. These texts therefore remove the necessity for students to attend institution based lectures at specific times and places, bringing a new freedom to study their chosen subject at the time they need and a pace and place to suit them. This same freedom is highly beneficial to industry since staff can receive training without spending significant periods away from the workplace attending lectures and courses, and without altering work patterns.

Contributors

AUTHORS

Mr J A M van Balken DSM Research, Geleen, The Netherlands

Dr T G Cartledge, Nottingham Trent University, Nottingham, UK

Dr J Green, Napier University, Edinburgh, UK

Dr J F Hall, De Montfort University, Leicester, UK

Dr R O Jenkins, De Montfort University, Leicester, UK

Dr Ing J Kamphuis, DSM Research, Geleen, The Netherlands

Dr C K Leach, De Montfort University, Leicester, UK

Dr G Mijnbeek, Bird Engineering bv, Schiedam, The Netherlands

Dr J Verweij, Gist Brocades bv, Delft, The Netherlands

Dr E de Vroom, Gist Brocades bv, Delft, The Netherlands

EDITOR

Dr R O Jenkins, De Montfort University, Leicester, UK

ADVISORY EDITOR

Dr G Mijnbeek, Bird Engineering bv, Schiedam, The Netherlands

SCIENTIFIC AND COURSE ADVISORS

Professor M E C van Dam-Mieras, Open Universiteit, Heerlen, The Netherlands

Dr C. K. Leach, De Montfort University, Leicester, UK

ACKNOWLEDGEMENTS

Grateful thanks are extended to authors, editors, advisors and production team. Especial thanks to Ms K Brown.

The development of this BIOTOL text has been funded by COMETT, The European Community Action Programme for Education and Training for Technology. Additional support was received from the Open Universiteit of The Netherlands and by the University of Greenwich (formerly Thames Polytechnic).

Contents

How to use an open learning text

An open learning text presents to you a very carefully thought out programme of study to achieve stated learning objectives, just as a lecturer does. Rather than just listening to a lecture once, and trying to make notes at the same time, you can with a BIOTOL text study it at your own pace, go back over bits you are unsure about and study wherever you choose. Of great importance are the self assessment questions (SAQs) which challenge your understanding and progress and the responses which provide some help if you have had difficulty. These SAQs are carefully thought out to check that you are indeed achieving the set objectives and therefore area very important part of your study. Every so often in the text you will find the symbol Π, our open door to learning, which indicates an activity for you to do. You will probably find that this participation is a great help to learning so it is important not to skip it.

Whilst you can, as an open learner, study where and when you want, do try to find a place where you can work without disturbance. Most students aim to study a certain number of hours each day or each weekend. If you decide to study for several hours at once, take short breaks of five to ten minutes regularly as it helps to maintain a higher level of overall concentration.

Before you begin a detailed reading of the text, familiarise yourself with the general layout of the material. Have a look at the contents of the various chapters and flip through the pages to get a general impression of the way the subject is dealt with. Forget the old taboo of not writing in books. There is room for your comments, notes and answers; use it and make the book your own personal study record for future revision and reference.

At intervals you will find a summary and list of objectives. The summary will emphasise the important points covered by the material that you have read and the objectives will give you a check list of the things you should then be able to achieve. There are notes in the left hand margin, to help orientate you and emphasise new and important messages.

BIOTOL will be used by universities, polytechnics and colleges as well as industrial training organisations and professional bodies. The texts will form a basis for flexible courses of all types leading to certificates, diplomas and degrees often through credit accumulation and transfer arrangements. In future there will be additional resources available including videos and computer based training programmes.

Preface

The BIOTOL series of texts are designed to provide a self-study resource concerning the principles, practices and applications of biotechnology. This text is designed to explain how biotechnology may be applied to solve problems encountered in the production and use of organic chemicals in a wide range of business sectors.

Although the success of the organic industry is undoubted, conventional industrial organic chemical enterprises are faced with two main problems. Firstly, there are limitations on the specificity of reactions that can be carried out by purely chemical processes. This reduces yields of desirable products and frequently generates harmful by-products. Secondly, many reactions are carried out under conditions, and via mechanisms, that are energy demanding and incompatible with living systems. Furthermore elements are often used in concentrations and in combinations with other elements in ways not encountered in living systems. The outcome is that purely chemical processes often require complex and expensive purification procedures, generate materials that are recalcitrant to degradation and, in many instances, are toxic to living systems. Although recalcitrance may, in some cases, be desirable, for example in producing insulation materials for electrical devices, it is often undesirable leading to accumulation of non-biodegradable end products and the removal of material from natural geocycling processes. It is for these reasons that the chemical technologies are often regarded as environmentally "dirty" technologies.

The advent of contemporary biotechnology offers the potential to solve many of these problems by providing new routes for producing traditional products or by enabling the production of new material to fulfil commercial and practical objectives hitherto either unattainable, or attainable only by the production of environmentally unsatisfactory products. The key to these is the utilisation of biologically mediated catalysis either by employing purified enzymes or by using whole organisms.

In this text, the potential of the biotechnological approach to chemical synthesis is described and the strategy and outcome of applying biotechnological processes and principles to the syntheses of organic products are explained. Inevitably, authors have had to be selective, it is impossible to include all of the possibilities in a single text. A careful selection of examples have been made to cover both bulk, intermediate volume products and low volume fine products. The style is to use a more-or-less case study approach to illustrate the principles involved and authors have taken the opportunity to emphasise different aspects of their study areas. Thus, some give quite extensive examinations of the economics of processes whilst others focus mainly on the purely technical areas. The text therefore provides opportunities to learn about a whole range of issues in an array of product areas. The product specific chapters are built upon a generic section dealing with the advantages and limitations of using biotechnological approaches for satisfying chemical objectives.

This text is targeted at a senior undergraduate/postgraduate level and should be of value to all engaged, or seek to be engaged, in the organic chemical industry in a wide variety of business sectors. It is built upon the assumption that readers have quite extensive biochemical and microbiological knowledge, although authors have incorporated many helpful reminders into relevant sections of their contributions.

Our thanks to the authors whose experience as researchers and as teachers is reflected in their quality of their contributions.

Scientific and Course Advisors: Professor M C E van Dam-Mieras
Dr C K Leach

An introduction to biotechnological innovations in the chemical industry

An introduction to biotechnological innovations in the chemical industry

1.1 Introduction

Although chemistry as an emperical fundamental dicipline has a long history, its application in industry gained importance after the introduction of the use of fossil energy sources during the industrial revolution. The chemical industry withdraws materials, mainly fossil oil, from the earth's reserves to use as an energy source and as a source of raw materials for production processes. The products so produced are rather different from naturally occurring materials and mankind has become heavily dependent on them. However, other energy and raw material sources have to be sought because fossil sources are non-renewable and will eventually become depleated.

non-renewable fossil oil

Nowadays, renewable biological materials such as starch, sugar, oils and molasses are used on a relatively small scale as energy and raw materials for the chemical industry. These biological sources are mainly derived from waste products and overproduction in agriculture. With these materials less drastic conversions are applied, in comparison with fossil oil, in order to make as much as possible of the chemical structure present in the raw material. Such conversions are therefore generally performed by micro-organisms or parts of them. This means that the production process includes at least one biological conversion step.

renewable biological material

1.2 Production processes

In production processes, raw material are converted into desired products using a series of unit operations. Such unit operations may be few in number and they are linked together in a logical sequence. Typical unit operations include such activities as the transport of solids and liquids, the transfer of heat, crysallisation, collection and drying.

unit operations

In a chemical production process at least one of the unit operations (the chemical reactor) is the place in which chemical conversion takes place. However, the chemical reactor is proceeded by a series of unit operations in which the new materials are prepared (the upstream operations). After conversion has taken place, the products are subjected to a further series of unit operations (the downstream operations). These downstream operations include product recovery and purification steps. A typical example of a production process is illustrated in Figure 1.1.

upstream& downstream operations

Π Which of the unit operations described in Figure 1.1 represents the chemical reactor?

You should have identified the "conversion step".

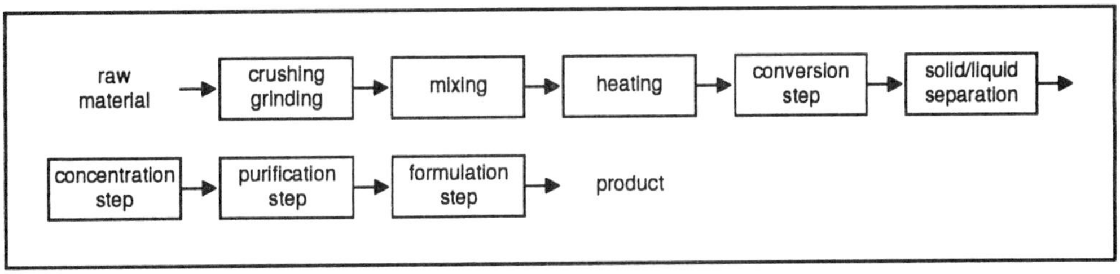

Figure 1.1 Example of a simple production process with eight unit operations.

byproducts

auxiliary material

In practice, production processes are usually rather more complex. Raw materials are usually impure and thus some pre-purification steps may be required. Obviously impurities in the raw materials will incresae the probability of impurities and byproducts occuring in the output stream from the chemical conversion step. Even using pure raw materials, most chemical conversion are incomplete and often lead to the formation of undesirable byproducts. Furthermore often additional (auxiliary) materials are used (for example catalysts, specific solvents), which have to be separated from the desired product, Thus, in typical production processes a large number of separation steps are required.

To improve the efficiency of the process, raw materials and auxiliary chemicals are recycled providing it is economically viable. Similarly ways are sought to find uses for byproducts and intermediates. This usually involves using them as feeds for further reactions. Invariably, production processes produce waste streams. These must be brought to an acceptable state before being disposed of. This is especially a concern relating to chemical production processes in which the compounds produced may be incompatable or toxic to living systems and can thus cause pollution problems. Increasing regulatory and technical burdens are being place on chemical process operators to ensure that such environmental problems do not arise from their operations.

biotechnological processes are more environmentally friendly

In biotechnological processes, the conversion of raw material to product is usually performed by micro-organisms, or parts of micro-organisms (eg enzymes) known as a fermentation or bioconversion processes respectively. On a large scale, the conversion is generally carried out in a so-called bioreactor. The conditions under which the conversion is done are generally very gentle with regards to temperature, pressure and pH, when compared to those in a chemical process. Other advantages of biotechnological production processes include high reaction specificity and selectivity (therefore fewer byproducts), and the need for relatively few reaction additives. Another important difference between a chemical and a biotechnological production process is that the latter type is closely related to naturally occurring processes: byproducts may only be carbon dioxide and water. The implementation of biotechnological production methods can, therefore, be seen as an environmentally friendly production strategy. We shall compare chemical and biotechnological catalysis in more detail in the next chapter.

mixed chemical/ biochemical processes

Within the chemical industry, micro-organisms and enzymes are often used as catalysts. It is possible for a unit operation in an essentially chemical production process to be a biochemically catalysed step: giving rise to a mixed chemical/biochemical production process. The products of these reactions include organic chemicals, solvents, polymers, pharmaceuticals, and purfumes. Mixed chemical/biochemical production processes are continuously innovated and optimised, mainly for economical reasons.

1.3 Choice of production process

economic considerations

It is possible to produce many chemical moieties (partly) by means of biotechnological production processes. For example, ethene (C_2H_4) commands a large market and is produced from fossil oil. This chemical can also be produced from ethanol, which in turn can be produced by micro-organisms using agricultural wastes. Ethene is a 'building unit' for the petrochemical industry from which several other intermediates and end products such as plastics are produced. The biotechnological production method is, however, for economical reasons still not used in practice. The costs of producing ethene from sugar via ethanol are relatively too high. This is partly due to the cost of the raw materials and the product yields on the two different substrates. The separation of ethanol from the aqueous fermentation liquid is also relatively expensive. Unfortunately micro-organisms that make large amounts of ethene directly from glucose have not been found. Nevertheless, the biotechnological production method may become the cost effective option as fossil energy sources become depleted and relatively more expensive.

specificity of reaction

Other chemicals such as gluconic acid cannot be produced by petrochemical production methods. Gluconic acid is used in the pharmaceutical industry and even as an addition to concrete. Gluconic acid can be produced from glucose, derived from potato starch, using the bacterium *Gluconobacter* or the fungus *Aspergillus*. These micro-organisms are able to modify glucose rapidly to gluconic acid, which is slowly consumed again. In this way gluconic acid is temporarily accumulated in the fermentation fluid. Technically, it would be possible to perform this process chemically starting from glucose, but in this case the biological method is preferred as the specifity of this reaction is very high and thus no undesired byproducts are formed. For the same reason biotechnological production processes are preferred if optically pure chiral compounds, such as L-configuration of a certain amino acid, has to be produced. The price of such products is, however, relatively more expensive. In later chapters of this text we consider the production of organic acids and amino acids in some detail.

biodegradable products

Another example of producing a chemical in bulk from sugar with the help of a micro-organism is the polymer polyhydroxybutyric acid. Many micro-organisms accumulate this compound as a reserve material. This polymer could be substitute for polyester or polypropene plastics. The big advantage of polyhydroxybutyric acid is that it can be degraded microbially. Products such as plastics that have been used for shorter or longer times and when they are not needed any more are brought back into the environment. However, when returned to the environment they are not readily biodegraded (they are recalcitrant) and thus accumulate. The accumulation of materials that are not readily returned to natural geocycling is of major concern. In a world where mankind has become aware that more sustainable environmental practice have to be used to prevent pollution, biotechnology will become more and more important.

integral life cycle management

A first obvious consequence of such considerations is that we should not only look at the costs of the product from an economic point of view, but that we must consider the costs of the production process in a broader sense. We must take into account the raw materials used, the amount of energy invested and the possibility to design alternatives, more environmentally friendly processes. In other words, we should not only look at the desired product, but we must consider the total life cycle of the product. The design of a production process taking into account these aspects is often referred to as integral life cycle management.

The application of the principles of integrated life cycle management, generally favours the replacement of products dependent upon conventional chemical and physical processes by biotechnological products and processes. As we described earlier, most biotechnological processes use biological (renewable) feedstocks and energy sources and the products are also compatable with biological (living) system. These products are readily biodegradable and returned to the natural geocycling and, as a consequence, do not pose the same intensity of pollution caused by the recalcitrant materials and byproducts generated by physico-chemical processes.

Biotechnology therefore offers a more environmentally friendly and sustainable approach to fulfilling the needs of society. It can achieve this by, for example, offering alternative routes to the manufacture of products hitherto made by potentially environmentally damaging routes. Alternative, it enables the production of novel products, which are less environmentally damaging than products made via conventional chemical routes. We will use two examples to illustrate these principles.

nitrogen fixation

Nitrogen fixation via the Haber-Bosch process is a well established chemical process in which dinitrogen gas (N_2) and hydrogen are combined to produce ammonia (NH_3). The major use of this produce is as a nitrogen fertiliser. Several million tonnes are produced annually. On the positive side, use of ammonia (usually as an ammonium salt) has undoubtedly increased the yields of crops. In strictly limited economic terms, the increase in crop yields achieved by the use of ammonia from the Haber-Bosch process more than outweighs the cost of producing the ammonia. With this limited perspective, the Haber-Bosch process is undoubtedly successful. If, however, we take an integrated life management approach, the issue is not so clear cut.

eutrophication

The reduction of dinitrogen is an energy expensive process. Energy is needed to split the stable N≡N bond. In the Haber-Bosch process, high temperatures (400°C) and pressures are used to achieve significant conversions. This energy input is invariably derived from non-renewable energy sources. However, the environmentally damaging effects of this activity is not limited to the production of ammonia. Much of the ammonia-based fertilisers applied to land is washed out (leached) from soils. This ends up in rivers and in impounded water, causing eutrophication (increase in organic content). The consequence of this, is these waters support greater 'blooms' of algae, which in time die and decompose. This decomposition is accompanied by the consumption of oxygen, which tends to lead to anoxia. Thus the waters lose amenity value because they no longer support fish life, are more difficult to treat to become potable; they become odourous and are no longer suitable for bathing. Thus, if one adds to the cost of the Haber-Bosch process the true environmental costs, then the virtue of this process is less than clear cut.

Biotechnology, however, offers an alternative approach to achieving the same objective as the Haber-Bosch process. It has long been known that bacteria capable of utilising atmospheric nitrogen can supply plants with nitrogen in a form that the plants can use and very little of this "fixed" nitrogen is leached from the soil. In essence, what biotechnology offers is the potential to widen the range of crops that can be supported using bioiogically generated nitrogen fertilisers. These biological nitrogen-fixers use biological energy sources (carbohydrates) to drive fixation and do not lead to the same levels of entrophication as does the application of chemically produced ammonia.

Even on the rather simplified arguments described here, it should be clear that biotechnological approaches are generally more environmentally friendly and that we can apply biotechnological strategies to inorganic, as well as organic chemicals. We could, for example, cite the use of micro-organisms in the mining of a range of metals from low grade ores, by processes generally referred to as "acid mining".

In the example above we have illustrated how biotechnology may, in integrated life management terms, offer environmentally better routes of manufacture. It may also lead to environmentally more acceptable products. As an example consider the production and use of pesticides. The majority of pesticides are made by synthetic organic chemisry leading to molecules that are distinctly non-biological. Often they contain functional groups (for example halogens) and are made by reaction mechanisms (for example using free radicles) that are, in general, incompatable with biodegradation. These products, although used in low concentration, tend to accumulate in the environment especially within biological systems. Of particular importance is the accumulation of these materials in relatively high concentrations in organisms at the end of food chains. Thus an insecticide may be present only in low levels in particular insects but when these are eaten by birds, the biological part of the insect is metabolised, while the recalcitrant insecticide remains. Thus the concentration of the insecticide becomes greater in the birds than in its food. This process, called biomagnification, may result in the concentrations of the insectide in the birds reaching toxic levels. Furthermore, chemically produced pesticides generally have wide ranging activities, killing both beneficial as well as pest species.

pesticides

biomagnification

biodegradable pesticides

The advent of contemporary biotechnology has enabled development of new strategies to achieve the same objectives: protecting crops using biologically-produced, biodegradable pesticides. A typical example is the production and use of proteinaceous insect toxins encoded by Baculo virus. These types of pesticides are readily biogradable and are target specific.

We will not enlarge on the environment potential of biotechnology any further at this stage. We will, however, raise some environmental issues in later section of this text. If you would like to learn more, we recommend the BIOTOL text "Biotechnological Innovations in Environmental and Energy Management".

competition in the market place

legislation

The development of new products based on cleaner production processes and alternative raw resources is not only a question of technological development. The products have to compete in the market place and have to be acceptable to potential customers. Also the introduction of new processes and products depends upon gaining both the confidence and the financial resources of potential investors. Regulations may also greatly influence (both positively and negatively) the adoption of these new processes. Restrictive regulations may deter investors and may, by raising the spectra of potential hazards, alienate the general public and reduce the accetability of the products. We site for example the EC Directives and National Legislation concerning the safe handling of genetically manipulated organisms. To many workers this is seen as inhibitory to the development and exploitation of genetically modified organisms. Such legislation is seen as a constraint on the development of new processes and products. To others, this legilation is a positive bonus to biotechnology because it reduces the prospect of there being a major bio-catastrophy from these activities and it reassures the public that the work is undertaken in a safe manner and leads to "safe" products, thereby making them more acceptable to the public. This in turn encourages investment and development. However the reverse effect may also be true. To some, if biotechnology is "safe" it would not need to be regulated in this way. In effect, for some individuals the introduction of legislation indicates that biotechnology is inherently "unsafe", as a result this legislation may in some circumstances, make biotechnology and its products less publically acceptable.

role of legislations, politicians, investors, pressure groups on biotechnological development

Thus we can see that legislators, politicians, investors and society as a whole are important influences on the development and adoption of biotechnology. Biotechnological development is not only dependant upon technological/scientific advance but also economic, political and sociological developments. We could, for example, envisage that specialist lobby groups (eg "Green" groups, animal rights activists) may, through influencing public opinion, greatly influence biotechnology. It is also incuberent on education to ensure that public opinion (and thus investment and legislation) is developed upon knowledge and not upon emotive and ill-founded claims.

In the following chapters, we predominantly use a case study approach to illustrate a range of issues that arise from using biologically-based approaches to the production of chemicals. There are such an enormous range to choose from that we have had to be selective. Our selection has been made predominantly to ensure that the reader develops an understanding of the range and potential of biotechnology in this area, and developes an appreciation of the major advantages and limitations of this approach.

In Chapter 2, we provide an overview of the types of chemical transformations that can be mediated by organisms or their constituent parts, particularly enzymes. This chapter provides a context for later chapters. In Chapter 3, we examine the cellular energetic consequences of metabolite overproduction by organisms. We use this chapter to consider the limitations on yield of products. Chapter 4 considers the production of single cell protein, particularly using fossil fuel (methane/methanol) as substrate. This chapter enables us to explain how changing market values and social acceptability greatly influences the success or otherwise of biotechnological processes. In the case of SCP from methane, we use economic data from the 1960s (when the process appeared to be profitable) and the 1970s (when it became uneconomic) to illustrate this point. We feel further justification for including this case study because it was this project that led to the commercial development of large scale air-lift bioreactors which find ever increasing use in a wide range of biotechnological processes.

We have included chapters on large volume organic acid production to illustrate how intermediary metabolism may be manipulated to achieve overproduction of metabolic intermediates. Chapters on antibiotics, amino acids, polysaccharides and lipids are used to illustrate the application of biological systems to achieve specific transformations. Each has been chosen to enlarge on particular aspects. Thus within the amino acids chapter you will for example compare the technology and the economics of fermentative and enzymological strategies to produce stereospecific forms of particular amino acids. In the chapter on antibiotics, we illustrate how biological systems can be manipulated or used to diversify the range and characteristics of particular groups of molecules. Similarily, in the chapter on lipids, we use sterol/steroid interconversions in the health care sector to enlarge on the concepts of biological specificity. Later in the same chapter we turn attention to the bulk lipid market associated with food manufacture and show how biological systems can be used to convert a low prices lipid into a higher value lipid with desirable organoleptic ("mouth feel") properties. The chapter is also used to introduce the reader to the problems posed by attempting to use biological systems (which are largely aqueously based) to carry out conversion with substances which have only limited compatability with water.

By the end of the text you should appreciate the enormous potential that biological systems have for making a wide range of products and to achieve a variety of objectives. You should also have knowledge and be able to cite specific examples, of how economic, social and political attitudes may impinge upon the adoption of the technology.

Biocatalysts in organic chemical synthesis

Biocatalysts in organic chemical synthesis

2.1 Introduction

traditional and new biotechnology

Biotechnology is often divided into two categories called 'traditional biotechnology' and 'new biotechnology'. The major products of the traditional biotechnology industry are industrial alcohol, food and flavour ingredients, antibiotics and citric acid. The new biotechnology involves the newer techniques of genetic engineering and cell fusion to produce organisms capable of making useful products. Products of the new biotechnology are extremely diverse and include steroid derivatives, antibiotics and special proteins for therapeutic use (eg human growth hormone, interferons and interleukins).

The market for products of traditional biotechnology is currently worth around 250 times more than those of the new biotechnology, although it is predicted that the new biotechnology will account for an increasingly larger fraction of the total biotechnology industry.

The purpose of this chapter is to compare and contrast various production strategies of the biotechnology industry and to consider some of the major decisions that have to be made during bioprocess development. Many of the areas touched upon will be developed in greater detail in other chapters of this book. The book is limited to the use of micro-organisms and enzymes as bioprocess catalysts and does not consider catalysis by plant and animal cells. As you will see later in this chapter, industrial microbiology is the major foundation of biotechnology and there are many reasons why micro-organisms dominate as production organisms in both traditional and new biotechnological processes.

2.2 Micro-organisms as catalysts of organic synthesis

Microbial cells are very attractive as a source of catalysts for the production of organic chemicals because of their broad range of enzymes capable of a wide variety of chemical reactions, some of which are illustrated in Table 2.1.

Esterase enzymes - cleavage of various ester bonds to yield an acid and an alcohol, ie

$$R_1COOR_2 + H_2O \longleftrightarrow R_1COOH + R_2OH$$

Lipase enzymes (a subgroup of esterase enzymes) - hydrolyse fats into glycerol and fatty acids, eg

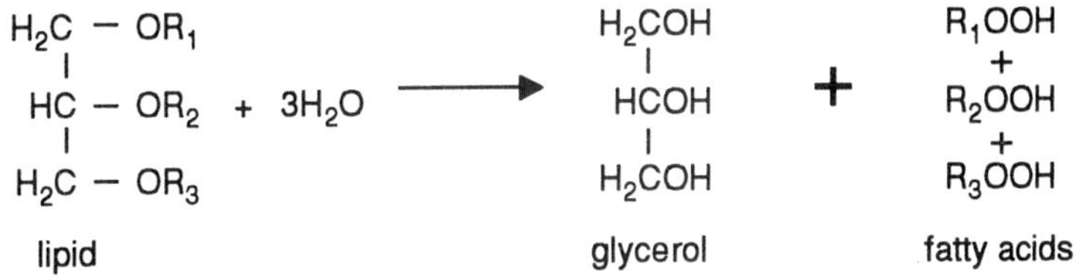

Proteolytic enzymes - hydrolyse proteins selectively, either on terminal groups (exopeptidases) or internal linkages (endopeptidases), eg

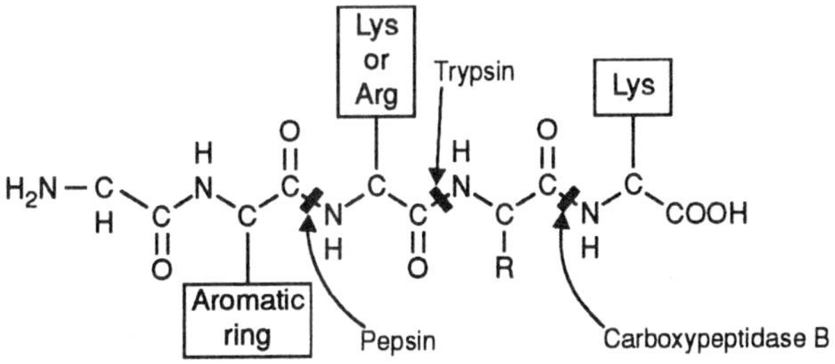

Oxidoreductases - catalyse oxidation-reduction reactions, eg

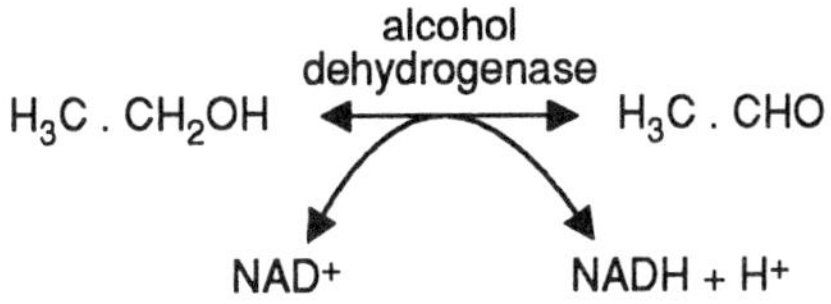

Oxygenases - add one (monooxygenases) or both (dioxygenases) atoms of molecular oxygen to molecules, eg

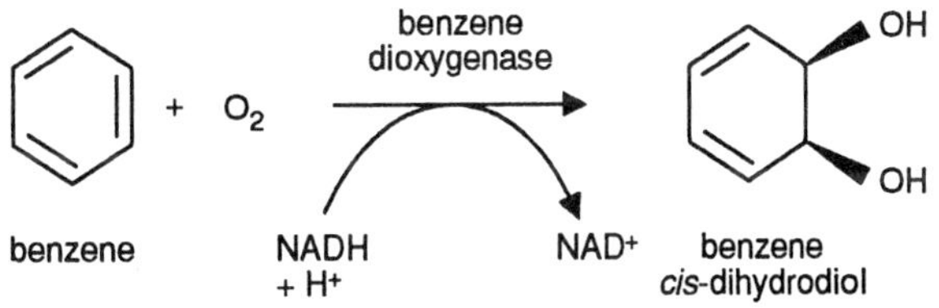

Table 2.1 Some reactions catalysed by microbial enzymes. In principle each enzyme catalyses the reverse process as well.

Lyases - catalyse the breakage of c-c bonds, eg

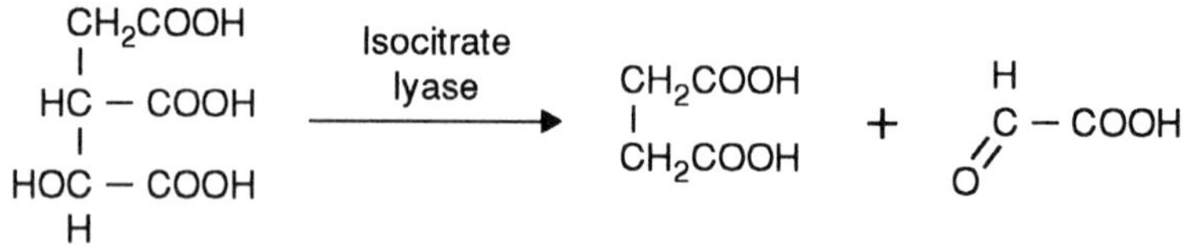

Isocitric acid | Succinic acid | Glyoxylic acid

Transferases - catalyse the transfer of functional groups, eg

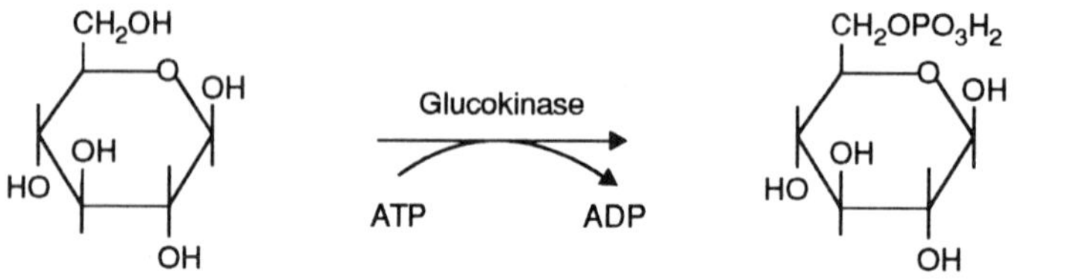

D-Glucose | D-Glucose-6-phosphoric acid

Isomerases - catalyse the inter-conversion of isomers, eg

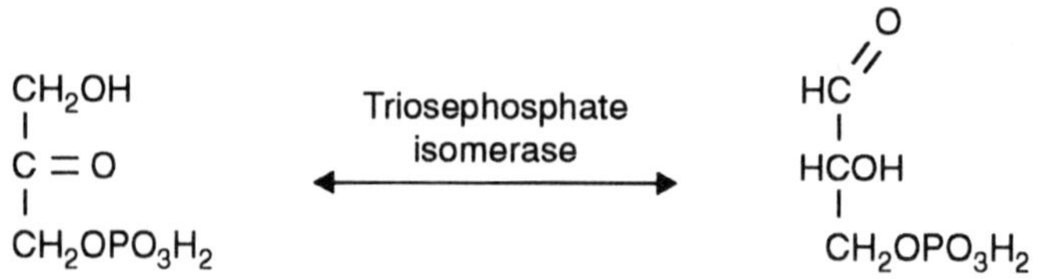

Dihydroxyacetone phosphate | Glyceraldehyde-3-phosphate

Table 2.1 ...... Continued

Microbial cells, rather than plant and animal cells, are generally preferred for the production of organic chemicals. There are several reasons for this.

Π Make a list of at least five commonly found features of micro-organisms that would benefit their use as catalysts for organic synthesis.

Your list could have included the following commonly found features:

advantages of using microbial cells

- high growth rates which allows the generation of large amounts of catalyst (micro-organism);
- substrates for growth are often cheap and include waste materials from other industrial processes;
- they are generally more robust and less fastideous than plant and animal cell cultures;

- there are many different types of microbes, each with unique nutritional and physiological features, which may be desirable for process development;
- collectively, micro-organisms have a broad complement of enzymes capable of a wide variety of chemical reactions;
- production plants involving micro-organisms are generally independent of climatic conditions and require little space (compared to crop plant production);
- for some microbes, such as *Escherichia coli*, the genome is well known and relatively easy to manipulate genetically;
- many microbes are single celled organisms that grow well in stirred tank bioreactors (fermentors).

Although it is possible to obtain cells from whole animals or plants and to cultivate them in suitable nutrient solutions, in general they are not as easy to handle as microbes. Nevertheless, plant and animal cells are a valuable genetic resource for biotechnology and many newly developed bioprocesses rely on transfer of their genes to micro-organisms.

biosynthetic processes and biotransform-ation

Microbial enzymes can be applied as catalysts for chemical synthesis in biosynthetic processes or in biotransformations (bioconversions). In a biosynthetic process the product is formed *de novo* by the microbial cell from substrates, such as monosaccharides, molasses, soybean and corn steep liquor. In a biotransformation, however, a precursor that is usually chemically synthesised is converted in one or several enzyme catalysed steps into the desired chemical. This chemical may be the end product or may serve as a precursor for further chemical modification.

2.3 Enzyme preparations versus whole cell processes

In designing a process we have the choice of using the whole organism or specific enzymes isolated from it. As always both options have pro's and cons. Broadly speaking we could say that biosynthetic processes mostly rely on whole cells, whereas biotransformations can be catalysed by whole cells and by enzyme preparations.

hydrolytic enzymes

Hydrolytic enzymes such as proteases, esterases and lipases (Table 2.1) account for more than half of all reported biotransformations. These enzymes are particularly easy to use because:

- they are available in large amounts from industrial sources;
- they are stable in non-aqueous solvent;
- they do not have cofactor requirements.

Π Why do you think many processes based on redox reactions involving dehydrogenase enzymes are still carried out using whole cells?

dehydrogenases

Dehydrogenase enzymes generally require NADH or NADPH, and although methods for recycling these cofactors are now available on a laboratory scale, little progress has been made in the scale-up to industrial level.

oxygenases

Similarly, many oxygenation reactions (Table 2.1), which also require cofactors, are usually performed using whole micro-organisms. Collectively, oxidoreductases and oxygenases account for around 30% of all reported biotransformations.

Enzymes such as lyases, transferases and isomerases (Table 2.1) account for most of the remainder of industrially applied biotransformations.

2.3.1 Enzyme catalysed processes

Enzymes isolated from micro-organisms have many desirable properties as catalysts for the synthesis of industrial chemicals, but there are associated problems:

associated problems

- their protein structure may not be stable under non physiological conditions which may be detrimental to their long term use, especially at elevated temperatures;
- the provision of enzyme cofactors can be expensive;
- most enzyme reactions are carried out in water and the enzymes must be separated from the product stream;
- the product stream is often very dilute, presenting problems of product concentration and recovery;
- preparation of a crude or purified cell-free enzyme preparation is necessary.

enzyme immobilisation

Advances in genetic and chemical enzyme modifications, enzyme immobilisation and enzymatic reactions in organic solvents, have increased the actual use and potential of enzymes in the production of industrial chemicals. Enzyme immobilisation, in particular, has proved to be a valuable approach to the use of enzymes in chemical synthesis. The term denotes enzymes that are physically confined or localised in a defined region in space with retention of their catalytic activities. A detailed consideration of immobilisation techniques is beyond the scope of this chapter; the subject is covered adequately in the BIOTOL text entitled 'Technological Applications of Biocatalysts'.

enzyme bioreactor design

Enzyme immobilisation allows the construction of enzyme reactors in which the enzyme can be reused. Furthermore, the process operates continuously and can be readily controlled. Enzyme reactors currently in use include those illustrated in Figure 2.1.

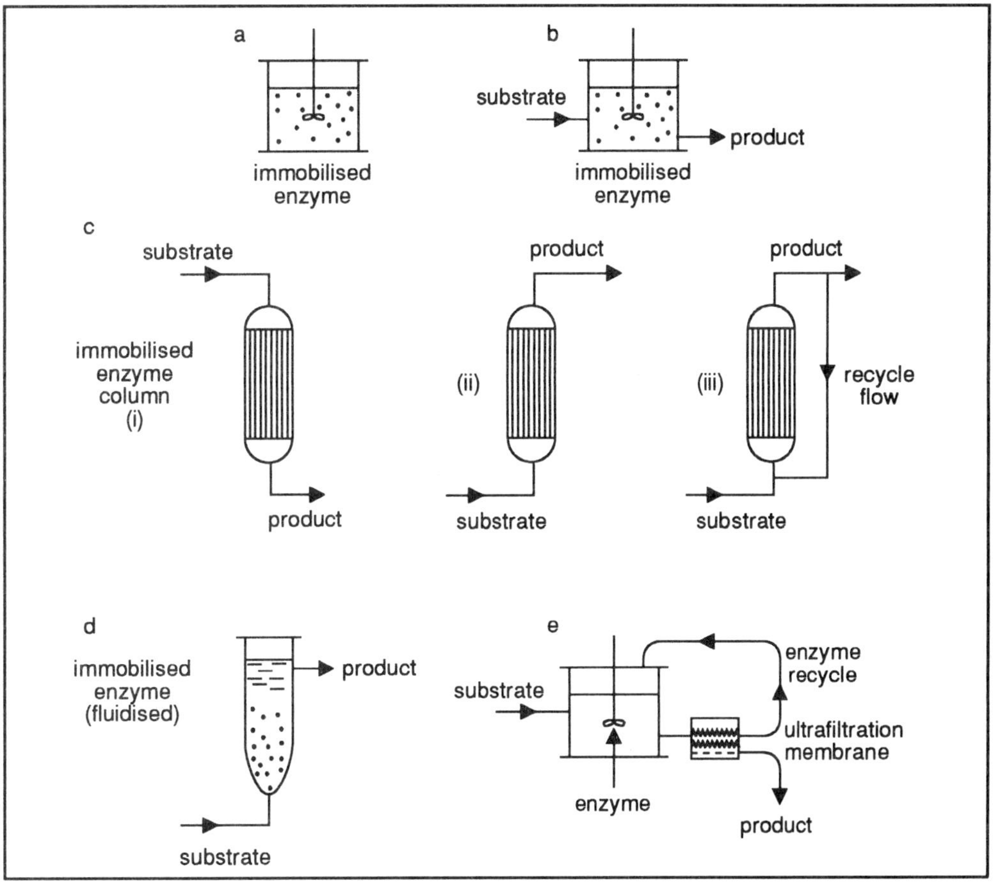

Figure 2.1 Examples of enzyme bioreactor design.
Bioreactors: a) batch stirred tank; b) continuous stirred tank; c) continuous packed-bed i) downward flow, ii) upward flow and iii) recycle; d) continuous fluidised-bed; e) continuous ultrafiltration. Redrawn from Katchalski - Katzir E. (1993) *Trends in Biotechnology II*, 471-477.

advantages of enzyme immobilisation

Some of the potential advantages of enzyme immobilisation are:

- high enzyme loads;
- prolonged enzyme activity;
- enzyme can often be regenerated by suitable treatment and used again;
- the ability to recycle products;
- high flow rates;
- reduction in cost (energy and labour), energy and waste products;
- easy to scale up to large systems;
- high yields of pure materials;
- higher substrate concentrations can be used.

2.3.2 Whole cell processes

We can also use microbial cells (fermentation) containing the desired catalytic activity without isolating the enzymes responsible.

Advantages of whole cell processes include:

advantages

- cell disruption not necessary;
- enzyme isolation not necessary;
- more suited to multiple step processes;
- cofactor regeneration not a problem;
- increased enzyme stability;
- reduced catalyst preparation costs.

Among the disadvantages of whole cell processes are:

disadvantages

- system not fully understood (black box situation);
- product contamination by cellular enzymes or other end products of metabolism;
- reduced catalytic specific activity;
- cell structures acting as diffusion barriers;
- contamination by other micro-organisms may be a problem.

Synthesis of industrial chemicals by microbial cells may be by fermentation (free, living cells), immobilised growing cells, immobilised resting cells or immobilised dead cells.

immobilised cells

Immobilised cells have all the advantages of immobilised enzymes. Cell immobilisation is preferred for reactions catalysed by intracellular enzymes because it avoids tedious and expensive extraction and purification procedures, which often result in preparations of low yield and stability.

SAQ 2.1

Identify which of the following statements are true for immobilised biocatalysts, when compared to free enzyme or free cell systems.

1) Conversions carried out by immobilised cells give higher yields than those carried out by growing and dividing cells.

2) Downstream processing can be much easier.

3) Smaller reactor volumes achieve similar rates of product formation.

4) The volume of effluent can be reduced.

SAQ 2.2

Benzene dioxygenase is a complex enzyme consisting of three protein components, that catalyse the conversion of benzene to benzene *cis*-dihydrodiol. Give two reasons why this biotransformation should be carried out using whole cells as opposed to using enzyme preparations.

2.4 Scale of production

The decision as to which approach - free enzyme, immobilised enzyme, fermentation, immobilised cells - is mainly dictated by economics. Commercial aspects of bioprocess development are considered in Section 2.6. The analysis of the various factors involved is a critical part of the decision-making process and involves inputs from scientists, engineers and marketing personnel. For products derived from micro-organisms, process design can be based mainly on biochemical engineering considerations or on microbial physiology considerations. When compared to the improvements achieved in bulk chemicals manufacture and in petroleum refining, the application of biochemical engineering principles to microbial processes has not been as successful over the past thirty years. It is now accepted that individual factors affecting overall optimisation of microbial processes are best handled by individual specialists - microbial biotechnologists and chemical engineers. The biotechnologist, in addition to having an in-depth knowledge in a particular field, must also have the appropriate skills and knowledge to communicate and interact effectively with chemical engineers. Such interaction is thought to be essential for technological innovation and commercial success of microbial processes of the future.

need for communication

The necessity for interaction between biotechnologists and chemical engineers increases with the scale of production. In chemical manufacture three categories of product can be defined according to the scale of production (Table 2.2).

Category	production range single plant	examples of biotechnology products
fine chemicals	100 kg/annum-100 tonnes/annum usually batch reactors	vitamins, vaccines, nucletides, amino acids) antibiotics) some enzymes)
intermediate volume chemicals	100-20,000 tonnes/annum batch or continuous reactors	glutamic acid) citric acid) food industry lactic acid) antibiotics for agriculture, enzymes for industry, many fermented foods and beverages
bulk chemicals	> 20,000 tonnes/annum usually continuous flow	single cell protein, ethanol for industry, biopolymers (for enhanced oil recovery), biogas (methane), sewage and wastewater treatment plants

Table 2.2 Product categories in chemical manufacture.

fine, intermediate volume and bulk chemicals

The design of production plants for the manufacture of the three categories of product varies considerably. Fine chemicals are usually produced in batch reactors, which may also be used for the production of a variety of similar products. Fine chemicals usually have demanding product quality specifications and, consequently, a significant fraction of the production costs are involved in product purification and testing. Intermediate volume chemicals have less rigorous quality specifications than fine chemicals and are usually manufactured in product-specific-plants, either as batch or continuous flow processes. Bulk chemical production plants usually operate continuous flow processes and the products do not have rigid quality specifications; rather they are marketed on the basis of overall product performance criteria.

2.5 Modes of operation of bioprocesses

batch mode

As you might have already gathered, the majority of industrial fermentations are batch processes. In closed batch systems, the growth medium is inoculated with cells and growth and product formation is allowed to proceed until the required amount of conversion has taken place. After harvesting the culture the vessel is cleaned, sterilised and filled with fresh medium prior to inoculation. For some processes, addition of all the feedstock prior to inoculation, as is done in closed batch fermentations, is undesirable and it is preferable to incrementally add the carbon source as the fermentation proceeds. Such a process is known as fed-batch culture and the approach is often used to extend the lifetime of batch cultures and thus product yields; fed-batch cultures are considered further in Section 2.7.4.

continuous mode

The alternative to batch mode operation is continuous operation. In the continuous mode there is a continuous flow of medium into the fermentor and of product stream out of the fermentor. Continuous bioprocesses often use homogenously mixed whole cell suspensions. However, immobilised cell or enzyme processes generally operate in continuous plug flow reactors, without mixing (see Figure 2.1, packed-bed reactors).

Π Complete the following statements by filling in the missing words.

Conditions change with _________ in batch fermentations but with distance along the reactor in ___________ reactors. ___________ conditions are maintained in continuous suspended fermentations.

The missing words are 'time', 'plug flow' and 'constant'.

advantages of batch mode

Advantages of the batch mode over the continuous mode of operation include:

- several different products can be made using the same bioreactor, thus providing operational flexibility;
- genetic stability of the process organisms is not as great a problem;
- the risk of contamination is relatively small;
- it is possible to identify all the material involved in making a particular batch of product, which may be important in quality control.

advantages of continuous mode

Advantages of the continuous mode over the batch mode of operation include:

- the proportion of the down-time (time spent preparing the bioreactor for a new run) to fermentation time is relatively small;
- they are relatively easy to operate and control when in steady state;
- the demand on services are relatively constant during operation;
- the product streams have a relatively constant composition (it may be difficult to prevent batch-to-batch variability);
- waste products of metabolism are unlikely to accumulate to inhibitory levels.

2.5.1 Productivity in batch and continuous systems

Productivity is an important parameter in evaluating the cost-effectiveness of a fermentation. It is defined as:

productivity = product concentration/fermentation time

total and maximal productivity

Typical units for productivity are kg m^{-3} h^{-1}. Factors that influence productivity include the production time of the fermentation, the time required to clean and set up the reactor, the sterilisation time and the length of the lag phase of growth. Figure 2.2 shows how total productivity and maximal productivity can be calculated for a batch fermentation. The decision as to when the fermentation is terminated (maximum or total productivity) depends on the operating costs, which include the capacity of the fermentation vessel, energy costs and labour costs.

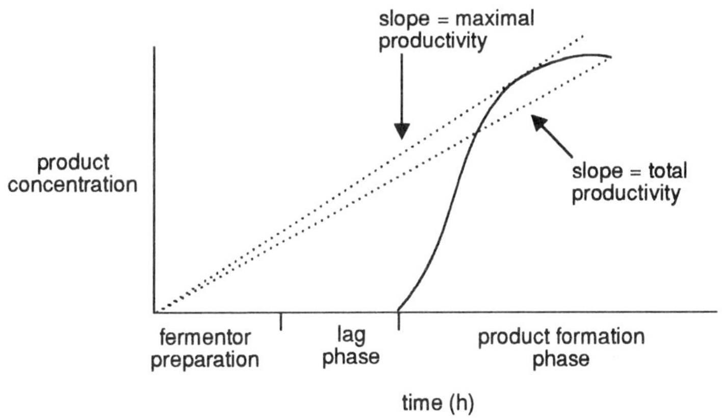

Figure 2.2 Types of productivities for batch fermentation.

Π Make a list of factors, other than those mentioned already, that could directly affect productivity.

Your list could be extensive since any factor that influences the rate and/or amount of product formed, ie the fermentation conditions or the characteristics of the process organism could influence productivity, eg pH, temperature, solubility of substrate.

In continuous fermentation:

$$\text{productivity} = D \cdot \bar{x}$$

where

D = dilution rate (flow rate/culture volume; units of h^{-1})

$\bar{x}$ = steady state product concentration (units kg m^{-3}).

In continuous fermentation, maximum productivity equals total productivity since the preparation time and the time in lag phase of growth are small relative to the total fermentation time.

Π Explain what is meant by a 'steady state' for a continuous fermentation.

A steady state is when the conditions in the bioreactor (biomass concentration, residual substrate concentration etc) remain constant over time.

Π In a continuous fermentation the preparation time is 20 hours and the fermentation is run at two dilution rates: 0.1 h^{-1} for 48 hours and 0.2 h^{-1} for 96 hours. What is the total productivity of the fermentation if the steady state product concentrations at the two dilution rates are 0.5 kg m^{-3} and 2.1 kg m^{-3} respectively? (Try to do this before looking at our solution below).

Time period (h)	Time (h)	D (h^{-1})	$\bar{x}$ (kg m^{-3})	$D.\bar{x}$ (kg m^{-3} h^{-1})
0-20	20	-	-	-
20-68	48	0.1	0.5	0.05
68-164	96	0.2	1.1	0.22

$$\text{Total productivity} = \frac{(48 \times 0.05) + (96 \times 0.22)}{164} = 0.14 \text{ kg m}^{-3} \text{ h}^{-1}$$

SAQ 2.3

Which of the following statements apply to 'batch mode' and which to 'continuous mode' of operation of fermentation? In each case give a reason for your decision.

1) Provides operational flexibility.

2) Genetic instability of the process organism is more likely to be a problem.

3) Loss of product as a result of microbial contamination is very much greater.

4) The 'down time' of the bioreactor is a significant portion of the total fermentation time.

5) It is possible that substances inhibitory to the fermentation may accumulate throughout the fermentation.

6) More suitable for fermentations in which the product is synthesised after growth.

7) The source of variation occurring in the product may be difficult to locate.

8) Relatively easy to operate and control.

2.6 Biotechnological processes verses chemical synthetic processes

Biotechnology has attracted enormous interest and high expectations over the past decade. However, the implementation of new technologies into industrial processes has been slower than initially predicted. Although biocatalytic methods hold great industrial potential, there are relatively few commercial applications of biocatalysts in organic chemical synthesis. The main factors that limit the application of biocatalysts are:

- the inherent disadvantages of biocatalysts;
- the existence of well-developed traditional technologies (economic considerations).

The disadvantages of biocatalysts will now be considered, followed by their advantages. You should note that the pros and cons of biotechnology versus chemical synthesis are very general and that exceptions may exist.

2.6.1 Disadvantages of biocatalysts

economic perspective

From an economic perspective, it is important for bioprocesses to be integrated into existing organic chemical infrastructure. Typical production sites for bulk chemicals manufacture, such as those of the petrochemical industry, are unlikely to provide all of the requirements of industrial microbiological processes. The requirements of industrial microbiological processes include:

- high quality process water;
- low temperature cooling water;
- clean air;
- clean steam;
- hygienic surroundings;
- waste water treatment facilities;
- storage facilities for feedstock and product;
- sophisticated quality control laboratories.

Satisfying these requirements invariably requires capital investment that can significantly affect overall economics of microbiological processes.

As far as fine chemicals are concerned, many biotransformations and some fermentations can be carried out in existing fine chemical reactors. In the absence of suitable existing facilities, significant investment in a state-of-the-art fermentation/biotransformation plant is required. However, existing chemical technology is well known and, in many cases, the investment in a factory plant has been paid for, so there is little economic incentive for implementing new processes. It is, therefore, unlikely that existing chemical technology will automatically be replaced by biotechnology.

mixed chemical/biochemical processes

At present most bioprocesses in the organic chemical industry are actually mixed chemical/biochemical processes. In such processes, chemically synthesised educts (chemical precursors) are biotransformed and then re-enter chemical synthesis. The main reason for this approach is that, in general, higher volumetric productivities can be achieved with chemical catalysts.

Apart from economic considerations, the inherent disadvantages of biocatalysts have also limited the transformation of new technologies into industrial processes. Table 2.3 lists the major drawbacks of bioprocesses.

Educt, substrate or product inhibition of reaction.
Poor understanding of reaction kinetics.
Educt or substrate poorly soluble in water.
High temperatures damage enzymes or cells (cooling required).
Contamination by unwanted micro-organisms.
Genetic instability of process micro-organism.
Need for purified substrates and water to avoid poisoning of biocatalysts.
Low concentration product stream.
Poor reliability and reproducibility of process.
High cost of state-of-the-art fermentation/biotransformation plant.

Table 2.3 Possible drawbacks of bioprocesses and mixed chemical/biochemical processes compared to purely chemical synthetic processes.

metabolic control

Several of the problems associated with whole cell bioprocesses are related to the highly effective metabolic control of microbial cells. Because cells are so well regulated, substrate or product inhibition often limits the concentration of desired product that can be achieved. This problem is often difficult to solve because of a poor understanding of the kinetic characteristics of the metabolic pathway leading to the desired product.

mild conditions

When compared to purely chemical synthesis, bioprocesses are operated under relatively mild conditions and in aqueous solvents: they are essentially low temperature processes with operating temperatures usually below 40°C. The pH of most bioprocesses is between 6 and 8 and the pressure is usually one atmosphere. Under these conditions, substrates (eg oxygen) can be poorly soluble in water, which may limit productivity. Since reactions can generate considerable amounts of heat, waste heat generated during bioprocesses often has to be adequately dissipated to ensure high temperatures do not damage enzymes or cells.

organic solvents

Many substrates currently produced in the chemical industry are immiscible with water, but are readily miscible with organic solvents. Most enzymes, however, will not operate efficiently, or not operate at all, in non-aqueous media. Some exceptions do exist, such as lipases and esterases, which can operate in non-aqueous environments. Currently, there is considerable interest in extending the range of enzymes that do work in organic solvents.

Π What possible advantages might enzymes operating in organic solvents have for a bioprocess?

Organic solvents may improve substrate solubility, specificity of reaction and equilibrium reaction. The last two benefits are considered further in Section 2.6.2.

contamination

genetic instability

Since most bioprocesses are monoseptic (one type of organism) operations where the production strain has been genetically modified (crippled) to overproduce the desired product, contamination by unwanted micro-organisms is a constant threat. The contamination may arise externally and aseptic procedures must be used to reduce the risk - such sterile engineering can be complex and costly. Genetic instability of the producing organism can give rise to contamination from within, where some change in the producing organism (reversion of a mutation; loss of a plasmid) affords a growth rate advantage that enables the unwanted organism to outcompete the producing strain.

chemical contamination

Cells and isolated enzymes are often susceptible to poisoning at low levels of chemical contamination. It is, therefore, necessary to carry out expensive purification of substrates (feedstocks) and water used in bioprocesses.

product purification

In whole cell bioprocesses, extracellular products are preferable because this removes the requirement for cell disruption and this reduces the level of impurities in the product solution. Nevertheless, product isolation and purification can be prohibitively expensive particularly for low concentration product streams, which is a feature of many bioprocesses.

reproducibility

When compared to traditional chemical synthesis, processes based on biocatalysts are generally less reliable. This is due, in part, to the fact that biological systems are inherently complex. In bioprocesses involving whole cells, it is essential to use the same strain from the same culture collection to minimise problems of reproducibility. If cell free enzymes are used the reliability can depend on the purity of the enzyme preparation, for example iso-enzyme composition or the presence of other proteins. It is, therefore, important to consider the commercial source of the enzyme and the precise specifications of the biocatalyst employed.

Π Give a possible drawback of bioprocesses, when compared to purely chemical synthesis, under each of the following headings (refer to Table 2.3 if you are unable to do this).

Reaction kinetics:
Economics:
Product stream:
Reliability:
Temperature:
Genetics:
Inhibition:
Solubility:

2.6.2 Advantages of biocatalysts

Despite the inherent disadvantages of bioprocesses, there are many advantages which can make a biotechnological approach to chemical synthesis the sole or desirable approach. The possible advantages of bioprocesses are shown in Table 2.4.

Stereospecificity
Regiospecificity
Reaction specificity
Reduced disposal costs
Waste products likely to be less environmentally damaging
Reduced number of synthesis steps
Reduced temperature and pressure costs
Mild reaction conditions and aqueous solvents
Mild reaction conditions give slower destruction of reaction vessel
Mild reaction conditions allow reactions with labile molecules
Safer reaction technology and working environment
Renewable feedstocks
Non-toxic, biodegradable and non flammable educts and substrates

Table 2.4 Potential advantages of mixed chemical/biochemical processes compared to purely chemical synthetic processes.

stereospecificity

One of the most important advantages of biocatalysts is their stereospecificity. An example of a stereospecific biotransformation is given in Figure 2.3. In 1992 the Food and Drug Administration (FDA) in the United States addressed the issue of whether it mattered, for drug approval purposes, that a preparation of a synthetic chiral molecule (one or more asymmetric centers) contains not one but two compounds (racemates) - the two mirror-image stereoisomers (individual enantiomers).

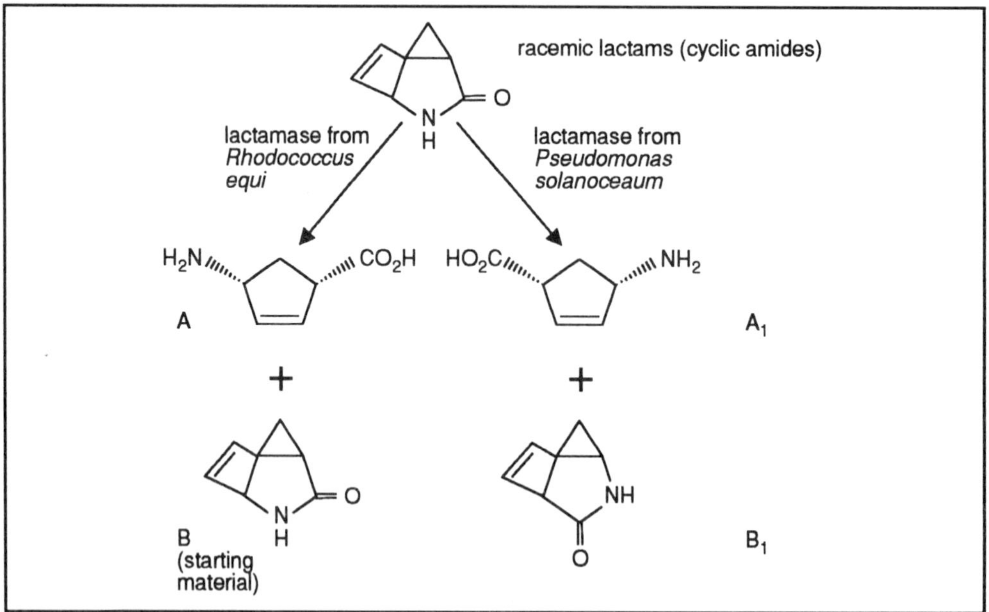

Figure 2.3 An example of a stereospecific biotransformation: resolution of bicyclic lactams. A and A_1 are an enantiomer pair as are B and B_1.

FDA guidelines

The FDA gave drug companies the choice of developing chiral drugs as racemates or as single enantiomers. In effect, this means that although the development of racemates is not prohibited, such drugs have to undergo rigorous justification before approval by the FDA. This justification includes putting each of the two forms of the molecule separately into animals and, possibly, humans. Consequently, the majority of drug companies have decided to develop single enantiomers, if feasible, in preference to racemates; some have decided to avoid the problem by switching to nonchiral molecules. The FDA's policy on chiral compounds has created, in effect, many opportunities for chemists in 'chirotechnology' and biotechnology is proving to be a powerful tool. The annual sales of the top ten best selling single enantiomeric drugs has been estimated to be around US $10 billion. These molecules include several penicillin and cephalosporin based antibiotics and market research has indicated that chiral molecules are likely to take an increasing share of the pharmaceutical drug market in the 1990's.

regiospecificity

The specificity of biocatalysts also extends to site specificity (regiospecificity). This means that if several functional groups of one type are present on the molecule, only one specific position will be affected. An example of this is the microbial oxidation of D-sorbitol to L-sorbose, a key step in the synthesis of vitamin C (Figure 2.4).

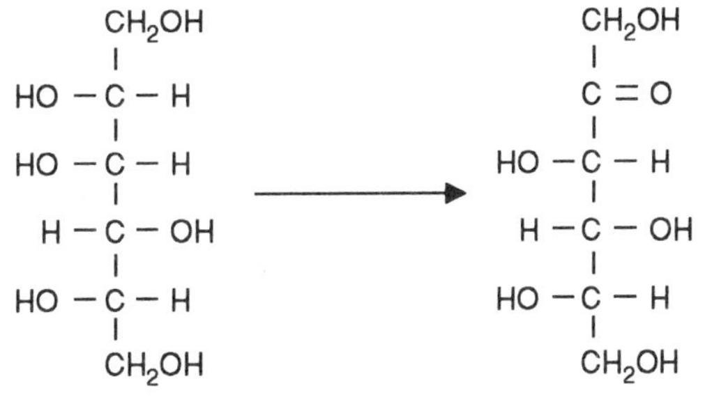

Figure 2.4 Microbial oxidation of D-sorbitol to L-sorbose.

medium engineering

The specificity of enzyme reactions can be altered by varying the solvent system. For example, the addition of water-miscible organic co-solvents may improve the selectivity of hydrolase enzymes. Medium engineering is also important for synthetic reactions performed in pure organic solvents. In such cases, the selectivity of the reaction may depend on the organic solvent used. In non-aqueous solvents, hydrolytic enzymes catalyse the reverse reaction, ie the synthesis of esters and amides. The problem here is the low activity (catalytic power) of many hydrolases in organic solvents, and the unpredictable effects of the amount of water and type of solvent on the rate and selectivity.

pollution control

The high specifity of biocatalysts also has the advantage of reducing disposal costs (pollution control costs) because relatively few useless and potentially harmful byproducts are generated. In addition, waste products that might be produced are, by their very nature, likely to be biodegradable and, therefore, less environmentally damaging compared to those produced in purely chemical synthetic processes.

multi-step synthesis

In processes involving whole cells the required product can often be formed in a single step, although the cells essentially carry out a multi-step synthesis. This means that only a single product purification is necessary. Conversely, in chemical synthesis of compounds, each step in the synthesis is usually carried out separately. Thus the product of one reaction must often be purified before it can be used in the next step in the synthetic sequence. This multi-step approach is expensive, time consuming and can require a complex process plant to handle the individual steps on an industrial scale.

non-corrosive conditions

The almost non-corrosive conditions (low temperature, neutral pH, atmospheric pressure) of most bioprocesses have the advantage that there is much slower destruction of the reaction vessels, which means financial depreciation over a much longer period of time. Indeed, it has been estimated that biotechnological reactors have a life span three times that of chemical reactors. The relatively mild conditions of bioprocesses are also favourable compared with chemical synthesis, where the high temperatures and pressures can incur considerable costs.

renewable feedstocks

In bioprocesses, the feedstocks required to grow the catalysts and produce the chemical are generally renewable resources, such as sugar from crops. Conversely, purely chemical synthesis relies largely on non-renewable resources such as oil, coal and natural gas. It follows that as non-renewable resources dwindle, it is likely that biotechnology will become increasingly important to the chemical industry.

Π Make a list of possible advantages of bioprocesses, when compared to purely chemical synthetic processes, under each of the following headings (refer to Table 2.4 if you are unable to do this).

Specificity:
Reaction conditions:
Educt/substrate/products:
Others:

SAQ 2.4

Briefly compare and contrast downstream processing for bioprocesses and for purely chemical synthetic processes, from an economic perspective.

2.7 Bioprocess development

commercial consideration

Micro-organisms have produced chemicals for industry for many years and their potential for the future in this respect is enormous. However, the development of new and improved bioprocesses depends on the realities of the commercial world:

- the product must generate a profit;
- the bioprocess must have a significant advantage over existing or potential chemical processes;
- the industry must be willing to risk substantial capital expenditure.

Other factors that are important for the development of bioprocesses include:

- the identification of new large-market products and/or high-profit margin products;
- the ability to protect patents or the confidentiality of new technical advances;
- the need to minimise the use of expensive capital equipment and labour-intensive processes;
- the need to minimise the requirements that new products have to meet to attain regulatory approval (because of the high costs, time-delay and risks involved).

research and development

Even when a successful process has been established, continuous improvement is required to meet the challenges of competitors. Ongoing commercial success will rely, therefore, on optimisation of the many diverse stages in the bioprocess through research and development. For a typical bioprocess, the stages may include:

- growth and improvement of cells for catalytic use;
- immobilisation of biocatalyst;
- preparation of chemical feedstocks to be modified or purification of growth substrates;
- design of bioreactor to allow contact with biocatalyst and feedstock;
- optimisation of conditions in bioreactor;
- separation of product from byproducts and waste products;
- recycling or disposal of byproducts and waste products;
- product isolation from dilute aqueous solution;
- product purification and formulation;
- scale-up of a laboratory scale process to a new and large scale, ie design, construction and operation.

Next, some of the decisions involved in bioprocess development will be considered.

2.7.1 Choice of process micro-organism

flexibility

Increased flexibility into the choice of organism will rely on knowledge of the genetics of species other than those presently used in genetic engineering. Bioreactors nowadays can be designed and constructed in such a way that simple modifications permit the growth of prokaryotes, filamentous fungi, plant cells and mammalian cells in the same reactor. Obviously genetic stability of the organism is extremely important but physical, chemical and nutritional factors also have to be considered.

shear resistance

In aerobic processes, mechanical agitation by rotating impellers is often used to disperse air in the form of fine bubbles throughout the growth medium and to promote adequate mixing. However, the shearing action of the impellers also tends to damage the cells employed. Such damage is particularly pronounced in the case of filamentous fungi, actinomycetes and bacteria with appendages (flagella and fimbriae). Fungal morphology can be important with respect to the formation of certain products, for example citric acid. For these products, culture conditions must be adapted or strains selected to minimise mycelium injury or alterations in morphology. As far as damage to bacterial cells is concerned, the main hydrodynamic effects on appendages, particularly flagella that can have lengths many times that of the cells, can lead to stripping of such appendages with subsequent release of cell contents. Selection of a more robust fungus or bacterium may enhance the commercial success of the process.

resistance to environmental stresses

Micro-organisms with resistance to environmental stresses such as solvents, extremes of pH, high salt concentration, and having broad temperature and dissolved oxygen optima are more suited to process applications. Improved process instrumentation and control reduces the disadvantages of fastidious microbes in this respect. However, some commercial-scale bioreactors, for example column and pressure-cycle systems, have rapidly changing physical and chemical conditions and robust micro-organisms are essential if good overall process economics are to be achieved. Process organisms also need to be resistant to inhibition by substrates and products and to inactivation by proteases. In addition, the bioprocess must use micro-organisms that are acceptable to the regulatory authorities, that can be grown easily on media, and that possess constitutive enzyme activities.

Π Make a list of desirable characteristics for a process micro-organism. If you are unable to do this re-read Section 2.7.1.

2.7.2 Strategies for improvement of process micro-organisms

genetic modification

Natural isolates usually produce commercially important products in very low concentrations and the potential productivity of the organism is controlled by the genome. In practice, the process of strain improvement involves the continual genetic modification of the culture, followed by reappraisals of its cultural requirements. Genetic modification may be achieved by:

- selecting natural variants (eg enrichment culture of soil organisms);
- selecting induced mutants (eg produced by use of UV light or chemical mutagens);
- selecting recombinants (eg produced by protoplast fusion or by *in vitro* genetic engineering).

An array of strategies and techniques is available for each approach. A detailed consideration of these is beyond the scope of this chapter, although you will encounter several strategies elsewhere in this BIOTOL text.

There is a small probability of a genetic change occurring each time a cell divides. Therefore, selection of natural variants may result in increased yields but it is not possible to rely on such improvement, and techniques must be employed to increase the chances of improving the culture.

genetic recombination

In its broadest sense recombination can be defined as any process which helps to generate new combinations of genes. The use of recombination mechanisms for improvement of industrial micro-organisms has been limited by the lack of basic knowledge of the genetics of industrial micro-organisms. The ease of applying mutation and selection techniques and the spectacular success of these approaches have also limited the use of recombination techniques for industrial micro-organisms.

2.7.3 Composition of the medium

ideal characteristics

The growth medium for a bioprocess should ideally provide all the requirements of the process organism in such a way that the organism grows rapidly, produces large amount of the desired product, does not degrade the final product, and is able to survive under harsh environmental conditions (such as shear forces, oxygen limitation etc).

Whenever possible the individual components of a medium should not be provided in excess because if they are not consumed or built into the product, they must be removed by costly purification processes.

Π Make a list of characteristics of an ideal medium for biotransformations using whole cells. (Try this on a piece of paper before reading on).

Your list could have included the following characteristics:

- allows fast growth and high yields of process micro-organism;
- does not allow growth of contaminants;
- no excess of any component (all substrates utilised by end of biotransformation);
- no colour, odour or byproduct at the end of the biotransformation;
- substrates cheap, stable, soluble and of constant quality;
- product stable;
- protects cells from adverse environmental conditions;
- prevents foaming.

2.7.4 Technical options

continuous product removal

In many cases, problems cannot be overcome by biological means. This is especially true for those related to inhibition by substrate or product. There may, however, be technical solutions to these problems. Nowadays, complicated feed strategies with different substrates can be achieved through the use of flow injection analysis, on-line sensors, mass flow meters and sophisticated computer control. Such control coupled to a fed-batch mode of operation (Figure 2.5) can often eleviate problems caused by substrate inhibition. For some processes, continuous product removal can avoid the problems associated with product inhibition; the various options include:

- vacuum fermentation (volatile products);
- solvent extraction;
- dialysis;
- ion exchange;
- cell recycling.

Some of the most important technological approaches to bioprocess problems are shown in Figure 2.5.

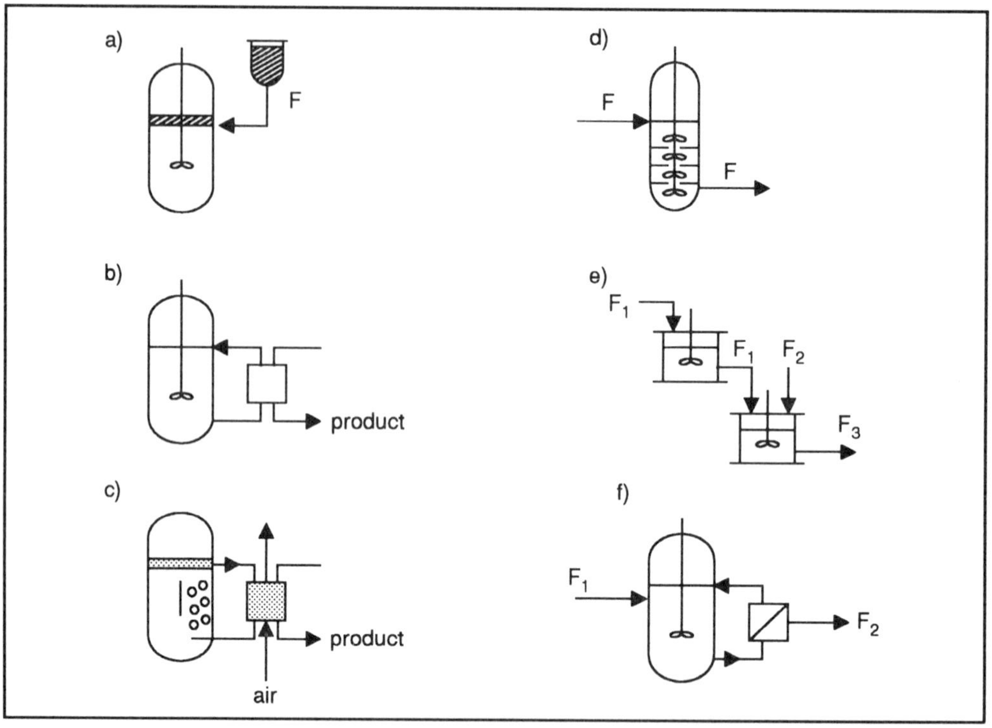

Figure 2.5 Possible technological solutions to bioprocess problems: a) Fed-batch culture; b) Continuous product removal (eg dialysis, vacuum fermentation, solvent extraction, ion exchange etc); c) Two-phase system combined with extractive fermentation (liquid-impelled loop reactor); d) Continuous culture, internal multi-stage reactor; e) Continuous culture, dual-stream multi-stage reactor; f) Continuous culture with biomass feedback (cell recycling). (See text for further details).

feb-batch culture

We have already seen that the term 'fed-batch culture' is used for batch cultures which are fed, continuously or intermittently, with fresh medium without removal of culture fluid. Benefits of such systems include:

- extending the lifetime of the product formation phase beyond that possible with batch culture;
- maintaining very low concentrations of the growth limiting nutrient.

Π In what circumstances would the maintenance of very low concentrations of the growth limiting nutrient be an important process design criterion?

When excess substrate interferes with growth and/or product formation. One example is the production of baker's yeast. It is known that relatively low concentrations of certain sugars repress respiration and this will make the yeast cells switch to fermentative metabolism, even under aerobic conditions. This, of course, has a negative effect on biomass yield. When maximum biomass production is aimed at, fed batch cultures are the best choice, since the concentration of limiting sugar remains low enough to avoid repression of respiration.

two-phase systems

In a two-phase system (Figure 2.5c), the organic (water immiscible) solvent may be used as product extractant. In addition, recirculation of the organic phase can serve to transfer oxygen and to mix the aqueous phase.

The advantages of the continuous mode of operation of fermentations were considered in Section 2.5. One of the main drawbacks is that in aerated well mixed continuous processes, high substrate concentrations can remain. This problem may be overcome using a continuous cascade (multi-stage) reactor, in which substrate concentration near the outflow of the bioreactor is reduced to the same low levels characteristic of fed-batch processes. In multi-stage continuous processes, conditions for growth and/or product formation may vary considerably between stages, which may have considerable benefits for the process. For example, in secondary product processes, where product formation takes place after growth has ceased, the first stage of the multi-stage process may be optimised for growth with subsequent stages optimised for product formation, in the absence of growth.

multi-stage continuous processes

single-stream multi-stage

With complex media (eg with more than one source of carbon) a single-stream multi-stage process may be necessary to achieve utilisation of all substrate. For example, such an approach is necessary if utilisation of one carbon source represses the expression of genes for utilisation of another. In the single-stream multi-stage process, utilisation of one substrate in the first stage allows utilisation of the other in the second stage. In general, the system provides a series of different environments.

multi-stream multi-stage

The multi-stream multi-stage system is a valuable means for obtaining steady-state growth when, in a simple chemostat, the steady-state is unstable eg when the growth-limiting substrate is also a growth inhibitor. This system can also be used to achieve stable conditions with maximum growth rate, an achievement that is impossible in a simple chemostat (substrate-limited continuous culture).

biomass feedback

Biomass feedback refers to increasing the concentration of biomass in the culture vessel. This is achieved by fitting some device, either internally or externally, to the continuous culture which retains or returns biomass to the vessel. The main advantage of biomass feedback is that the maximum output rate of biomass (and products) in the vessel with a given medium can be increased. This is particularly useful when the growth-limiting substrate is unavoidably dilute, for example if substrate has low solubility or has to be limited because of the formation of an inhibitory product.

Combinations of different technological approaches are, of course, possible. Several methods have not been mentioned in this chapter, but you will encounter these in a specific context elsewhere in the text.

SAQ 2.5

Select an appropriate combination of process design factors for each of the processes.

Process problems	Design factors
Process A:	Mode of operation:
Genetic instability	Batch
Substrate repression	Fed-batch
Multi-step synthesis	Continuous
Product (volatile) inhibition	
Process B:	Biocatalyst:
Genetic instability	Free enzyme
Poor enzyme stability	Free cells
Cofactor requirement	Immobilised enzyme
Product (non-polar) inhibition	Immobilised cells
Process C:	Reactor technology:
Product purification complicated	Vacuum fermentation
by presence of low molecular components	Biomass feedback (cell recycle)
Single-step biotransformation;	Two phase system
Enzyme stable and readily available;	Ultrafiltration with enzyme recycling
No cofactor requirement;	Single-stream multi-stage dual-stream
Low molecular weight product	Multi-stage
	Solvent extraction

Summary and objectives

Biotechnology has the potential to become the most important tool in organic chemical industry in the early part of the next century. However, this will rely on successful integration of biology, chemistry and engineering. Micro-organisms have tremendous potential as biocatalysts because of their flexibility, but they can be unpredictable. All bioprocesses require active biocatalysis, which may require improvement by mutagenic procedures or by recombinant DNA technology. Downstream processing is invariably a decisive cost factor and its minimisation is an important prerequisite of a successful bioprocess in organic chemical synthesis. Rapid progress currently being made in reactor design and process control strategies, together with advances in enzyme and recombinant DNA technologies, will ensure that biotransformations of increasing complexity and that novel products will continue to be realised.

Now that you have completed this chapter you should be able to:

- list advantages of using micro-organisms as catalysts of organic synthesis;
- compare and contrast the use of enzyme preparations and whole cells as catalysts of organic synthesis;
- categorise products of the chemical industry according to scale of production;
- compare and contrast batch and continuous fermentations;
- list advantages and disadvantages of bioprocesses and mixed chemical/biochemical processes compared to purely chemical synthetic processes;
- list factors that are of major importance for the development of bioprocesses;
- describe broadly the major decisions that have to be made in bioprocess development.

Efficiency of growth and product formation

Efficiency of growth and product formation

3.0 Introduction

In order to develop a rational approach to improving rates of metabolite production, it is necessary to consider the fate of the nutrients that are required for its synthesis. However, overcoming the major flux control points within a metabolic pathway may not lead to metabolite overproduction if the energetic consequences of the alteration are unfavourable to the organism.

energetics perspective

In this chapter, we will consider growth and product formation from an energetics perspective. In the first part of the chapter, the stoichiometry of growth is considered in some detail. The relationships between product formation and growth are then described, together with approaches to determining key parameters of growth efficiency. Finally, classes of metabolites are defined, according to the relationships between energy and metabolite synthesis. Examples of commercially significant products in each class are also discussed.

Since process technologists rely on quantitative relationships, many seemingly complex equations describing growth and product formation stoichiometry are presented in this chapter. These are intended to illustrate a quantitative approach to the study of efficiency of growth and product formation. You are not expected to recall all details of the equations, rather the factors that need to be considered in such a quantitative analysis. If this approach is entirely new to you, we recommend Chapters 8-10 of the BIOTOL text 'Bioprocess Technology: Modelling and Transport Phenomena', which deals with the modelling of growth and product formation.

3.1 Growth stoichiometry

3.1.1 Yield coefficients

yield coefficient

In any quantitative assessment of growth and/or product formation, it is essential to link formation of microbial biomass and products with the utilisation of substrate and nutrients. In the case of microbial biomass production, the total amount of cell mass formed is often proportional to the mass of substrate utilised. Mathematically this is expressed as the corresponding ratio, or yield coefficient:

$$Y_{x/s} = \frac{\Delta X}{\Delta S}$$

where

ΔX = amount of biomass produced
ΔS = amount of substrate consumed

Yield coefficients may be defined for different substrates in the medium and are usually based upon substrate change in units of mass or mol of substrate eg, Y_{X/O_2} is used to relate the amount of biomass formed to the amount of oxygen consumed, so:

Organism	**Substrate**	$\mathbf{Y_{x/s}}$	
		$g\,g^{-1}$	$g\,mol^{-1}$
	glucose	0.38	68.4
Pseudomonas fluorescens		Y_{X/O_2}	
	oxygen	0.85	27.2

stoichiometric relationships

Where yield coefficients are constant for a particular cell cultivation system, knowledge of how one variable changes can be used to determine changes in the other. Such stoichiometric relationships can be useful in monitoring fermentations. For example, some product concentrations, such as CO_2 leaving an aerobic bioreactor, are often the most convenient to measure in practice and give information on substrate consumption rates, biomass formation rates and product formation rates.

In practice, variations in yield factors are often observed for a given organism in a given medium. For example, yield coefficients often vary with growth rate. An explanation for these variations comes from a consideration of the fate of substrate in the cell, which can be divided into three parts:

- assimilation into cell mass;
- energy for growth;
- energy for maintenance.

energy for maintenance

Energy for maintenance is the energy required for survival, or non-growth related purposes. It includes activities such as active transport across membranes and turnover (replacement synthesis) of macromolecules.

Where a single substrate serves both as carbon and energy source, which is the case for chemoheterotrophic organisms used for biomass production, we can write:

$$\Delta S = \Delta S_{\text{assimilation}} + \Delta S_{\text{growth energy}} + \Delta S_{\text{maintenance energy}}$$

where

ΔS = the total amount of substrate consumed
$\Delta S_{\text{assimilation}}$ = amount of substrate assimilated
$\Delta S_{\text{growth energy}}$ = amount of substrate consumed to provide energy for grwoth
$\Delta S_{\text{maintenance energy}}$ = amount of substrate consumed to provide energy for maintenance

Or, expressed as yield coefficients:

$$Y_{x/s} = \Delta X/\Delta S_{\text{assimilation}} + \Delta X/\Delta S_{\text{growth energy}} + \Delta X/\Delta S_{\text{maintenance energy}}$$

where ΔX is amount of biomass produced
$Y_{x/s}$ is the yield coefficient

growth rate dependence

Whereas the amount of substrate assimilated per unit of biomass formed ($\Delta X/\Delta S_{assimilation}$) is constant regardless of the growth rate, the overall yield coefficient ($Y_{x/s}$) is variable and dependent upon the environmental conditions within the culture. To illustrate this growth rate dependence, consider two extremes: a culture growing at its maximum specific growth rate will use most of its substrate for assimilation and growth energy, whereas a stationary phase culture (non growing) will consume substrate for maintenance without any growth. From the biotechnological process point of view, yield coefficient variability is extremely important and yield coefficients must, of course, be optimised.

Later in this chapter (section 3.2.1) we shall consider yield coefficients with respect to product formation.

3.1.2 Elemental material balances for growth

The stoichiometry of growth and metabolism can also be described by elemental material balances. This approach can provide an insight into the potential of the organism for biomass or product production, and thus the scope for process improvement.

empirical formula for dry biomass

An elemental material balance approach to growth stoichiometry requires an empirical formula for dry weight material:

$C_\theta H_\alpha O_\beta N_\delta$

The ratios of subscripts in the formula can be determined if the elemental composition of an organism growing under particular conditions is known. A unique cell formula can then be established by relating elemental composition to one gram-atom of carbon, ie $\theta = 1$, then α, β, and δ are set so that the formula is consistent with known relative elemental weight content of the cells. The formula can be extended to include other macro-elements, such as phosphate and sulphur, if elemental analysis shows these elements to be a significant proportion of cell material.

Π Complete the following statements:

1) One C-mol of cells is the quantity of ________ containing one gram-atom of ________.

2) One C-mol of cells corresponds to the cell _____ weight with the carbon subscript (θ) taken as ________.

The missing words are: 1) 'cells' and 'carbon'; 2) 'dry' and 'unity'.

Π Use the data on elemental composition shown below to determine the empirical chemical formula and the formula weight for the yeast.

	composition (% by weight)						Empirical chemical formula	Formula weight (C-mole of cells)
	C	H	N	O	P	S		
Bacterium	47.1	7.8	13.7	31.3			$CH_{1.666}N_{0.20}O_{0.27}$	20.8
Yeast	44.7	6.2	8.5	31.2	1.08	0.6	?	?

Atomic weights: C, 12.011; H, 1.008; N, 14.008; O, 16.000; P, 30.98; S, 32.06.

The empirical chemical formula is $CH_{1.65}N_{0.16}O_{0.52}P_{0.01}S_{0.005}$ and the formula weight is 24.6. For example, the subscript for O in the chemical formula is determined as follows: (31.2/16)/(44.7/12.011).

The formula weight is then calculated by multiplying the coefficients by the atomic weights and summing them.

Thus (1 x 12.011) + (1.008 x 1.65) + (14.008 x 0.16) + (16 x 0.52) + (30.98 x 0.01) + (32.06 x 0.005) = 24.6.

Now lets consider the elemental approach to stoichiometry for a relatively simple situation: aerobic growth where the only products formed are cells, carbon dioxide and water. The following formulas can be used if we consider the four main elements:

Cell material $CH_{\alpha}O_{\beta}N_{\delta}$

Carbon source CH_xO_y

Nitrogen source $H_lO_mN_n$

stoichiometric coefficients

We can now write the reaction equation by introducing stoichiometric coefficients for all elements of the equation.

$$a'\,CH_x\,O_y + b'\,O_2 + c'\,H_l\,O_m\,N_n \longrightarrow CH_{\alpha}\,O_{\beta}\,N_{\delta} + d'\,H_2O + e'\,CO_2 \qquad \text{E - 3.1}$$

You should note that there are only five unkown stoichiometric coefficients (a', b', c', d', e') since the coefficient of cells is taken as unity. The reaction equation can be used to establish relationships between the unknown coefficients by considering balances on the four elements, as follows:

C: $a' = 1 + e'$

H: $a'x + c'l = \alpha + 2d'$

Π Write material balances for the remaining two elements in the reaction equation (E - 3.1).

The material balances are:

O: $a'y + 2b' + c'm = \beta + d' + 2e'$

N: $c'n = \delta$

SAQ 3.1

A bacterium is grown aerobically with glucose as sole source of carbon and ammonium ions as nitrogen source. Experimental analysis shows that six moles of glucose are utilised for each mole of biomass produced. Write the reaction equation for growth if the elemental composition of the cells is $CH_{1.666}O_{0.27}N_{0.20}$.

(Hint: commence with the 'abstract' equation E - 3.1).

For a more detailed stoichiometric representation for aerobic growth of a chemoheterotrophic organism we must consider:

- the generation and utilisation of ATP;
- the oxidation-reduction balance of substrates and products.

This allows an assessment of the relative efficiencies of the biochemical pathways involved in microbial growth and metabolism.

It may be assumed that neither ATP nor NADH accumulates, ie formation must be balanced by utilisation. Let us first consider the formation and utilisation of ATP; we may write:

energy balance

ATP formation	=	ATP utilisation
(substrate level phosphorylation)		(biosynthesis)
(oxidative phosphorylation)		(maintenance and dissipation)

For substrate level phosphorylation we can write:

$$\varepsilon_s a\,(ADP + P_i) \rightarrow \varepsilon_s a\,(ATP + H_2O) \qquad \text{E - 3.2}$$

Where: ε_s = number of substrate-level phosphorylations per mole of carbon utilised.

For oxidative phosphorylation we can write:

$$2b\,(P/O)\,(ADP + P_i) \rightarrow 2b\,(P/O)\,(ATP + H_2O) \qquad \text{E - 3.3}$$

Where: P/O = is the number of ADP phosphorylations per atom of oxygen consumed.

For biosynthesis (ATP utilisation) we can write:

$$\frac{MW_B}{Y^{max}_{ATP}}(ATP + H_2O) \rightarrow \frac{MW_B}{Y^{max}_{ATP}}(ADP + P_i) \qquad \text{E - 3.4}$$

Where:

MW_B = molecular weight of biomass;

Y^{max}_{ATP} = mass of cells formed per mol of ATP utilised in biosynthesis.

Y_{ATP}^{max}

The Y_{ATP}^{max} in the equation can be determined from growth yields and known routes of ATP synthesis. For growth of *Escherichia coli* on glucose and mineral salts the Y_{ATP}^{max} value, estimated from known cell composition and known biosynthetic pathways, is 28.8 g dry weight mol^{-1} ATP. However, the Y_{ATP}^{max} determined experimentally from yield measurements is often around 50% of the theoretical (12 to 14 g dry weight mol^{-1} ATP).

Π Explain the discrepancy between theoretical and experimentally derived values for Y_{ATP}^{max}.

The discrepancy arises because ATP is used to drive processes which are not directly related to growth, eg membrane transport processes, protein turnover. These are called the 'maintenance and dissipation' demands for ATP.

For maintenance and dissipation we can write, simply:

$$c(ATP + H_2O) \rightarrow c(ADP + P_i) \qquad \text{E - 3.5}$$

balance for NADH

Now lets consider the balance for NADH, ie:

NADH formation	=	NADH utilisation
(energy source dissimilation)		(biosynthesis)
		(oxidative phosphorylation)

To represent the balance for NADH using quantitative relationships, we must consider the degrees of reductance of substrate and products.

degree of reductance

The degree of reductance of material is the number of available electrons per atom of carbon and is determined using C(+4), H(+1), 0(-2) and N(-3). So, for biomass with an empirical formula of $CH_{1.666}N_{0.20}O_{0.27}$, the degree of reductance (γ) is:

$$(4) + (1 \times 1.666) - (3 \times 0.2) - (2 \times 0.27) = 4.526$$

Π What are the degrees of reductance of (1) CO_2, (2) NH_3 and (3) $C_{55.2}H_{86.2}O_{45.1}$?

The degrees of reductance are (1) 0, (2) 0 and (3) 3.93. The answer for (3) was determined as follows:

$$4 + (86.2/55.2 \times 1) - (45.1/55.2 \times 2)$$

reduction balance

For energy source dissimilation we can then write:

$$a\,CH_xO_y + H_2O + \frac{\gamma_s}{2}\,a\,NAD^+ \longrightarrow a\,CO_2 + \frac{\gamma_s}{2}\,a\,(NADH + H^+) \qquad \text{E - 3.6}$$

Where: γ_s = the degree of reductance of carbon substrate.

For biosynthesis we can write:

$$(1 + \sigma)\, CH_x O_y + \frac{\delta}{n} H_l O_m N_n + \frac{1}{2} (\gamma_B - \gamma_S (1 + \sigma) - \frac{\delta}{n} \gamma_N) (NADH + H^+)$$

$$\longrightarrow CH_\alpha O_\beta N_\delta + \sigma\, CO_2 + H_2O \qquad \text{E - 3.7}$$

Where:

γ_B = the degree of reductance of biomass;

γ_N = the degree of reductance of nitrogen source;

γ = the degree of reductance of compound.

For oxidative phosphorylation we can write:

$$2b\,(NADH + H^+) + bO_2 \longrightarrow 2b\, NAD^+ + 2bH_2O \qquad \text{E - 3.8}$$

The coefficients *a*, *b* and *c*, which appear throughout these balance equations describe the extent to which these reactions occur relative to the growth reaction (ie $1 + \sigma$) and are written taken into account elemental balances for each reaction.

Π How should the balance reactions already described for aerobic metabolism be adapted for anaerobic fermentations?

1) By deleting O_2 everywhere.

2) By dropping the oxidative phosphorylation reaction.

SAQ 3.2

Use the equations already given to predict how you might expect $Y^{max}_{X/S}$ to be influenced by:

1) a decrease in the degree of reductance of substrate;

2) an increase in the efficiency of oxidative phosphorylation;

3) a decrease in energy demand for biomass synthesis?

Give reasons for your responses.

3.2 Relationships between product formation and growth

process kinetics

When considering product formation stoichiometries it is essential to define the relationship between product formation and growth. Essentially, this classification divides microbial product production processes into four types:

1) The main product appears as a result of primary energy metabolism. Examples: production of biomass, ethanol and gluconic acid.

2) The main product arises indirectly from energy metabolism. Examples: citric acid and some amino acids.

3) The main product is independently elaborated by the organism and does not arise directly from energy metabolism (the product is a secondary metabolite). Example: antibiotics such as penicillin and streptomycin.

4) Biotransformation, in which the main product is formed from substrate through one or more reactions catalysed by enzymes in the cells. Examples: steroid hydroxylation.

Figure 3.1 illustrates the main patterns for batch fermentation process kinetics for type 1, 2 and 3 processes.

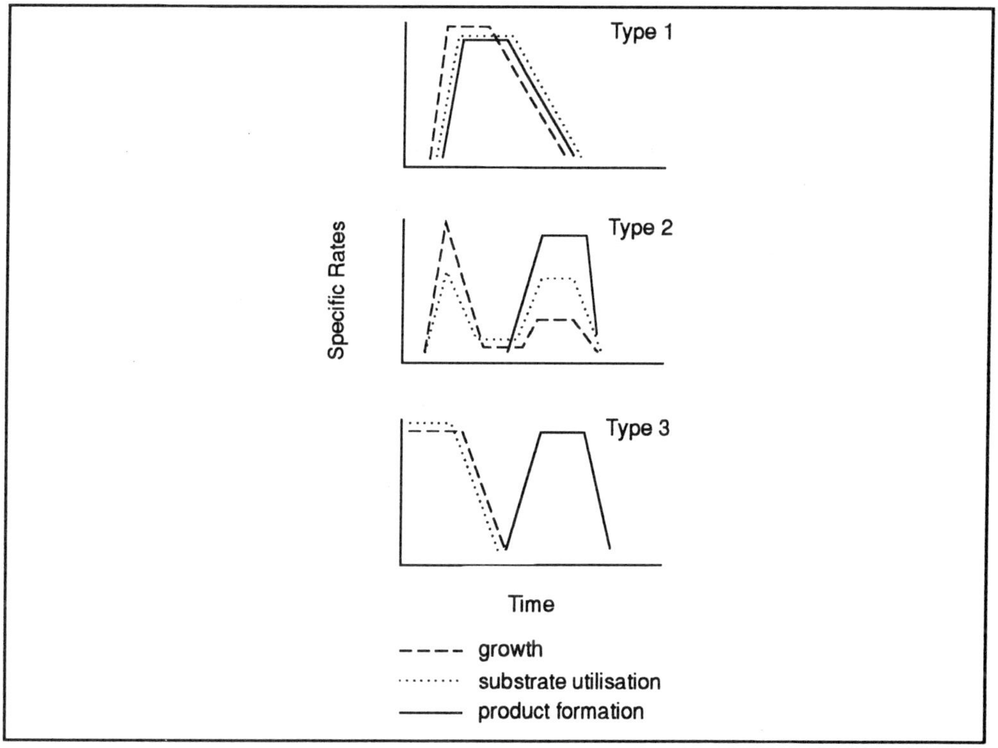

Figure 3.1 General patterns for batch fermentation process kinetics.

type 1 In type 1 processes, substrate utilisation, biomass formation and product formation are linked in a simple chemical reaction. For example, if the product contains C, H, O only, we simply extend the biomass material balance equation used earlier:

$$a'\,CH_x O_y + b'\,O_2 + c'H_l O_m N_n \longrightarrow$$

$$CH_\alpha O_\beta N_\delta + d'\,H_2O + e'\,CO_2 + f'\,C\,H_v O_w \qquad \text{E - 3.9}$$

biomass product

The product yield coefficient can then be calculated, taking into account the relative numbers of carbons in the substrate and product. The molar yield coefficient is then written as

$$Y_{p/s} = \frac{\text{product formed}}{\text{substrate used}} = \frac{f' n_s}{a' n_p}$$

(mols product formed/mols substrate consumed). E - 3.10

Where:

n_s = number of carbon atoms in the substrate molecule;

n_p = number of carbon atoms in the product molecule.

Π Determine the molar yield coefficient for exopolysaccharide production if 9.55 mols of glucose ($C_6H_{12}O_6$) are required to produce one mol of $C_{55.2}H_{86.2}O_{45.1}$.

$Y_{p/s} = (1 \times 6) / (9.55 \times 55.2) = 0.011$

type 2 For type 2 processes, the simple stoichiometry (E - 3.9) does not apply. Here, product formation is not necessarily proportional to substrate utilisation or biomass formation. In these cases, we need to consider a product formation step in addition to the growth reactions considered earlier, ie the formation (or utilistion) of both NADH and ATP relative to product formation.

$$d\, CH_xO_y + H_2O + \frac{1}{2}(\gamma_s - Z\gamma_p)NAD^+ \longrightarrow$$

$$d\, CH_yO_w + d\,(1 - Z)CO_2 + d\frac{1}{2}(\gamma_s - Z\gamma_p)\,(NADH + H^+) \qquad \text{E - 3.11}$$

$$\varepsilon_p\, d\,(ADP + P_i) \longrightarrow \varepsilon_p\, d\,(ATP + H_2O) \qquad \text{E - 3.12}$$

Where:

Z = the fraction of carbon substrate used for convertion to product;

ε_p = the number of ATPs generated by product formation.

Π If the coefficient ε_p in the product formation equation (E - 3.12) was negative, what would this indicate?

The ATP is utilised rather than generated in connection with product formation.

Π What does the coefficient d and the parameter γ_p denote in the product formation equation (E - 3.12)?

The coefficient d denotes the extent to which the reaction occurs relative to the growth reaction. The parameter γ_p denotes the degree of reductance of product.

type 3 and 4

For type 3 processes, growth and metabolic activity reach a maximum early in the batch process cycle (Figure 3.1) and it is not until a later stage, when oxidative activity is low, that maximum desired product formation occurs. The stoichiometric descriptions for both type 3 and 4 processes depend upon the particular substrates and products involved. In the main, product formation in these processes is completely uncoupled from cell growth and dictated by kinetic regulation and activity of cells.

Product formation stoichiometry can be used to estimate the upper bounds for product yields in processes. A relatively simple example is the anaerobic fermentation of glucose by yeast. Here, carbon dioxide and ethanol are the only products. Modification of (E - 3.9) then becomes:

$$CH_2O \longrightarrow \frac{1}{3} CO_2 + \frac{2}{3} CH_3 O_{0.5} \qquad \text{E - 3.13}$$

Π For the ethanolic fermentation described by (E - 3.13), determine the upper bound on the molar yield factor, ie $Y^{max}_{P/S}$.

$$\text{From (E - 3.10), } Y^{max}_{P/S} = \frac{\frac{2}{3} \times 6}{1 \times 2} = 2$$

Yields much lower than the upper bound value indicate that there is significant substrate utilisation to support growth, maintenance or synthesis of other products.

3.2.1 Product yield considerations

We can see that for type 1 processes, high growth rate is obligately linked to a high rate of product formation. Indeed, this is the case for all products produced by a fermentative mode of metabolism, eg ethanol, lactic acid, acetone. Chemostat studies have shown that for most aerobic processes when growth is limited by some nutrient other than the carbon source, the yield of product decreases with increase in specific growth rate (μ or D; μ = dilution rate (D) in chemostat culture). Conversely, both the biomass yield and the specific rate of substrate utilisation (q_S; g substrate g biomass^{-1} h^{-1}) increase with specific growth rate.

growth rates and product yields

The relationships between specific rate of substrate consumption and dilution rate and between yield coefficients and dilution rates are shown in Figure 3.2.

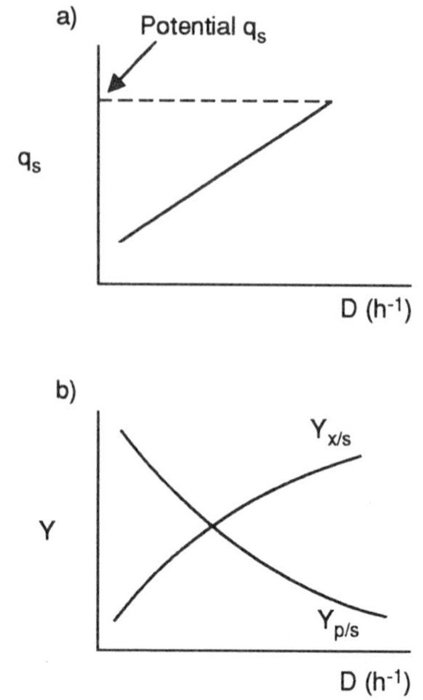

Figure 3.2 Theoretical relationships for (a) q_s against dilution rate and for (b) $Y_{p/s}$ and $Y_{x/s}$ against dilution rate. The micro-organism is grown aerobically in a nitrogen limited chemostat culture.

These observations can be explained by considering the substrate uptake rates. At high growth rates there is relatively little difference between the organisms potential substrate uptake rate and the substrate requirement for growth. However, at low growth rates relatively little substrate is required for growth but the potential substrate uptake rate is unaltered - this means that the organism can channel more of its substrate carbon into product.

Π Can you think of a disadvantage of operating a process at low dilution rate?

The main disadvantage is that, although high yields of product might be achieved, it is difficult to achieve high productivities (kg product m^{-3} h^{-1}).

High productivities at low dilution rate would require:

- expression of substrate uptake rates near their potential maximum;
- tight control of respiratory activity to ensure that substrate is converted to product and not to CO_2.

Ideally, we would wish for high substrate uptake in the absence of growth and in the absence of maintenance energy requirements. Since aerobic micro-organisms control their rates of substrate uptake when growth is slow or absent, manipulation of substrate uptake may be necessary.

These comments also apply to batch cultures where product formation occurs after growth has ceased.

maintenance requirement

The relationships between energy metabolism and product formation in an organism can be used to predict the influence of maintenance requirements on product yields. In aerobic fermentation, where type 3 and 4 product formation is often the goal, increased maintenance requirements decrease product yields by increasing substrate utilisation for energy production. However, the converse is true for some type 1 processes, for example the production of ethanol by anaerobic fermentation in a medium containing glucose and ammonia (as a nitrogen source). Here, substrate-level phosphorylation is the only source of ATP and a relatively high maintenance requirement means a relatively large amount of glucose will be metabolised. This in turn generates relatively more NADH, which is in excess to that used in biosynthesis. The NADH has to be oxidised to NAD^+ to ensure continued catabolism of glucose. This is achieved by increases in product formation (ethanol), ie the stored electrons in NADH are transferred to a product which is more reduced than the substrate and so the oxidation-reductance balance is maintained.

P/O quotient

energy dissipation

Where biosynthesis of a product requires the net input of energy, the theoretical yield will be influenced by the P/O quotient of the process organism. Furthermore, where the formation of a product is linked to the net production of ATP and/or NADH, the P/O quotient will influence the rate of product formation. It follows that to estimate the potential for yield improvement for a given primary or secondary metabolite, it is necessary to determine the P/O quotient of the producing organism.

We have seen that both the maintenance energy requirement and the P/O quotient of the process micro-organism influences the rate of product formation. In the following sections we will consider how these two factors can be determined, together with the maximum biomass yield.

SAQ 3.3

To which of the four types of processes do each of the following statements apply?

1) High growth rate is obligately linked to a high rate of product formation.

2) Product formation is completely uncoupled from cell growth.

3) The main product is a secondary metabolite.

4) Growth, substrate utilisation and product formation time courses exhibit coincident maxima.

5) The main product arises indirectly from energy metabolism.

SAQ 3.4

A 1 m^3 aerobic bioprocess was operated in a continuous mode with nitrogen as the growth limiting nutrient. The steady state biomass concentration ($\bar{x}$), the biomass yield coefficient ($Y_{x/s}$) and the product yield coefficient ($Y_{p/s}$) were determined at a low and at a high dilution rate (D).

$D = 0.1\ h^{-1}$	$D = 0.6\ h^{-1}$
$\bar{x} = 10$ kg biomass m^{-3}	$\bar{x} = 8$ kg biomass m^{-3}
$Y_{x/s} = 0.25$ kg biomass (kg substrate)$^{-1}$	$Y_{x/s} = 0.5$ kg biomass (kg substrate)$^{-1}$
$Y_{p/s} = 0.6$ kg product (kg substrate)$^{-1}$	$Y_{p/s} = 0.2$ kg product (kg substrate)$^{-1}$

Which of the two dilution rates should the process be operated at? (Hint: compare productivities for the product). Note that biomass productivity = $D.\bar{x}$).

3.3 Determination of maintenance energy requirement and maximum biomass yield

Cells need a certain amount of energy for maintenance. The maintenance energy is, for instance, needed for maintaining the proton motive force which is, among other purposes, used for maintaining the ion gradients across the cell membrane. Furthermore, energy is needed for the turnover of proteins and mRNA, for repair and for movement (if mobile).

Maintenance energy requirements can be defined in terms of rate of substrate consumption per unit of biomass for maintenance: this is known as the maintenance coefficient (m).

In the absence of product formation, the relation between the observed growth yield ($Y_{x/s}$) and the maximum growth yield ($Y^{max}_{x/s}$) can be written as:

$$\frac{1}{Y_{x/s}} = \frac{1}{Y^{max}_{x/s}} + \frac{m}{\mu} \qquad \text{E - 3.14}$$

where:

μ = specific growth rate
m = maintenance coefficient.

intercept is $\frac{1}{Y^{max}_{x/s}}$

This relationship is very useful experimentally because it can be used to determine both $Y^{max}_{x/s}$ and m. In practice, carbon limited chemostats are used and $Y_{x/s}$ is measured at different dilution rates (D). In a plot of $\frac{1}{Y_{x/s}}$ against $1/\mu$, the intercept is $\frac{1}{Y^{max}_{x/s}}$ and the slope is m. (Note that at steady state μ = D).

slope is m

The observed values of $Y_{x/s}$ at the different dilution rates can also be used to determine Y_{x/O_2} values. This is achieved by defining $Y_{x/s}$ and Y_{x/O_2} in terms of their respective rates of production and taking into account the degrees of reductance. So:

$$Y_{x/s} = \frac{\text{rate of biomass production}}{\text{rate of substrate utilisation}}$$

$$Y_{x/O_2} = \frac{\text{rate of biomass production}}{\text{rate of oxygen consumption}}$$

Combining these two rate equations and taking into account the degree of reductance, we have:

$$Y_{x/O_2} = \frac{-\gamma_o}{\gamma_x} \cdot \frac{Y_{x/s} \cdot \frac{\gamma_x}{\gamma_s}}{1 - Y_{x/s} \cdot \frac{\gamma_x}{\gamma_s}} \qquad \text{E - 3.15}$$

(where γ signify the degree of reduction for O_2, γ_x for biomass, γ_s for substrate)

SAQ 3.5

A bacterium was grown as a glucose-limited chemostat culture and steady state growth yield ($Y_{x/s}$) was measured at different dilution rates.

D (h^{-1})	$Y_{x/s}$ (C-mol biomass/C-mol substrate)
0.9	0.475
0.4	0.470
0.2	0.485
0.1	0.455
0.05	0.426

Use the data to determine the maintenance coefficient (m) and the maximum growth yield $Y_{x/s}^{max}$.

Determine the Y_{x/O_2} value for $\mu = 0.9\ h^{-1}$ (the empirical formula for biomass is $CH_{1.666}N_{0.20}O_{0.27}$).

3.4 Determination of P/O quotients

As we noted earlier, P/O is the number of ADP phosphorylations per atom of oxygen consumed, ie the amount of ATP produced per 0.5 moles O_2.

The quotient can be derived from the maximum yield for oxygen ($Y_{O_2}^{max}$) using the equation:

$$Y_O^{max} = Y_{ATP}^{max} \cdot P/O \qquad \text{E - 3.16}$$

carbon limited cultures

In practice, carbon limited chemostat cultures are used to estimate the P/O quotient. These conditions are used because they favour the most efficient conversion of the carbon substrate into cellular material, ie the highest efficiency of energy conservation. The steady state respiration rate (q_{O_2}) is measured as a function of dilution rate (specific growth rate) and $Y_{O_2}^{max}$ can be obtained from the reciprocal of the slope of the plot. q_{O_2} is also known as the metabolic quotient for oxygen or the specific rate of oxygen consumption.

substrate level phosphorylation

You will note that the equation used to determine P/O does not take into account ATP synthesis via substrate level phosphorylation, which is a limitation of the P/O estimation.

Π How would you expect substrate level phosphorylation to affect the P/O estimation?

Substrate level phosphorylation leads to an overestimation of P/O by about 7% and 30% at P/O quotients of 3 and 1 respectively.

The P/O quotient obtained in this way is therefore only an approximation. Nevertheless, the values can be used as a comparitive measure of growth efficiency, provided organisms are grown under carbon limited conditions (difference in Y_{ATP}^{max} is minimised). Such a comparison of the energetic efficiency would otherwise not be possible.

SAQ 3.6

A bacterium was grown as a glucose-limited chemostat culture and steady state respiration rate (q_{O_2}) was measured at different dilution rates:

D (h^{-1})	q_{O_2} (mmol g^{-1} h^{-1})
0.6	9.2
0.5	8.7
0.4	6.8
0.3	5.3
0.2	3.5
0.1	2.8

Determine the P/O quotient for the bacterium if Y_{ATP}^{max} during growth on glucose is 13.9 g dry wt mol^{-1}.

3.5 Metabolite overproduction and growth efficiency

The theoretical yield for a given metabolite can be estimated provided the following are known:

- the biosynthetic pathway for synthesis of the metabolite;
- the P/O quotient of the producing organism.

The extent to which the yield of metabolite can be improved is indicated by the difference between the theoretical and observed yields. The latter must, of course, be corrected for substrates requirements of growth and maintenance. Clearly, the influence of the P/O quotient on the theoretical yield will depend on the relationship between energy and metabolite synthesis. Three classes of metabolite can be distinguished in this respect.

classes of metabolites

1) Metabolites whose biosynthesis is energy requiring, for example exopolysaccharides using certain substrates. Here, part of the substrate has to be oxidised to provide ATP for biosynthesis and thus the P/O quotient of the producing organism influences the theoretical yield.

2) Metabolites whose biosynthesis leads to the net production of ATP and/or reducing equivalents, for example organic acids and certain secondary metabolites. In these cases, the P/O quotient influences the extent to which energy can be dissipated.

3) Metabolites that are composed of structures of quite different oxidation states. Certain secondary metabolites and biosurfactants fall into this class since they have both carbohydrates and fatty acids in their structures.

It should be noted that, for some metabolites, the class to which they belong depends on the fermentation conditions, for example the class to which certain antibiotics belong depends on the substrate(s) used for their production.

It is obvious that rapid metabolite production requires high fluxes of carbon through the metabolic systems responsible for its synthesis. The rate of metabolite production, for a wide range of micro-organisms, has been shown to increase with decrease in growth efficiency ($Y_{O_2}^{max}$). In addition, micro-organisms with low growth efficiency have a far greater capacity to dissipate energy (turnover ATP) than those with high growth efficiencies. We will now consider the three classes of metabolites in relation to growth efficiency.

dissipitation of energy

growth efficiency

ATP turnover

- In class 1 (ATP requiring), the rate of metabolite production is limited by the micro-organisms capacity to dissipate energy.
- In class 2 (ATP generating), the rate of metabolite production, and oxidation state, are inversely related to the growth efficiency.
- In class 3, the rate of metabolite production from a single substrate may be limited by the rate of ATP turnover. Provision of ready made precursors can increase both the metabolite yield (final concentration) and rate of production by decreasing the requirement for ATP turnover during biosynthesis.

In the following sections we will consider the production of each class of metabolite separately.

SAQ 3.7

1) Which of the following statements are applicable to class 2 metabolites?

a) The P/O quotient influences the extent to which energy can be dissipated.

b) The rate of metabolite production is inversely related to $Y_{O_2}^{max}$.

c) Biosynthesis of the metabolite leads to the net production of energy.

d) The rate of metabolite production increases with increase in growth efficiency.

3.5.1 Exopolysaccharide production

nature of substrate

In this section we will consider the energetics of exopolysaccharide production in some detail. We will see how chemostat (substrate limited) derived yield coefficients and elemental balances can be used to determine how the nature of the substrate influences rates of metabolite production, and to give an indication of the scope for improvement of the producing micro-organism. You should note that for most industrial bioprocesses, the unavailability of data in the primary literature would prevent such an analysis. Further aspects of exopolysaccharide production are covered in Chapter 7 of this text.

The energetic requirements of exopolysaccharide production from various carbon sources can be calculated if the P/O quotient during growth on the carbon substrate is known. Table 3.1 shows molar growth yields measured during carbon limited growth in chemostat culture.

Π Use the data given in Table 3.1 to calculate the P/O quotient with succinate as limiting substrate.

$$Y_O^{max} = Y_{ATP}^{max} \cdot P/O$$

$$P/O = (29/2)/10/1 = 1.4$$

Now enter this value for P/O quotient into Table 3.1.

constant P/O quotient

We can see from Table 3.1 that the P/O quotient is virtually independent of the carbon source. We can therefore assume a constant P/O quotient when calculating the energetic consequences of exopolysaccharide production from different carbon sources. It is also reasonable to assume that the rate of ATP turnover is similar on different carbon sources.

Carbon Limitation	Carbon Source						
	Glucose $C_6H_{12}O_6$	Gluconate $C_6H_{12}O_7$	Sorbitol $C_6H_{14}O_6$	Xylose $C_5H_{10}O_5$	Glycerol $C_3H_8O_3$	Succinate $C_4H_6O_4$	Ethanol C_2H_6O
μ_{max} (h^{-1})	0.41	0.25	0.18	0.21	0.16	0.32	0.16
Biomass (g dry wt l^{-1})	3.2	2.17	2.88	2.46	2.69	2.46	3.43
$Y_{x/s}$ (g dry wt mol^{-1})	80.0	69.4	73.3	52.0	41.8	41.9	30.4
Y_{O_2} (g dry wt mol^{-1})	31.0	27.3	23.6	22.0	29.2	29.0	16.6
Y_{ATP}^{max} [g dry wt (mol ATP)$^{-1}$]	13.9	13.9	13.9	12.7	12.7	10.1	7.17
P/O quotient (mol ATP per 0.5 mol O_2)	1.1	1.0	0.9	0.9	1.1		1.2
Nitrogen Limitation							
Biomass (g dry wt l^{-1})	1.16	1.66	2.09	1.53	1.59	1.49	1.28
q_s (m mol g^{-1} h^{-1})	2.72	2.10	1.62	2.37	2.12	4.03	1.5
q_{O_2} (m mol g^{-1} h^{-1})	2.66	2.40	2.86	2.27	3.37	4.40	3.3
Exopolysaccharide (g l^{-1})	7.86	5.2	6.38	5.86	2.07	6.57	0.09
q_p (g g^{-1} h^{-1})	0.24	0.15	0.12	0.15	0.06	0.18	0.002

Table 3.1 Parameters of growth and exopolysaccharide production for *Agrobacterium radiobacter* grown in chemostat culture on various carbon sources. Data obtained from Linton J. D. *et al* (1987) *Journal of General Microbiology 133*, 2979-2987.

Π How would you expect the growth efficiency to be influenced by the P/O quotient?

The growth efficiency ($Y_{O_2}^{max}$) is proportional to the P/O quotient; this is expected from the equation used to calculate P/O (section 3.4, E - 3.16).

We already know that product formation is enhanced when substrate other than the main carbon source limits growth. Further, during nitrogen limited growth in chemostat culture, exopolysaccharide has been shown to be the only product excreted. We can see from Table 3.1 that, under nitrogen limited growth, the specific rate of exopolysaccharide production (q_p) is highest with glucose as substrate. During exopolysaccharide production on glucose, gluconate and xylose the respiratory rates (q_{O_2}) are similar. However, respiratory activity increases during growth on carbon sources either more reduced or oxidised than glucose. An explanation for this can be obtained from a consideration of the energy requirements of exopolysaccharide production.

respiratory activity

SAQ 3.8

1) Determine the degrees of reductance of the carbon substrates shown in Table 3.1. Which substrates are more reduced and which are more oxidised than glucose?

2) Examine Table 3.1. What is the relationship between specific rate of exopolysaccharide production and growth efficiency?

3) Use the data in Table 3.1 for glucose limited growth to calculate q_s, at a dilution rate of 0.2 h^{-1}, in units of g g^{-1} h^{-1}. (Molecular weight of glucose = 180).

The ATP requirement for exopolysaccharide production can be twice that required for cell biosynthesis. High rates of ATP synthesis are therefore required to support high rates of exopolysaccharide production. However, whereas synthesis of the sugar backbone of exopolysaccharide is energy requiring, the production of the oxidised parts of the molecule (eg acetate and uronic acids) is energy generating. Optimal yield of exopolysaccharide occur, therefore, when carbon and energy fluxes are integrated. The extent of such integration depends on:

- the degree of reductance of the exopolysaccharide;
- the degree of reductance of the carbon source;
- the P/O quotient of the producing micro-organism.

degree of reductance

For substrates (glucose, gluconate and xylose) that are around the same degree of reductance as exopolysaccharide (ie around 3.9), virtually all of the energy generated above that required for growth goes into exopolysaccharide production. However, the rate of exopolysaccharide production decreases with more oxidised (succinate) or more reduced (sorbitol, glycerol, ethanol) substrates because carbon and energy fluxes are not favourable. For example, in the case of succinoglycan biosynthesis form ethanol (Table 3.2), there is a massive overproduction of ATP and high rates of exopolysaccharide production would require some form of energy dissipation.

Carbohydrate	Net ATP produced/unit succinoglycan	Rate of succinoglycan production (g g^{-1} h^{-1})
glucose ($C_6H_{12}O_6$)	-18.25	0.24
gluconate ($C_6H_{12}O_7$)	-0.2	0.15
sorbitol ($C_6H_{14}O_6$)	+5.91	0.12
glycerol ($C_3H_8O_3$)	+12.95	0.06
ethanol (C_2H_6O)	+91.3	0.002

Table 3.2 Rates of succinoglycan synthesis from carbohydrates and net ATP production during synthesis. Data obtained from Linton J. D. (1990) *FEMS Microbiology Reviews 75*, 1-18.

scope for improvement

In order to quantify the scope for improvement of exopolysaccharide production, it is first necessary to correct the observed yields of exopolysaccharide for the amount of carbon substrate and oxygen required for cell production. The corrected yields are then compared with the theoretical calculated from the P/O quotient for the producing micro-organism. Such a comparison is made in Table 3.3.

Carbon source	Experimental stoichiometry												
	Carbon source	+ NH_4^+	+ O_2	=	cells ($C_{0.937}H_{1.79}N_{0.2}O_{0.573}$)	+	intracellular polysaccharide ($C_6H_{10}O_5$)	+	CO_2	+	exopolysaccharide ($C_6H_{9.37}O_{4.9}$)	+	H_2O
Glucose	$6C_6H_{12}O_6$	+ 0.565	+ 6	=	2.83	+	0.196	+	6.20	+	4.06	+	14.58
Gluconate	$6C_6H_{12}O_7$	+ 1.06	+ 8.4	=	6.16	+	0.141	+	11.4	+	3.38	+	22.35
Sorbitol	$6C_6H_{14}O6$	+ 0.90	+ 10.53	=	4.52	+	0.30	+	8.23	+	3.80	+	20.40

	Yield of exopolysaccharide					
	observed		Corrected for cell production*		Theoretical**	
	Y_s	Y_{O_2}	Y_s	Y_{O_2}	Y_s	Y_{O_2}
Glucose	0.6	3.37	0.74	5.7	0.83	5.3
Gluconate	0.46	2.0	0.72	6.0	0.65	5.56
Sorbitol	0.56	1.8	0.80	3.35	0.84	5.6

Table 3.3 Experimental stoichiometries for cell growth and exopolysaccharide production from various carbon sources by *Agrobacterium radiobacter* under nitrogen-limiting conditions.
* calculated using $Y_{x/s}$ and Y_{x/O_2} of corresponding carbon-limited cultures

** calculated using the mean observed P/O quotient of 1.1 of carbon-limited cultures.
Data obtained from Linton J. D. *et al* (1987) *Journal of General Microbiology 133*, 2979-1987.

nitrogen-limited cultures

The requirement for oxygen and carbon source for cell biosynthesis are calculated using mass balance equations for growth during exopolysaccharide production (nitrogen-limited cultures). These balances are derived from experimentally determined values of:

- substrate and oxygen consumption;
- exopolysaccharide production;
- intracellular polyglucan production;
- bacterial biomass production;
- CO_2 production;
- cellular composition of C, H, O and N.

carbon-limited cultures

yield comparison

The mass balance equation and $Y_{x/s}$ and Y_{O_2} values of corresponding carbon-limited cultures are then used to determine the carbon substrate and oxygen requirements for cell production. The yield of exopolysaccharide (corrected for cell production) is then compared to the theoretical yield. The latter being calculated from the mean observed P/O quotient of carbon-limited cultures.

Experimentally determined yields of exopolysaccharide have been found to be 70% of the theoretical. This suggests that exopolysaccharide production is an efficient process with little scope for major yield improvements.

SAQ 3.9

To which class(es) of metabolite, based on the relationship between energy and metabolite synthesis, would you expect exopolysaccharides to belong? Explain your reasoning.

3.5.2 Citric acid production (class 2 metabolite)

The production of organic acids by micro-organisms, and especially citric acid, is considered in detail in Chapter 4. In this section therefore we will only briefly consider citric acid production, from an energetics perspective.

energy dissipation

additional electron transport chain

During citric acid production, ATP and NADH are generated via glycolysis and the tricarboxylic acid cycle. Reoxidation of NADH occurs via the respiratory chain, which can lead to the synthesis of ATP by oxidative phosphorylation. The major allosteric control enzyme in glycolysis is phosphofructokinase, which is inhibited by both ATP and NADH. Clearly, for metabolite production to proceed rapidly there must be a means of dissipating energy (ATP) and NADH. It has been found that organisms with low P/O quotients have a greater capacity to dissipate energy than those with high P/O quotients.

high carbon fluxes

In the case of industrial citric acid production, the process micro-organism (*Aspergillus niger*) has a relatively high P/O quotient when not in citric acid production mode. However, during citric acid production the organism is able to reoxidise NADH via an additional electron transport chain, which is not coupled to oxidative phosphorylation. In essence this allows dissipation of energy and ensures that high carbon fluxes through glycolysis can be maintained leading to citric acid production.

Π Which one of the following organisms is likely to produce gluconic acid at the fastest rate?

Organism	$Y_{O_2}^{max}$
A	2.7
B	2.1
C	2.8
D	2.3

Organism B because it has the lowest $Y_{O_2}^{max}$, ie lowest growth efficiency or highest capacity to dissipate energy.

3.5.3 Sophorolipid production (class 3 metabolite)

high ATP demand

Sophorolipid is a glycolipid, ie it is composed of carbohydrate and lipid. It therefore contains moieties of widely different oxidation levels and its synthesis from single carbon sources has a high ATP demand. However, the demand for ATP is reduced if a mixture of glucose and C-18 alkane is used. If glucose and fatty acid is used the ATP demand is reduced further and relatively high specific production rates can be achieved.

Substrate	ATP demand per molecule of sophorolipid
Glucose	25.5
C-18 alkane	19.11
Glucose + C-18 fatty acid	6

Π Why is there a high ATP demand for sophorolipid synthesis from 1) alkane substrate and from 2) glucose substrate?

1) The need to produce the carbohydrate moiety from alkane.

2) The need to produce the alkane moiety from glucose.

SAQ 3.10

1) Explain why the use of *A. niger* to produce citric acid might be regarded to be far from ideal.

2) Is a low or a high growth efficiency for the producing organism desirable for sophorolipid production? Explain your reasoning.

Summary and objectives

Overcoming the major flux control points within a given metabolic pathway may not, by itself, lead to rapid metabolite overproduction if the energetic consequences of the alteration are unfavourable. We have seen that in order to improve the yield of a given metabolite in a rational way, it is necessary to describe growth and metabolism by elemental material balances. Further, when considering product formation stoichiometries it is essential to define the relationship (four types) between product formation and growth. The theoretical yield of a given metabolite can be estimated provided 1) the biosynthetic pathway for its synthesis and 2) the P/O quotient of the producing organism is known. The extent to which the yield of metabolite can be improved is then indicated by the difference between the theoretical and observed yields. Three main classes of metabolite can be distinguished according to the relationship between energy and metabolite synthesis. We have also seen that the capacity to sustain large fluxes of carbon and energy required for rapid metabolite production is inversely related to the growth efficiency of micro-organisms.

Now that you have completed this chapter you should be able to:

- use stoichiometric data and knowledge of the elemental composition of the cell to write reaction equations for growth;
- interpret stoichiometric representations of aerobic growth;
- classify bioprocesses according to process kinetics;
- use yield coefficients and productivities to determine the best operating conditions for bioprocesses;
- determine P/O quotients, maintenance coefficients and maximum biomass yield coefficients from continuous culture data;
- determine the degrees of reductance of carbon substrates;
- classify metabolites according to the relationships between energy and metabolite synthesis;
- appreciate how growth efficiency of organisms influence product formation rates.

Single cell protein

Single cell protein

4.1 Introduction

In this chapter we examine the processes that have been developed to produce micro-organisms as a source of food protein. We will examine the reasons why micro-organisms have been considered as alternative protein sources, the substrates on which they have been grown, the various process technologies developed and the comparative economics of these processes. One process will be examined in depth, to illustrate how a team composed of such diverse people as microbiologists, process engineers, patent lawyers and cost analysts work together to develop a marketable product.

The driving forces behind the development of many single cell protein projects emerged from global economic conditions and social concerns of the 1960s. In the 1970s and early 1980s, there were considerable technological advancements associated with single cell protein process developments and many types of processes were operated commercially. In this chapter we present technological and economic data derived from these early developments to provide a historical context for single cell protein as a food and animal feed source. You will see that many important principles underpinning modern process technology are based on the experiences gained in the development of single cell protein processes.

4.2 Conventional protein sources

essential amino acids

Animals, including humans, cannot synthesise all the different amino acids they need and thus require them in their diet. These amino acids are called the essential amino acids. Proteins in food are hydrolysed in the digestive tract and the resulting amino acids are reassembled into proteins within the animal's cells. All animals are ultimately dependent on plants for protein, as it is plants that create protein by combining inorganic nitrogen from the soil (as nitrate) with organic molecules derived from carbon from the atmosphere (as CO_2).

organoleptic properties

For us to remain perfectly healthy, the protein in our diet must supply sufficient quantities of amino acids. We prefer to eat our protein in particular forms, that is in foods having particular textures, tastes and smells (these are called organoleptic properties). Conventional sources of protein are plants, mainly as cereals and pulses, and animals, mainly as meat, eggs and milk. The proportions of such proteins eaten in various parts of the world differ widely (Figure 4.1).

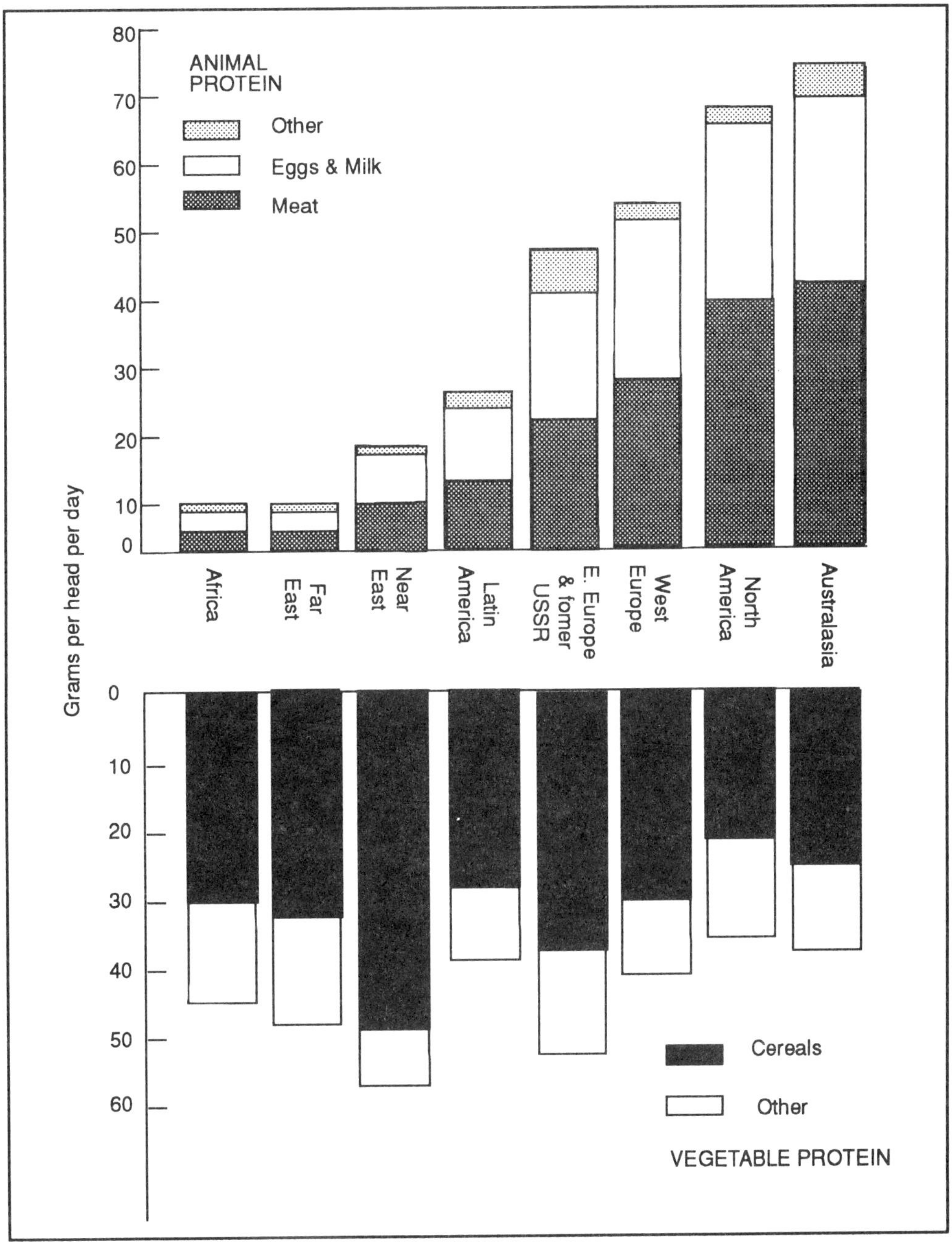

Figure 4.1 World protein consumption

Π List three factors you think account for such variations in the sources of proteins between various parts of the world?

Essentially the answer is history, climate, culture and money! Historically, people had to eat the food available locally, and this would be controlled by the local natural environment. Cultural influences have also led to preferences for certain food types. In more affluent countries foods such as meat, or high-protein feedstuffs on which to rear animals, can be produced or be imported. In less affluent countries such luxuries cannot be afforded. Increasing populations in some countries have overstretched food supplies, and so limited the availability of foods.

changing demands for dietary protein

There are problems, however, with these conventional sources of protein. Crop production is dependent upon a suitable climate and in most countries available arable land is already fully farmed. Fish stocks in the oceans are in danger of becoming depleted. In countries where animal meat forms a high proportion of dietary protein, there are controversies such as whether or not the fats eaten with the protein are healthy, whether or not we are justified in keeping animals in the unnatural conditions of some farms, or whether or not we are justified in killing animals for food at all. Such controversies are leading an increasing number of people to become vegetarian. It is likely that the world's population will double in the next few decades, yet the United Nations estimate that about one thousand million people are already suffering protein deficiency. It is estimated that between 1980 and 2000 the annual demand for protein as food for humans will increase from 50×10^6 tonnes to 79×10^6 tonnes, and the demand for protein as feed for animals will increase from 44×10^6 tonnes to 108×10^6 tonnes. Biotechnology is being applied to the rapid improvement of conventional food sources, both plant and animal, in an effort to meet the increased demand in food. Interest has also been shown in growing micro-organisms as a source of protein and it is developments in this area that we are going to examine here in detail.

4.3 Single cell protein

single cell protein/food and feed

Single cell protein, normally called simply SCP, is the term used to describe microbial cells, or proteins from them, which are used as food (food for humans) or feed (food for farm animals or fish). Although the term micro-organisms covers viruses, bacteria, fungi, algae and protozoa, viruses and protozoa are not considered suitable for SCP production.

Π Why do you think viruses and protozoa are not suitable for SCP production?

Both viruses and protozoa are difficult to grow in culture. Viruses need living cells to grow in and their small size makes them difficult to deal with. Protozoa need complex diets of organic materials. Bacteria, fungi and algae are relatively easy to grow in culture.

The term SCP is not exactly appropriate, as some filamentous organisms are used as SCP and these organisms are multicellular not unicellular.

You may be wondering why anyone should ever have considered using micro-organisms as a protein source. Let us consider why this should have been.

4.3.1 The advantage of micro-organisms as a protein source

filamentous blue-green bacteria

Eating micro-organisms is nothing new. You might not have been aware that some foods traditionally eaten by man are in fact micro-organisms. Filamentous blue-green bacteria (often called blue-green algae, or cyanobacteria) were collected from lakes and rivers and eaten by the Aztecs in Mexico, and people inhabiting the shores of Lake Chad in Africa still do so. Edible fungi have been collected from the wild for centuries and

yeasts

farmed throughout the last two. During the two World Wars this century, yeasts (unicellular fungi) were grown on a large scale in Germany and used as food and feed.

Micro-organisms are rich in protein. Microbial cells can contain as much protein as conventional foods. Bacteria can contain 60-65% (as a % of dry weight) protein whereas fungi and algae contain about 40%. In addition, microbial cells can be a rich source of fibre, unsaturated fats, minerals and vitamins. They are low in saturated fats and sodium.

protein from inorganic nitrogen

Micro-organisms create protein. Like plants, many micro-organisms can use inorganic nitrogen and can thus be used as an alternative to plants to create protein. In SCP processes inorganic nitrogen is usually supplied as ammonia (or as ammonium salts), which is readily available and is renewable, as it can be manufactured from atmospheric nitrogen and can be recycled through the nitrogen cycle.

autotrophs/ heterotrophs

Micro-organisms can use alternative carbon sources. Algae are autotrophs using atmospheric CO_2 (think of them as plants growing in water instead of soil). They compete with plants for CO_2 but there is not a shortage of CO_2 in the atmosphere and it is renewable by recycling through the carbon cycle. Other micro-organisms are heterotrophs (organisms which use organic sources of carbon) and can use a wide range of organic carbon sources. These can be materials unsuitable as food sources for animals (for example methanol). Others are waste products from industries or agriculture and have limited uses and can be a problem to dispose of by other means.

saves agricultural space

SCP processes are efficient on space. SCP production plants can be built on land unsuitable for agriculture and so need not compete for space with conventional sources of protein. Also they are much more efficient in terms of amount of protein produced per unit area (figures are quoted later for some processes).

rapid growth

Micro-organisms grow rapidly. Micro-organisms grow much more rapidly than plants or animals. Bacteria can grow with mean generation times (doubling times) as short as 20-30 minutes. The mean generation times of unicellular algae and fungi are about 1-3 hours, whereas those of multicellular algae and fungi may be longer. This means that micro-organisms have the potential to produce protein far more rapidly than is possible by rearing plants or animals.

Π By completing the following calculation you will be able to demonstrate the amazing potential for micro-organism to rapidly produce protein for food.

In batch culture, when growth is exponential, the number or organisms produced from one organism is given by 2^n, where n is the number of generations. So after one generation there are 2^1 (ie 2), after two generations 2^2 (ie 4) and after three generations 2^3 (ie 8) and so on.

Starting from a single bacterial cell with a mean generation time (doubling time) of 1 hour, and assuming exponential growth throughout, how many organisms would you have after 48 hours? As the dry weight of a bacterial cell is about 10^{-10}g, what would the dry weight of these cells be? Assuming these cells to be 50% protein, how much protein would there be? Assuming you are an average person, you require about 70 g of protein in your diet per day. How long would this protein last you? (Do not cheat, try the calculation before reading on). Now repeat the calculation to find out how much protein you would have after 72h.

After 48 hours there would be 2^{48} or 2.8×10^{14} cells.

This represents $2.8 \times 10^{14} \times 10^{-10} = 2.8 \times 10^4$g dry weight cells.

This represents $2.8 \times 10^4 \times 50\% = 1.4 \times 10^4$g protein.

This represents $1.4 \times 10^4/70 = 200$ days worth of protein for one person.

After 72 hours there would be 2^{72} or 4.7×10^{21} cells.

This represents $4.7 \times 10^{21} \times 10^{-10} = 4.7 \times 10^{11}$g dry weight cells.

This represents $4.7 \times 10^{11} \times 50\% = 2.35 \times 10^{11}$g protein.

This represents $2.35 \times 10^{11}/70 =$ ca. 3×10^9 meals, or enough protein to feed the entire population of China for 3 days!

You have now demonstrated the capability of exponentially increasing microbial populations to rapidly produce protein. However, although such outputs of protein are possible in theory, they cannot be achieved in practice, since exponential growth cannot be maintained for such periods because bioreactors are limited in size.

4.3.2 The disadvantages of micro-organisms as a protein source

Let us consider some of the disadvantages of micro-organisms compared to conventional sources of protein.

amino acid profiles

Are they nutritious? Guidelines on the nutritional quality of SCP have been given by the Protein Advisory Group (PAG) of the United Nations, and are based on amino acid profiles and feeding trials in animals. While most microbial cells are rich in protein, many do not contain sufficient quantities of essential amino acids. For instance, algal and fungal cells tend to lack methionine. Microbial cells may not be as easy to digest as conventional protein sources, for instance algae have cellulose cell walls which must be broken up if the proteins within the cell are to be easily digested by humans. The requirements for food are more strict than those for feed.

digestibility

importance of organoleptic properties

What are they like to eat? Humans are particular about the organoleptic properties of their food. Microbial cells may have little taste or smell, or even smell or taste unpleasantly to some people. The texture may not be the same as in conventional foods, particularly with unicellular organisms. These draw-backs can be overcome by adding a proportion of SCP to manufactured foods. However, even when SCP is incorporated into manufactured foods it may not have suitable characteristics such as stability, ability to bind water or fats, or ability to form gels, emulsions or foams. SCP for feed does not have to meet such strict requirements.

adverse effects

What happens after you have eaten them? Even if a micro-organism is palatable it may not necessarily be acceptable to the human digestive system, and if eaten in quantity can produce indigestion, flatulence, nausea, vomiting or diarrhoea. As little as 15 g yeast cells per day can produce such effects in humans.

cost

How much do they cost? SCP must compete in price with conventional protein foods and feeds. In countries where protein foods are readily available they can be relatively cheap. It has not always been possible to produce SCP at competitive prices.

safety

Are they safe to eat? Micro-organisms which are pathogenic or toxic obviously can not be used as SCP sources. In addition most microbial cells have a higher content of nucleic acid, particularly RNA, than conventional foods. When such cells are digested by animals these nucleic acids are metabolised to uric acid. Unlike most other mammals, humans do not possess uricase, which oxidises uric acid to soluble allantoid for

uric acid

excretion, and so uric acid can build up in the blood and may deposit as crystals in the joints, causing gout and arthritis. Thus, SCP used as food is usually processed to reduce the RNA content. PAG guidelines recommend that for humans the daily intake of nucleic acid should not be more than 4g, of which not more that 2 g should be obtained from SCP.

Π If SCP has a nucleic acid content of 15%, how much of that SCP could be safely ingested per day? If the SCP contains 50% protein, what proportion of the recommended human daily requirement (of 70 g protein) does this represent? Try to work these out for yourself before reading our answers.

Our calculation:

2g nucleic acid would be present in $2 \times \frac{100}{15}$ g =13.3 g SCP

At 50% protein this represents $13.3 \times \frac{50}{100} = 6.65$ g protein

This corresponds to $\frac{6.65}{70} \times 100 = 9.5$ % of the daily requirement.

SAQ 4.1

Which of the following factors supports the use of micro-organisms rather than higher plants for the production of protein food?

1) Plants are more difficult to digest than micro-organisms.

2) Micro-organisms can be used to convert organic wastes into proteins.

3) Micro-organisms grow more quickly than plants.

4) Higher plants need CO_2 as a carbon source.

5) Micro-organisms can use inorganic nitrogen.

SAQ 4.2

Suggest ways of overcoming or bypassing the following disadvantages of SCP as food.

1) Unpalatability

2) Indigestibility

3) Poor amino acid profile

4) Toxicity

We have seen that only certain micro-organisms that conform to nutritional and safety requirements are suitable for food or feed, and that food has more strict requirements than feed. In addition, for use as food, SCP should have a reduced nucleic acid content and should be palatable. Most often this means that its use is limited to processed foods, in which food technologists can produce acceptable tastes, smells and textures.

4.4 Substrates for SCP production

For a micro-organism to grow it must be supplied with all the nutrients required for cell material and energy production.

pectinases cellulases and hemicellulases

The physiological types of organism used in SCP production and their corresponding substrates are shown in Table 4.1. Photosynthetic bacteria utilise CO_2 from the atmosphere and nitrate in inorganic salts or natural ground water media. Algae are similar, growing on nitrate, ammonia or ammonium salt as nitrogen source. Some can also be grown as heterotrophs, in the dark, using sugars as sources of carbon and energy. Heterotrophic bacteria and fungi for SCP are grown on a variety of organic substrates, serving as both carbon and energy sources. Some organisms have additional requirement for growth factors, such as vitamins. For yeast, the substrate is in the form of sugars, as yeast cells cannot break down polysaccharides, whereas filamentous fungi may in addition be able to use starch (by secreting amylases), pectin (by secreting pectinases) and cellulosic material (by secreting cellulases and hemicellulases). Waste containing cellulosic material is in solid rather than in liquid form. Processes have also been developed with yeasts growing on *n*-paraffins or ethanol, and with bacteria growing on methanol. Inorganic nitrogen is supplied in such processes as ammonia, or as ammonium salt.

Organisms	**Physiological type**	**Carbon source**	**Energy source**	**Nitrogen source**
Blue-green bacteria	Autotroph	Atmospheric CO_2	Sunlight	NO_3^-
Algae	Autotroph	Atmospheric CO_2	Sunlight	NH_3, NH_4^+ NO_3^-
	Heterotroph	Carbohydrate (sugars)	As carbon source	NH_3, NH_4^+ NO_3^-
Fungi	Heterotroph	Carbohydrate (sugars, starch, pectin, cellulosics)	As carbon source	NH_3, NH_4^+
		Hydrocarbons and derivatives (*n*-paraffins, ethanol)	As carbon source	NH_3, NH_4^+
Bacteria	Heterotroph (methylotroph)	Hydrocarbon derivatives (methanol)	As carbon source	NH_3, NH_4^+

Table 4.1 Organisms and substrates in SCP production

Π In a culture medium for the growth of heterotrophs what do you think the carbon : nitrogen (C:N) ratio should be?

- 1 : 1
- 10 : 1
- 100 : 1
- 1000 : 1

C:N ratio

The C:N ratio should be about 10:1. The organic carbon in the medium provides both a source of energy and a source of carbon. Cells contain more carbon that nitrogen. The correct ratio of C:N is about 10:1, although this differs slightly between organisms. This means that a medium containing 3% w/v sugar should be supplied with ammonia at about 0.3% w/v. At a C:N ratio of 1:1 most of the ammonia would not be incorporated into cells (it is present in excess) and would be wasted. At C:N ratios more than 10:1 the ammonia would be completely used up before all the sugar, reducing the biomass output and wasting the sugar.

solid-substrate fermentations

The cost of substrates used in SCP production may represent 40-75% of the total production cost. Ammonia contributes 5-15% of the substrate cost but the major portion is the carbon source. Atmospheric CO_2 is free, but costly energy is needed for agitation to dissolve it into dense algal cultures. Wastes from agriculture and industry can be plentiful and relatively cheap, but may still represent 20-30% of the total production costs. Solid agricultural wastes, especially cellulosic ones, may also need expensive pre-treatment before they can be used in solid-substrate fermentations. Industrial wastes in the form of effluents can have high levels of BOD (Biological Oxygen Demand), which means they could cause pollution if disposed of in water without treatment. Using them as substrates for SCP production can reduce the BOD by as much as 70-80% and so save on treatment costs. Such agricultural and industrial wastes are derived from biomass (plant material) which is renewable and likely to remain plentiful and relatively cheap. Hydrocarbons and their derivatives can represent from 30-70% of total production costs. They are derived from oil or natural gas which are non-renewable, will not remain as plentiful as at present and will become increasingly expensive. They also have alternative uses as fuels and petrochemicals and their availability is often influenced by political issues.

renewable

4.5 Micro-organisms for SCP production

The physiological groups of organisms used in SCP production have been given in Table 4.1. We have examined the characteristics an organism should and should not have in order to be suitable as food or feed in Section 4.3. When selecting an organism for a particular production process, factors relating to growth of the organism also need to be considered.

Π Listed below are characteristics in culture of an organism you are intending to use in an SCP process you are developing. Consider whether each characteristic is an advantage or disadvantage to you. Tick the appropriate box, or if you think the characteristic is an advantage on the one hand but a disadvantage on the other, tick both boxes.

	advantage	disadvantage
i) Low growth rate.	☐	☐
ii) High biomass yield coefficient.	☐	☐
iii) Filamentous growth.	☐	☐
iv) Tolerance to broad range of temperatures.	☐	☐
v) Tolerance to broad range of pH.	☐	☐
vi) High spontaneous mutation rate.	☐	☐
vii) Low aeration requirement.	☐	☐

output

i) Disadvantage. High growth rate is needed for high output (weight of biomass produced per unit of time). The only advantage could be that as the RNA content of cells is generally proportional to the growth rate, growth at low growth rate could result in a product with lower nucleic acid content.

ii) Advantage. The biomass yield coefficient (weight of cells produced per unit of substrate consumed) should be high in order to give a high output. It also ensures efficient utilisation of the (expensive) substrate.

wall growth

iii) Advantage/Disadvantage. Compared to unicellular organisms, filamentous ones are easier (and cheaper) to recover from fermentation media (by sieving or rotary vacuum filtration) and have a more fibrous texture. However, dense broths of filamentous organisms can be difficult to aerate and wall growth can cause problems such as clogging of pipes and valves.

iv) Advantage. Temperature increases can occur during fermentations, as growth processes are exothermic. The ability of an organism to tolerate raised temperature would reduce the need for cooling. The ability of an organism to grow at ambient temperatures also overcomes the need for heating and cooling. The broader the temperature range tolerated, the less the need for temperature control.

non-aseptic processes

v) Advantage. The pH of a medium tends to change during fermentation. Most often media are buffered, and the fermentor is fitted with pH control. However, the ability to tolerate a wide range of pH can overcome the need for pH control. Fungi generally grow at lower pH than bacteria. Use can sometimes be made of this by operating fungal processes at very low pH, preventing bacterial growth. This means that an aseptic process (using sterilising procedures to maintain a pure culture) will be less prone to contamination if aseptic procedures fail. In some circumstances non-aseptic (non-sterile) processes can be operated, saving sterilisation costs.

vi) Disadvantage. Organisms for SCP production require a high degree of genetic stability. We have been considering the characteristics an organism must have for SCP production. These characteristics are under genetic control and any mutation could lead to an undesirable change in them. This is particularly important in continuous culture, which is often used for SCP production. Here the long growth period (in principle indefinite but in practice several weeks) can be long enough for mutants to arise, compete with the parent organism and predominate. In batch culture there is not enough time for this to occur.

vii) Advantage. For most SCP fermentation processes the running costs (costs of operating the fermentation unit) is 10-20% of the total production cost. Aeration costs contribute 30-60% to running costs. In other words aeration costs can be as much as 12% of the production costs of the SCP. The lower the aeration costs the better. Production costs for various SCP processes are discussed in more detail later on.

You are now familiar with the major characteristics of organisms that are useful for SCP production, and the types of substrates on which they can be grown. We are now going to consider in detail the processes that have been developed. Some of these processes have been developed only as far as the pilot scale, and have not reached commercial operation. Others have reached full production scale but have subsequently failed, for a variety of reasons. These have been included as well as the successes, as they show you the variety in the technology of SCP production, and also show how economic and political factors influence the success and failure of processes. These processes might also become useful and economic some time in the future. Emphasis will be put on the technology involved in the fermentation and down-stream processing of each process.

4.6 SCP from carbon dioxide

4.6.1 Spirulina

open lagoon system

Blue-green bacteria (cyanobacteria) of the genus *Spirulina* have been produced as SCP in Mexico, using natural bicarbonate-rich ground-water (into which atmospheric CO_2 readily dissolves). A flow diagram of the process is given in Figure 4.2. The single 10 hectare (1 ha = 10,000 m^2)open lagoon is about 0.6 m in depth and unmixed. The system is operated as a batch culture or as a semi-continuous culture (in which a proportion of the medium is removed and replaced by fresh medium intermittently). Nitrate is added as a nitrogen source, and other minerals are present in the water. The long filaments are raked mechanically from the pond onto screens (sieves), where water is drained and either recycled or disposed of. The biomass is then de-watered by rotary vacuum filtration, dried by vacuum drying, then dried in a drum drier and ground to a powder (to make the product more appealing). The product contains 56% protein and is sold as food.

Π The plant operates with an output of 10 g dry weight per square metre per day. What is the output of the 10 hectare lagoon per year? (Note 1 hectare = 10,000 m^2).

10 g m^{-2} equates to 10 x 10^4 = 10^5g $ha.^{-1}$

The daily output is thus: 10^2 kg $ha.^{-1}$ day^{-1}.

The 10ha lagoon produces 10^3 kg $ha.^{-1}$ day^{-1}.

This equates to 10^3 x 365 = 3.6 x 10^5 kg $year^{-1}$.

The output of the lagoon is thus 360 tonnes $year^{-1}$.

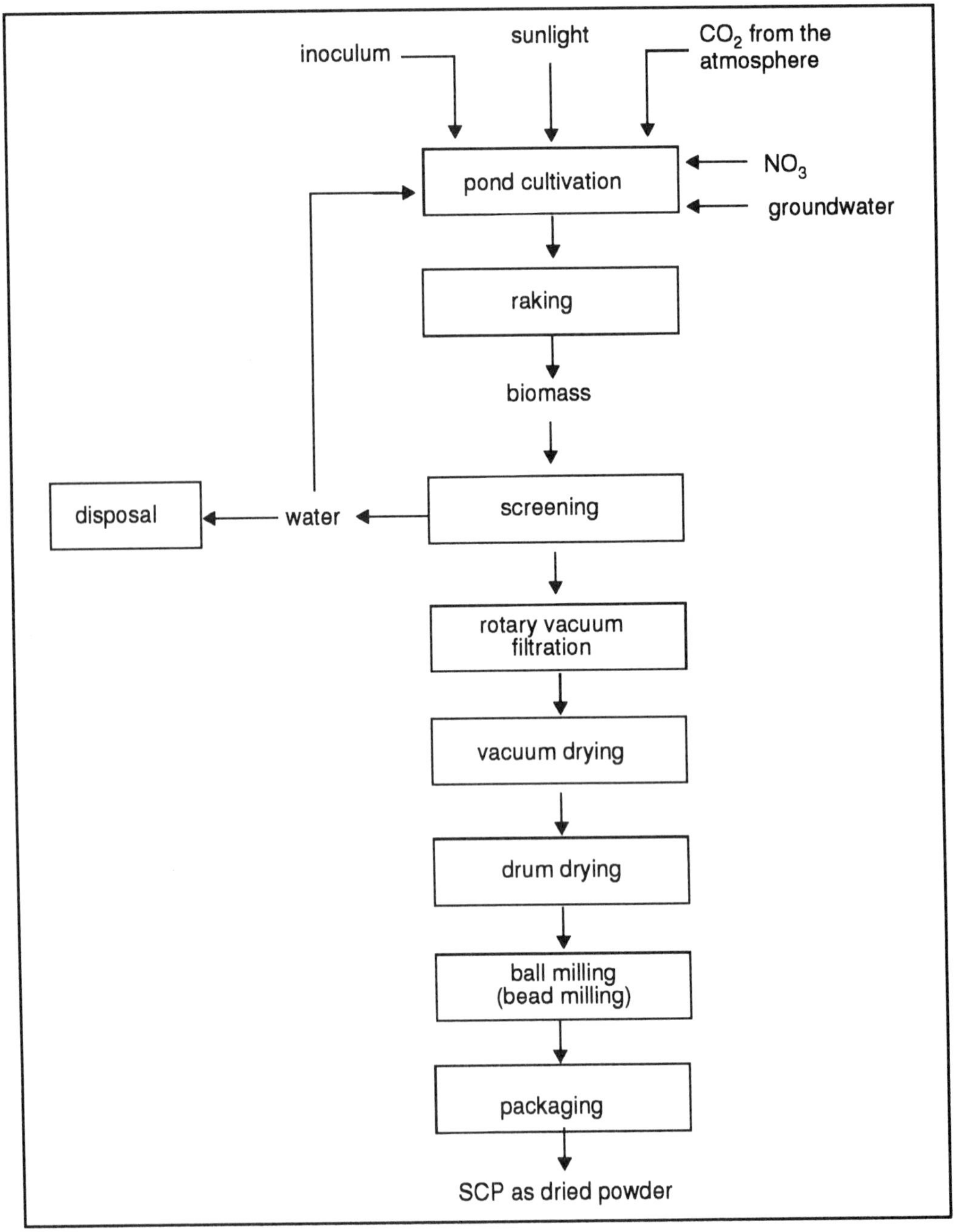

Figure 4.2 The production of *Spirulina maxima* for SCP

4.6.2 Algae

Eukaryotic algae (Chlorophyceae) of the genera *Chlorella* and *Scenedesmus* have been used for SCP production. Several types of cultivation systems have been considered, depending on the substrate used and whether the SCP is intended for use as food or feed.

CO_2 enrichment/

For use as food or feed, algae are grown in pure or mixed culture in a mineral salts medium containing NH_4^+ or NO_3^- and supplied with air or gaseous CO_2 (air contains only 0.03% CO_2, so CO_2 enrichment is required for high output). A flow diagram of such a process is essentially the same as that shown in Figure 4.2. Open systems have been developed with organisms growing as continuous cultures in open lagoons or circulation ditches, similar in design to oxidation ponds and ditches used in sewage treatment. The ponds can be lined with clay, concrete, brick or plastic sheeting and are 20-50 cm in depth. Mixing can be mechanical, using motor driven paddles, or can be manual, and it is necessary to prevent sedimentation of cells and uneven exposure to sunlight. Capital costs (excluding land costs) are lower than with other SCP systems, at $(US)20,000-30,000 per hectare (1989 prices). The organisms grow as phototrophs using sunlight as an energy source and atmospheric CO_2 as carbon source. Such systems are relatively simple (low-tech) but, as they are open, they are liable to contamination by wild algae and bacteria. Heterotrophic bacteria can grow in the ponds using organic materials released into the water by the algal cells.

phototrophs

Π Write down three reasons why the requirement for sunlight limits the use of algal ponds in SCP production?

Firstly, the lagoons need to be open to allow the sunlight to penetrate. This means that pond culture is non-aseptic and contamination usually limits the use of SCP to feed, where a higher degree of contamination is acceptable. The micro-biological standards of SCP as food and feed have been defined by the PAG and are based on comparisons with conventional foods and feeds. Secondly, the technology can only be applied between latitudes 35° North and South where there is sufficient sunlight to give high outputs. Thirdly, in dense algal cultures sunlight does not penetrate more than 40-50 cm, so lagoons are limited to such depths. This limits output, ie you cannot produce more algae by making the lagoon deeper.

centrifugation
flocculation
drum drying

Cell recovery presents a problem in algal culture, as centrifugation (the most effective method) can be prohibitively expensive (Table 4.2). Methods such as flocculation with calcium hydroxide and sedimentation can be employed for feed purposes. Waste water is recycled and recovered cells are dried preferably by drum drying, which breaks up cell walls making the product easier to digest by humans. The product contains 40-50% protein and 4-6% nucleic acid, and has been produced in relatively small quantities at a selling price \$4-10 kg^{-1} (1990 prices). This compares to soya protein concentrate (70% protein) and milk powder (36% protein) at about \$3 kg^{-1}.

Process	Cost (US $ per thousand m^3 of culture)	Concentration factor
Centrifugation	380	40-100
Flocculation/Flotation	340	85
Flocculation/Sedimentation	320	50

Table 4.2 Costs of algal harvesting processes (1989 prices)

Π In algal pond cultures cell concentrations of 2 g dry wt l^{-1} are possible, corresponding to outputs of 15 g dry wt per square metre per day. What would the protein output of such a system be (as kg protein per hectare per year), assuming the dry algal biomass contains 45% protein? (Use a piece of paper to do this calculation and then compare it with ours).

15 g dry wt $m^{-2}d^{-1}$ corresponds to $15 \times 10^4 = 1.5 \times 10^5$ g $ha^{-1}d^{-1}$.

The annual output is thus $1.5 \times 10^5 \times 365 = 5.5 \times 10^7$ g ha^{-1} yr^{-1} or 5.5×10^4 kg $ha^{-1}yr^{-1}$.

At 45% protein this corresponds to $5.5 \times 10^4 \times 45/100 = 2.5 \times 10^4$ kg protein ha^{-1} yr^{-1}.

With increased CO_2 concentrations, by the injection of CO_2 gas into ponds, outputs of 3×10^4 kg dry protein per hectare per year are possible. This compares very favourable with conventional sources of protein (Table 4.3), note the low yields of meat and milk.

Sources	kg protein per hectare per year
Algal ponds	30,000
Farmed fish	1,000
Potatoes	800
Rice	600
Peanuts	450
Wheat	360
Milk	120
Meat	80

Table 4.3 Output of protein from algal ponds and conventional sources

tubular loop reactor

A novel fermentation system, a tubular loop reactor (Figure 4.3), has been developed at laboratory scale which is capable of producing algal cultures with densities of 20 g dry wt l^{-1}. This system converts 18% of incident solar energy, far in excess of the 7% in algal ponds and 1-2% in agriculture. Such a system gives a theoretical output in European climates of 100,000-150,000 kg protein ha^{-1} year $^{-1}$, and could be operated in arid regions as the water is conserved and can be recycled.

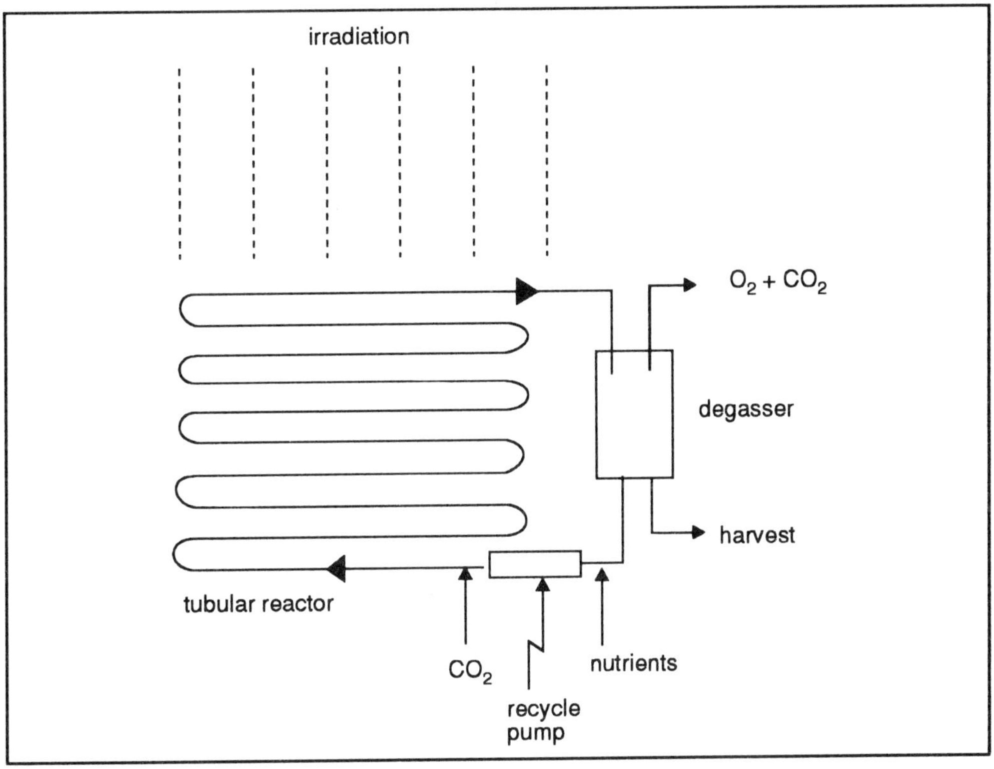

Figure 4.3 Tubular-loop reactor for the production of algal biomass

The photosynthetic production of *Chlorella* sp can be written as:

$$CO_2 + H_2O + NH_3 \xrightarrow{\text{light}} \text{Biomass} + O_2$$

high-rate algal ponds

Development work is being carried out growing algae for feed on municipal effluents and animal slurries (from intensive animal farms). This is carried out in high-rate algal ponds (shallow aerated lagoons operated at high dilution rates). Aerobic bacteria oxidise organic materials in the effluents, producing CO_2 which is used by the algae growing as photoautotrophs (using CO_2 and sunlight). The algae in turn produce O_2 which further stimulates the aerobic bacteria. Such systems are able to produce feed on the one hand and to reduce BOD, nitrate and phosphate (ie pollutants) from effluents on the other hand.

In Japan *Chlorella* spp has been produced for food in continuous aseptic systems in conventional bioreactors. The organisms are grown in the dark as heterotrophs using sucrose (in the form of molasses) or glucose as carbon and energy source. Production has been 2,000-3,000 tonnes per year at a selling price of \$(US)10-22 kg^{-1} (1990 prices). This product is sold as a high-value health food.

SAQ 4.3

Choose the correct completion to the following statement.

In the production of *Chlorella* spp for SCP from molasses, bioreactors rather than lagoons are used because:

1) molasses-based medium is darkly coloured and would not allow sunlight to penetrate lagoons;

2) the CO_2 required for growth can be more efficiently dissolved in fermentors;

3) lagoon systems containing molasses-based medium would become heavily contaminated by bacteria;

4) bioreactors give higher biomass yields than lagoons.

4.7 SCP from carbohydrates

Carbohydrates derived from plant biomass are plentiful and are renewable. They thus form an excellent substrate for SCP production by heterotrophs. Such carbohydrates can be in the form of sugars, which are readily used by microbial cells, or starch, pectin and cellulosic material, which require hydrolysis to sugars before they are available for cell uptake and use. Hydrolysis of starch can be carried out by chemical or enzymatic treatment to produce sugars for a fermentation medium, or by enzymes produced *in situ* by an amylase-producing organism growing in or on the starch substrate. Cellulosic material is solid and is normally used in solid-substrate fermentations. The estimated quantities of such materials available for SCP production (or other fermentation processes) are given in Table 4.4. Note the importance of grain-derived products.

Type	Source	Quantity (x 10^6 tonnes per year)
Sugars:	Molasses	9.3
	Whey	1.5 (USA)
	Sulphite waste liquor	12-2
	Glucose (from starch)	
	Wastes from fruit processing	
Starch:	Wastes from vegetable processing	
Cellulosics:	Bagasse (sugar cane fibre)	106
	Wheat bran	58
	Wheat straw	864
	Rice straw	599
	Corn (maize) wastes	193
	Urban refuse (paper/cardboard)	152
	Wood/forestry wastes	61

Table 4.4 Approximate world (or USA) availability of carbohydrate substrates

4.7.1 Sugars

yeasts

Yeasts are the product of many of the SCP processes based on sugars. This is because yeasts can use many different sugars, they have been used traditionally in foods and they have most of the desirable characteristics for SCP. Each process differs slightly according to the nature of the substrate and the organism used, but in principle all the systems are similar. The experience you gained in earlier sections should enable you to answer the following SAQ.

SAQ 4.4

The diagram below is a flow diagram of an SCP process from carbohydrate, with the letters A-K representing various operations, inputs and outputs. Using the list given, fit in appropriate operations, inputs or outputs at points A-K to complete the flow diagram.

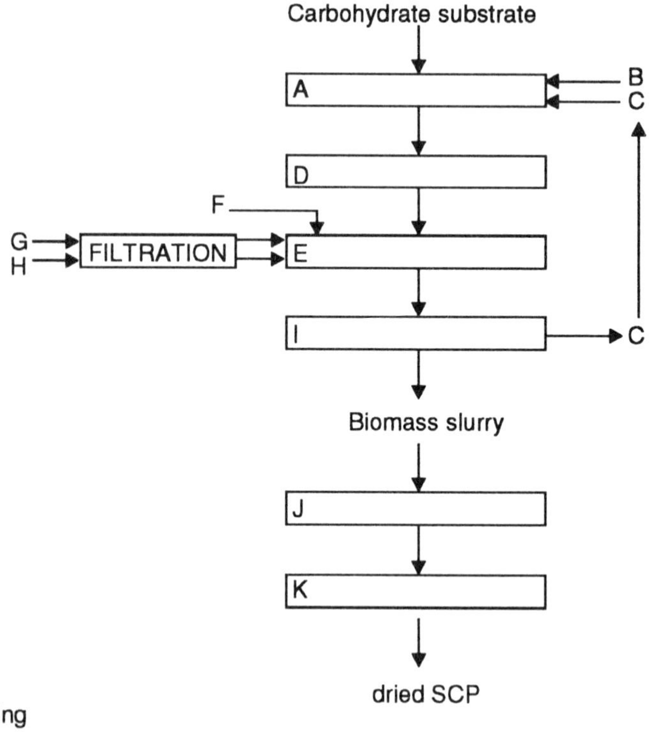

water
packaging
air
cell separation
medium blending
inoculum
continuous heat sterilisation
drying
NH_3
continuous fermentation
minerals (phosphate/sulphate)

molasses

Molasses is a by-product of sugar refining, mostly from beet or cane, and contains 35-50% sucrose and small quantities of nitrogen. It is used as sweeteners in the food industry and as a fermentation medium for the production of bakers yeast, ethanol and other products. For SCP production, the molasses is diluted to 4-6% sucrose, supplemented with phosphate and sterilised by continuous heat sterilisation.

continuous culture

Continuous processes are run in aerated fermentors with ammonia addition, producing food-grade *Candida utilis* and *Saccharomyces cerevisiae*. Systems operate at dilution rates 0.2-0.3 h^{-1} at pH 3.5-4.5 at 25-35°C. Yeasts are recovered by centrifugation, washed, dried by drum or spray drying and packaged. The product contains about 45% protein and is used as a high protein food supplement, particularly in Taiwan, South Africa and the former USSR.

Π Presented below are four flow diagrams representing processes for washing the concentrated yeast suspension (the suspension concentrated from the fermentation medium by centrifugation). Which of these processes do you think would be suitable for washing the yeast suspension?

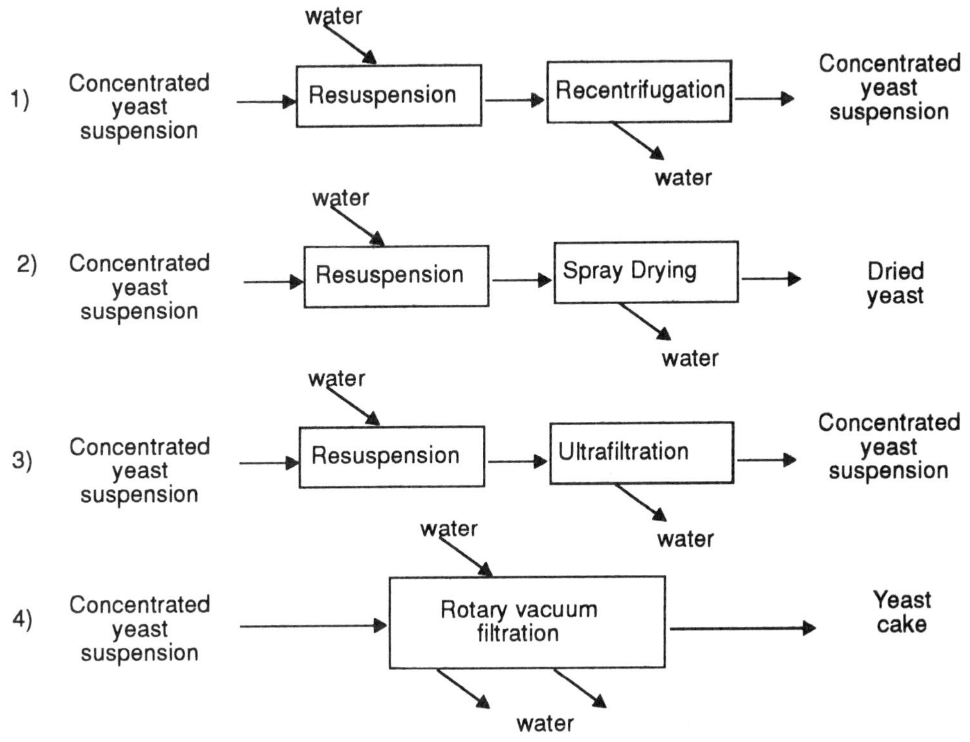

Methods 1 and 4 are appropriate methods of washing yeast cells. Yeast cells can be concentrated by centrifugation, so method 1 achieves washing. Method 4 makes use of the fact that concentrated yeast cells can be de-watered using a rotary vacuum filter which can incorporate washing. In order to filter yeast cells the filter would first have to be coated (pre-coated) with a layer of material such as starch granules, to decrease the effective size of the filter. Method 2 does not achieve washing, as medium components would be dried with the yeast and not removed in the water. Method 3 is not suitable for concentrating large volumes of yeast cells as ultrafilters would become blocked by dense yeast suspension.

rotary vacuum filter

whey

Whey is the effluent from cheese manufacture, and contains about 5% lactose and about 1% protein. About half of the global production is used as a feed supplement but the rest is unused. The BOD can be as high as 70,000 mg O_2 l^{-1}, which requires reduction by biological treatment (which is costly) prior to disposal. For SCP processes the valuable protein from the whey is first recovered by precipitation or ultrafiltration, and the deproteinised whey used in processes such as described for molasses. The yeasts used

deprotination

are food-grade organisms capable of using lactose as a carbon source, namely *Saccharomyces lactis*, *Candida utilis* and *Kluyveromyces* (previously *Saccharomyces*) *fragilis*. Feed-grade SCP is often prepared by spray drying the complete fermentation broth. A flow diagram of a process producing SCP from whey is shown in Figure 4.4.

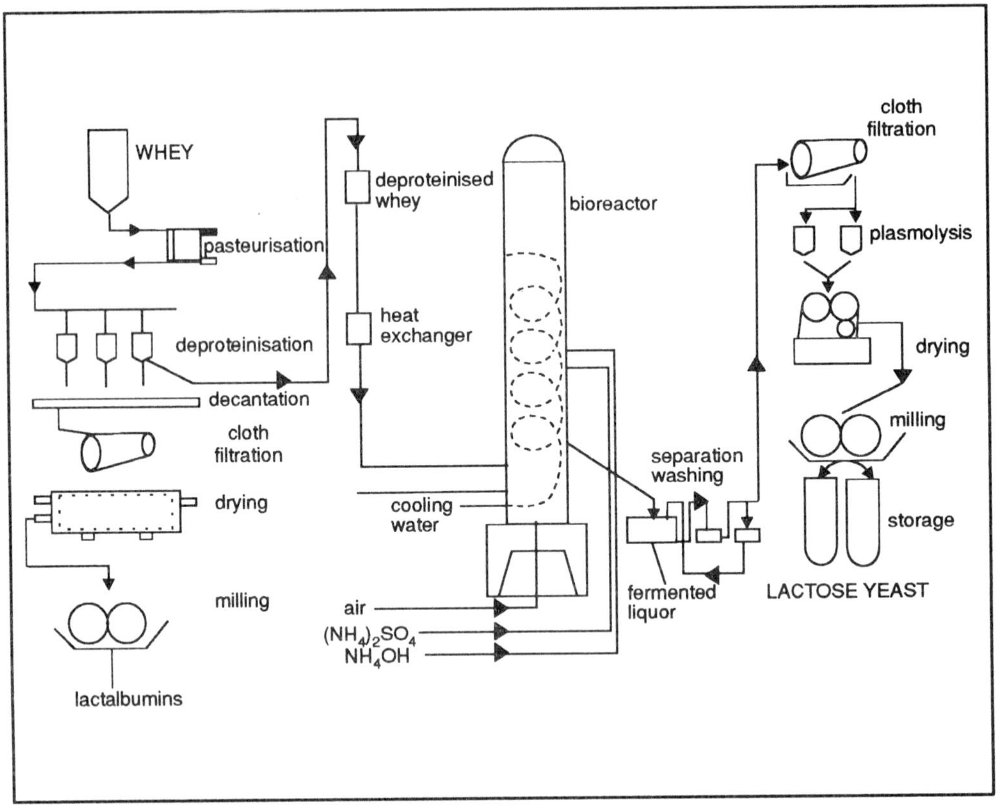

Figure 4.4 The production of whey protein and *Kluyveromyces lactis* from whey.

sulphite waste liquor

Sulphite waste liquor is a waste product of the sulphite wood pulping process. This process has now been replaced in many countries and the availability of sulphite waste liquor is now less than that indicated in Table 4.4. The liquor contains about 20% w/v sugars, in the form of both hexoses (6-carbon sugars) and pentoses (5-carbon sugars), and 6% w/v acetic acid and has a BOD of up to 50,000 mg $O_2\ l^{-1}$. Processes have been operated growing *Candida utilis* for food or feed but the most effective has produced feed from *Paecilomyces varioti* (the Pekilo process). This organism more effectively uses pentose sugars and the acetic acid present in the substrate compared to *Candida* spp, and so gives higher output and greater BOD reduction. The organism grows as short filaments (making cell recovery less costly) and has a protein content of 55-60%.

Any similar carbohydrate source from food processing (such as coconut milk or effluents from fruit canning) can be treated in this way, and there are many examples of development studies producing SCP from them.

Π The conversion of carbohydrate to biomass can be represented as follows:

$$C_6H_{12}O_6 + O_2 + NH_3 \longrightarrow Biomass + CO_2 + H_2O$$

Studies show that the production of 1kg dry biomass requires 2.0 kg sugar, 0.7 kg oxygen, 0.1 kg ammonia, with the liberation of 12,300 k Joules heat. A typical continuous fermentation operates at a dilution rate (D) = 0.2 h^{-1}, with sugar concentration of 3% (w/v) in the incoming medium. With a fermentor of 50 m^3 capacity and 90% utilisation of carbohydrate [ie 0.3% (w/v) sugar in the outgoing medium] what would be:

1) the biomass concentration (quantity biomass m^{-3});

2) the productivity (quantity biomass $m^{-3}h^{-1}$);

3) the output rate (quantity biomass from the system h^{-1});

4) the minimum oxygen-transfer rate (OTR) (quantity O_2 transferred $m^{-3}h^{-1}$);

5) the heat evolution rate (quantity of heat $m^{-3}h^{-1}$)?

Again we urge you to do the calculations on a piece of paper before checking your answer with ours.

1) Biomass concentration. 1 kg biomass is produced from 2 kg sugar, so the growth yield is 1/2 = 0.5 The incoming medium contains 3% sugar (30 g l^{-1} or 30 kg m^{-3}) and the outgoing medium 0.3% sugar (3 g l^{-1} or 3 kg m^{-3}). Sugar utilisation is thus 30-3 = 27 kg m^{-3}. With a yield factor of 0.5 the biomass concentration would be 27 x 0.5 = 13.5 kg m^{-3}.

2) Productivity. The continuous system operates such that the biomass concentration remains constant (specific growth rate = dilution rate). Productivity is therefore given by the biomass concentration x the dilution rate.

   The dilution rate is 0.2 h^{-1}, so the effect of dilution or growth is 13.5 x 0.2 = 2.7 kg $m^{-3}h^{-1}$. The productivity is therefore 2.7 kg m^{-3} h^{-1}.

3) Output. Output = productivity x reactor volume = 2.7 x 50 = 135 kg h^{-1}.

4) Minimum OTR. The production of 1 kg biomass requires 0.7 kg O_2. The productivity is 2.7 kg $m^{-3}h^{-1}$ (from 2 above). 2.7 kg biomass thus requires 2.7 x 0.7 = 1.89 kg O_2. The minimum OTR is thus 1.89 kg O_2 $m^{-3}h^{-1}$.

5) Heat Evolution. The production of 1kg biomass liberates 12,300 k Joules. The productivity is 2.7 kg $m^{-3}h^{-1}$. 2.7 kg biomass involves the liberation of 2.7 x 12,3000 = 33,210 k Joules. The heat evolution is therefore 33,210 k Joules $m^{-3}h^{-1}$.

   If you were unable to do these calculations, we would suggest that you read the BIOTOL text entitled '*In vitro* Cultivation of Micro-organisms' or other suitable texts of the fundamental principles of cell cultivation.

Π For SCP production from carbohydrates, aerobic rather than anaerobic growth is preferred, as aerobic growth leads to higher biomass yield coefficients (kg biomass produced per kg substrate used). Choose the correct reason for such higher yield coefficient from the following.

1) Anaerobic utilisation of carbohydrates is strongly inhibited by the end product ethanol, giving low yield coefficients compared to aerobic utilisation.

2) Aerobic growth utilises the glycolic pathway (for converting carbohydrate to pyruvate) yielding more energy than anaerobic growth, which utilises the pentose phosphate pathway, thus giving higher yield coefficients.

3) Aerobic growth (converting pyruvate to CO_2 and H_2O) yields more energy (ATP) than anaerobic growth (converting pyruvate to ethanol and CO_2), thus giving higher yield coefficients.

4) Aeration required for aerobic growth removes ethanol (produced by yeast growth) from the medium, keeping ethanol levels below inhibitory concentrations, thus giving a higher yield.

Response:

1) Incorrect - Ethanol can inhibit the growth of yeast at high concentrations (about 7% v/v). This could lead to lower productivity and output, and lead to higher residual sugar concentrations (lower degree of conversion). However this would not influence yield coefficients, which relate to the amount of substrate actually used.

2) Incorrect - The two pathways are alternative pathways for converting carbohydrate to pyruvate, and are found in various organisms. They operate in aerobic or anaerobic conditions.

3) Correct - Aerobic growth is more energy efficient (produces more ATP) than anaerobic growth, resulting in higher yield coefficients.

4) Incorrect - Aerated cultures do not produce alcohol. Even if they did, point 1) above applies, and aeration would not necessarily reduce ethanol levels significantly.

If you were unable to work out the answer to this it probably means you need to refresh knowledge of cell metabolism. BIOTOL texts on cell metabolism or other good introductory biochemistry text should prove adequate.

Two types of bioreactors have been used to produce SCP from sugar sources, namely stirred/baffled type and the air-lift with draught tube type (Figure 4.5), with various capacities up to 400 m^3.

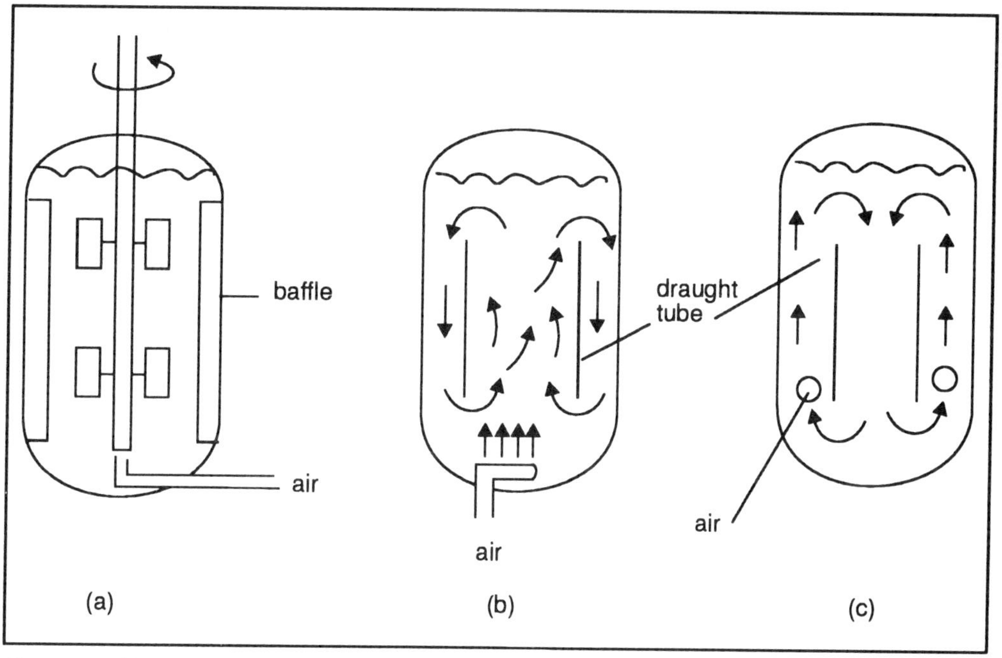

Figure 4.5 Fermentor types used in the production of SCP from sugar sources, (a) Stirred baffled system, (b) and (c) air lift systems

myco-protein

The patented 'Myco-protein' process, operating in the UK involves growing the filamentous fungus *Fusarium graminearum* on food-grade glucose syrup, produced by the enzymatic hydrolysis of wheat starch, a byproduct of wheat gluten production. The medium is composed of diluted glucose syrup, ammonia, mineral salts, choline and biotin. Ammonia gas is added to continuous aseptic systems via the air inlet and serves as a nitrogen source and to maintain pH at 9.0. The system used to develop the process was an air-lift fermentor of 36 m^3 capacity. For full-sale production a pressure-cycle bioreactor of larger volume is used. This technology was developed for SCP production from methanol (see Section 4.9) and was acquired under licence for the Myco-protein process. Operation is at dilution rates of about 0.2 h^{-1} with dry weights of 15-17 $g\ l^{-1}$. These conditions are selected partly to enhance the filamentous (fibrous) structure of the biomass. After concentration from the medium, by separation in a cyclone (whirlpool device), the cell suspension is heated to 64°C for about 20 minutes, during which time heat-stable RNAses degrade cellular RNA, the components of which are released from the cells. This reduces the RNA content from 10% to 2%. The heating process in addition inactivates proteases, which makes the produce more stable during storage. Biomass recovery is by filtration (de-watering) by horizontal belt vacuum filtration. The recovered biomass is usually blast chilled for storage by deep freezing, although it can be dried for storage. The product is given the trade name Myco-protein. This product contains 45% protein on a dry weight basis.

reduction of RNA content

SAQ 4.5

Use the description given above to draw a flow diagram to show the production of Myco-protein.

changing market circumstances

When the Myco-protein project was planned in the 1960s it was assumed the product would form a valuable protein source, needed to meet expected protein shortages. By the early 1980s, when the project had reached production, the protein shortage had not developed (at least in Europe). However, dietary habits had changed, and there was a move away from meat consumption to other 'healthier' foods. There was also a move towards 'convenience' foods such as deep-frozen prepared meals. It was to these markets that Myco-protein was directed.

diversification of product

The de-watered fungal biomass has a natural fibrous texture (being composed of fungal hyphae). The hyphae can be partially aligned by rolling the biomass between rollers. If the biomass is premixed with egg albumin, rolled, then steam-heated to set the albumin, a material with meat-like texture is produced. By altering the processing conditions, material with textures of various meats (such as beef, chicken or fish) can be produced. The product itself does not have a strong flavour and so will take up the flavour of dishes to which it is added.

nutritional quality

safety checks

Myco-protein was successfully tested for nutritional quality and safety in feeding trials in animals over a ten-year period. This included trials on human volunteers, carried out in the UK and USA. In 1985 the product was approved by the Ministry of Agriculture Fisheries and Food for sale as food in the UK. The product is marketed under the tradename Quorn, and is promoted on the basis of having satisfying meat-like texture with high protein and dietary fibre content while being without animal fat, low in calories and suitable for vegetarians. The nutritional quality of Quorn, compared to conventional protein sources is given in the Tables 4.5 and 4.6.

	Cheddar cheese	**Raw chicken**	**Raw lean beef**	**Stew steak**	**Fresh fish (cod)**	**Raw pork (leg)**	**Raw beef sausage**	**Quorn**
Protein (% w/w)	26.0	20.5	20.3	20.2	17.4	16.6	9.6	12.2
Dietary fibre (% w/w)	0	0	0	0	0	0	0	5.1
Fats (% w/w)	33.5	4.3	4.6	10.6	0.7	22.5	24.1	2.9
Ratio polyunsaturated fatty acids: saturated fatty acid	0.2	0.5	0.1	0.1	2.2	0.2	0.1	2.5
Cholesterol (mg/100 g)	70	69	59	65	50	72	40	0
Energy (k Joules/100g)	1665	496	504	721	311	1103	1226	328

Table 4.5 The nutritional quality of Quorn and some conventional foods

	Ideal Diet (WHO)	Average Diet (UK)	Quorn	Beef	Chicken	Soya
Isoleucine	4.0	4.7	4.4	5.0	4.6	4.5
Leucine	7.0	8.0	7.3	7.7	7.5	7.1
Lysine	5.5	6.3	6.8	8.8	9.0	7.4
Methionine and cysteine	3.5	3.8	2.6	3.9	3.7	3.0
Phenylalanine and tyrosine	6.0	8.2	7.9	8.3	8.0	8.0
Threonine	4.0	4.1	4.9	4.3	4.1	4.0
Tryptophan	1.0	1.3	1.4	1.3	1.1	1.3
Valine	5.0	5.6	5.3	5.1	4.8	5.0

Table 4.6 Essential amino acids content of a variety of protein sources

Π Does Quorn have a high ratio of polyunsaturated to saturated fatty acids compared with meat products? How does its cholesterol and essential amino acid content compare with other sources?

Quorn has a higher ratio of polyunsaturated to saturated fatty acids than the other foods listed. It contains no cholesterol. The amino acid profile is similar to that of other protein foods. It is slightly lower in methionine and cysteine content.

4.7.2 Starches

symbiotic culture

Effluents from the processing of starchy vegetables such as potatoes, cassava, rice or corn (maize) have been the substrates for commercially operated SCP processes. The patented Symba process, developed in Sweden, is a two-stage continuous process based on a symbiotic culture of the yeasts *Endomycopsis fibuligena* (which is amylolytic but of low value as SCP) and *Candida utilis*. The effluent is supplemented with ammonia, sterilised and fed to the first bioreactor containing *E. fibuligera*. This organism secretes amylase which hydrolyses the starch. The broth feeds a second bioreactor inoculated with *Candida utilis*, which grows faster and predominates. The process reduces the BOD of the effluent by 90%, to a level of 1000 mg O_2 per litre.

Development work has been carried out on processes to produce feed from starch effluents in non-aseptic waste treatment systems (such as oxidation ditches and aeration ponds) using amylolytic filamentous fungi. These often belonging to the genera *Aspergillus* and *Rhizopus*. In such processes, the levels of contaminating bacteria are depressed by operating the systems at low pH.

solid-substrate fermentations

Such fungi have also been used in development studies on increasing the protein content of solid starch wastes (vegetable rind or peel) or conventional feed (grains) by using solid-substrate fermentations. This involves chopping up the substrate, if necessary, and steaming it for about 1 hour (for hydration, and pasteurisation to reduce numbers of competing micro-organisms). After steaming, the substrate is cooled and the moisture content is adjusted, if necessary, to 50-75%. Ammonium salt and other necessary minerals are added, followed by inoculation and incubation. This type of process can be carried out on a small scale with manual aeration (by 'turning the pile'). On a larger scale, inoculated material can be incubated in tanks or bins with mechanical raking and forced aeration, or laid out in rows and turned mechanically. Once fungal growth is complete (after 24-48 h), the material can be fed directly to animals or dried for storage. In this way starchy material low in protein can be converted into a more balanced feed material containing 10-15% protein. The product is called 'protein-enriched feed'. This technology has been developed with a view to utilising cellulosic wastes, which are produced on a much larger scale. This is discussed below.

protein-enriched feed

SAQ 4.6

Draw a flow diagram to show the production of protein-enriched feed from solid starch wastes.

4.7.3 Cellulosics

hydrolysis to constituent sugars

You have seen from Table 4.4 that cellulosic material is produced in vast quantities as a waste. If an economic process can be found to hydrolyse cellulosic material to its constituent sugars, by chemical and/or enzymic means, then vast quantities of sugar could be made available for fermentation processes, including SCP production. Despite long and hard research effort, success has been limited. However, projects have been developed to enhance protein content of cellulosic wastes by growing cellulolytic fungi (such as *Trichoderma* spp) on them in solid-substrate fermentations (using the technology described for solid starch wastes). In this way the low-protein substrate (for example wheat straw or rice straw) can be converted into protein-enriched feed, with a protein content of up of 15%. This can be used as feed for ruminant animals. However, for high conversion rates, cellulosic material requires expensive physical or chemical pre-treatment. As yet no large-scale commercial processes have been developed.

Π What problems do you think might be caused by unwanted contaminating fungi in solid substrate fermentation processes for protein-enriched feed?

The cells of many fungi are toxic if eaten. In addition several fungi can produce toxins (called mycotoxins) which cause food poisoning to man and animals (for example *Aspergillus flavus* produces a toxin called aflatoxin). Other fungi can produce spores which can cause respiratory disorders in man and animals (for example *Asperigillus fumigatus*). These problems are minimised by using a heavy inoculum and so reducing the incubation period.

Processes have been developed in North America to pilot scale growing the bacterium *Cellulomonas* or the fungus *Trichoderma* on pre-treated milled cellulosic material in conventional fermentors, ie in a liquid medium. However, preparation costs are considerably higher than with solid-substrate fermentations.

4.8 SCP from hydrocarbons and derivatives

In countries fortunate enough to have oil and gas deposits, or having access to such deposits, hydrocarbons are available as SCP substrates. This is particularly true of non-industrialised countries in warm climates where hydrocarbons have limited uses as fuels or petrochemicals.

4.8.1 *n*-alkanes

n-alkanes

Processes have been developed to production scale growing yeasts, for feed, on purified C_{10} -C_{23} *n*-alkanes (such *n*-paraffins being liquid at normal ambient temperatures). A flow diagram of the process producing *Candida lipolytica* is given in Figure 4.6.

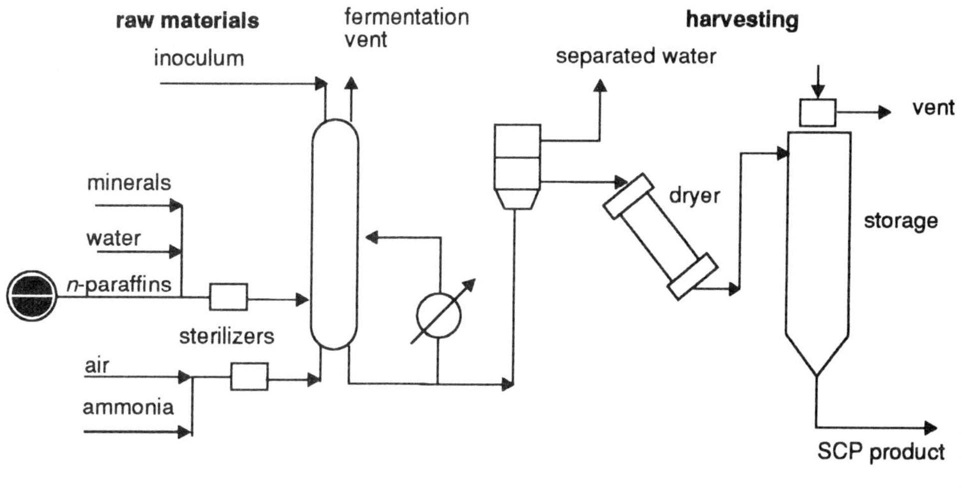

Figure 4.6 The production of *Candida lipolytica* from *n*-alkanes

You should now be completely familiar with this type of flow diagram as it is similar to that shown for the production of SCP from whey. However, the characteristics of the fermentation differ in several respects.

low solubility

n-Alkanes are only slightly soluble in water and form droplets in stirred aerated cultures. Direct contact with droplets is necessary if yeasts are to use the substrate, so steps are taken to maximise the surface:volume ratio of droplets. This is achieved by controlling agitation and also sometimes by the addition of surfactants (detergents) to produce droplets of 1-100 μm diameter.

oxygen transfer rate

Compared to carbohydrates, hydrocarbons are more highly reduced, and so require more oxygen for their oxidative degradation. To produce 1 kg biomass from *n*-alkanes requires 1-1.2 kg *n*-alkanes and 2.2 kg O_2. This requires high oxygen-transfer rates (OTRs) by (expensive) vigorous agitation and use of over-pressures (operating fermentors at positive pressures to enhance gas dissolution). This adds significantly to the cost of the product. To produce high OTRs, processes have made use of fermentor designs other than conventional stirred or air-lift systems, such as loop and tubular loop fermentors, and various stirring devices incorporating air-sparing (Figure 4.7).

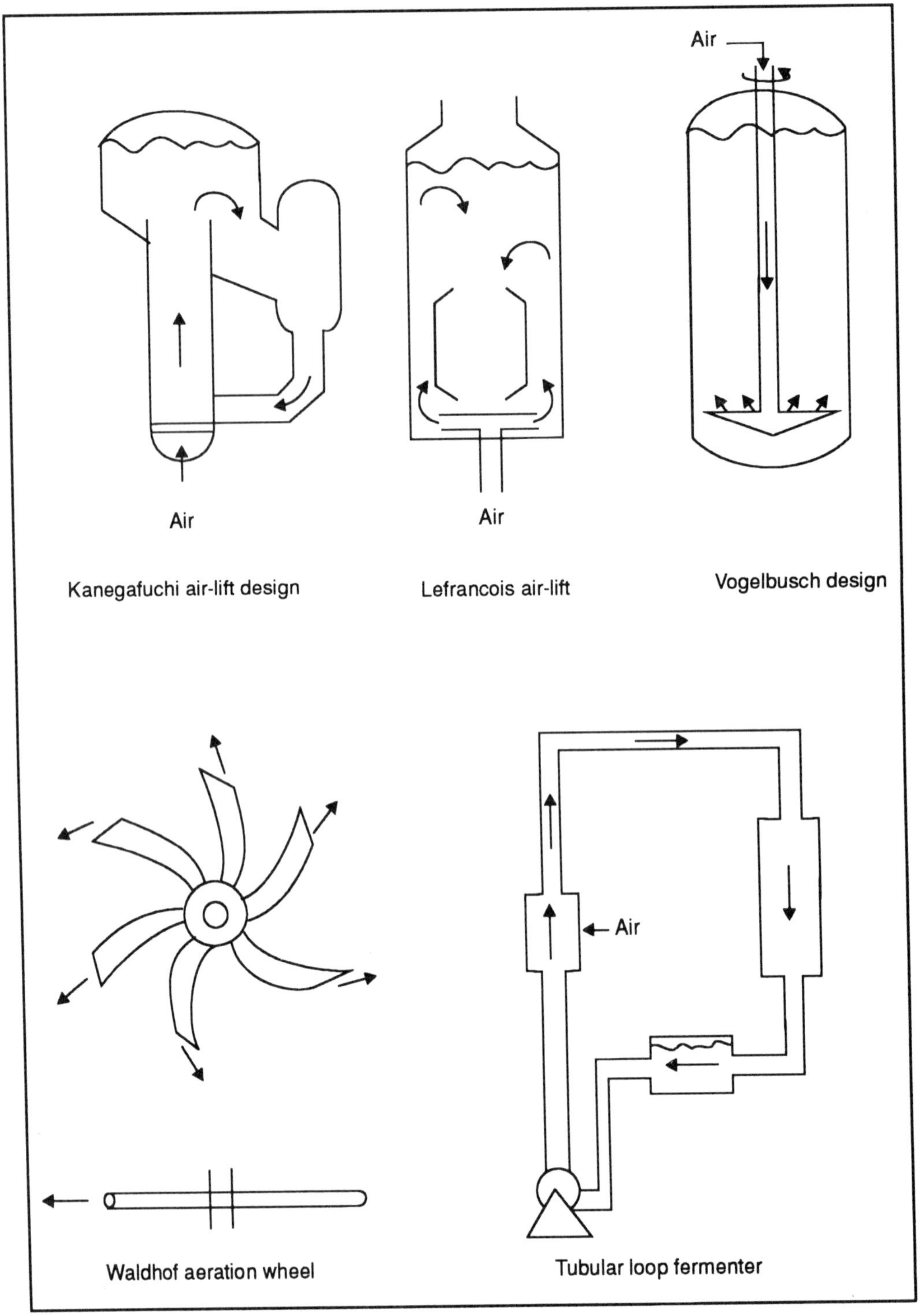

Figure 4.7 fermentor types used in the production of SCP from *n*-paraffins

heat generation

Oxidation of n-alkanes is strongly exothermic, the production of 1 kg biomass liberating 27,100 k Joules. Cooling costs to maintain temperatures at about 30°C are considerable.

SAQ 4.7

Compare a) the minimum OTR and b) heat evolution rate of a continuous fermentation system based on n-alkanes operating at a dilution rate of 0.2 h^{-1} and a biomass concentration of 13.5 kg m^{-3}, to a similar system based on carbohydrate. You may need to look back a few pages to get the relevant information concerning carbohydrate utilisation (Section 4.7).

substrate costs

The process illustrated in Figure 4.6 was developed to production scale with a capacity of 200,000 tonnes per year. This process, developed by British Petroleum, was one of several in Europe and Japan that, although fully developed, was never operated commercially. This was due to sharply increased substrate costs in 1973 and political and social pressures against the use of petroleum-based substrates (possibly contaminated with carcinogenic or toxic compounds). Such systems do operate in the former USSR, producing *Candida guilliermondii* as feed.

SAQ 4.8

Various wastes available as carbon substrates for SCP productions are listed in column A. Match each waste with a suitable organism (perhaps more than one) from column B. For each organism select the most appropriate production system from column C.

A	B	C
i) Waste paper	i) *Candida utilis*	i) Aqueous culture in open ponds or lagoons
ii) Exhaust gas emissions from a coal-fired power station	ii) *Chlorella regularis*	ii) Solid-substrate fermentation
iii) Molasses	iii) *Aspergillus niger*	iii) Aqueous culture in bioreactors
iv) Effluent from fruit and vegetable processing factory containing 0.5% sugar and 0.5% starch	iv) *Trichoderma viride*	

4.8.2 Ethanol

Ethanol can be produced relatively cheaply as a bulk petrochemical by the hydration to ethylene. A few processes operate in the USA using ethanol to produce food-grade *Candida utilis*, with capacity of about 7,000 tonnes per year. The technology is similar to that already described for SCP from sugars. On ethanol, the yield is about 0.7 kg dry wt per kg ethanol used.

4.8.3 Methanol

Methanol can be produced relatively cheaply as a bulk chemical by the oxidation of methane. Several processes have been developed to produce feed-grade SCP using methanol as a substrate. We will now examine one such process in depth, to show how a process is developed, from inception to production scale, and how the many problems encountered can be tackled and overcome.

4.9 The Pruteen process - a case study

4.9.1 The planning stage

historical perspective

In order to fully appreciate the driving forces behind the development of SCP projects you must imagine you have been transported back in time to the 1960s. It was then that the projects we have been looking at, and the one we are now going to look at in more detail, were planned. The world's population is booming and it is expected that a severe world protein shortage will develop before the year 2,000. European economies are recovering from the effects of war, and increasing prosperity is leading to increasing consumption of meat, which had been rationed throughout much of Europe during the war years and for some time after. Meat rationing is therefore still very much in peoples memories. Europe has never been self sufficient in animal feed, and relies on imports of soya from the USA and fish meal from South America.

This, then, was the scene in which SCP projects were planned in Europe. The need for alternative foods and feedstuffs was clear and, in the UK and elsewhere, oil and gas seemed a plentiful and cheap resource from which to produce them. North Sea gas fields were being exploited and research had shown that natural gas or its derivatives could be used to produce SCP feed of superior protein content to conventional feedstuffs. The economics of such processes seemed very promising. In 1971 the European prices for fish meal and soya meal were $200 and $100 per ton respectively. In 1973 oil price rises and a failure in the Peruvian fish meal supply pushed these prices up to $550 and $300 respectively. With such prices for the major feedstuffs it was considered that SCP feeds could be produced competitively.

4.9.2 Methane as a substrate

In the late 1960s, Imperial Chemical Industries (ICI) in the UK were interested in developing an SCP process using abundant and cheap methane from newly developed sources in the North Sea. However, it soon became apparent that methane was unsuitable as a substrate for fermentation.

Π Answer true or false to each of the following statements. Methane is unsuitable as a carbon source for SCP production because:

1) micro-organisms have not been isolated that are capable of using it as a sole source of carbon and energy;

2) at normal temperatures and pressures it has low solubility in water, limited productivity and output;

3) it is potentially explosive when mixed with air;

4) it is a gas, relatively difficult to store and transport;

5) it is difficult to obtain in a pure form;

6) it is toxic to micro-organisms at high concentration.

1) False. The ability to utilise methane as a sole source of carbon and energy (methanotrophy) is found in several genera of bacteria, such as *Methylomonas*, *Pseudomonas* and *Methylococcus*.

2) True. Even at high overpressures the low solubility of methane limits productivity and output.

3) True. The explosive nature of methane:air mixture requires careful operation of large-scale systems and incurs high insurance premiums.

4) True. Methane is expensive to liquefy and, as a gas, is more difficult to store and transport than a liquid.

5) False. Methane, as a gas, is much easier to obtain in pure form from oil and gas deposits than, of example, *n*-alkanes.

6) False. Methane is non-toxic to micro-organisms.

biomass yields

Although from theoretical considerations biomass yields from methane could be as high as 1.4, in laboratory-scale cultures values of about 1.0 were obtained, and in larger scale systems values were around 0.3-0.6. Methane fermentation also incurs high aeration and cooling costs.

biochemical pathway

The biochemistry of methanotrophic bacteria was studied to show how energy, reducing equivalents and C_3 skeletons were produced from the C_1 substrate. Methane is oxidised to methanol then formaldehyde. Formaldehyde can be assimilated, by various mechanisms, with C_5 sugars, to form C_6 sugars. The C_6 sugars are either assimilated, recycled to C_5 sugars or converted to C_3 skeletons such as pyruvate or glyceraldehyde (see Figure 4.9, section 4.9.6).

methane mono-oxygenase

The lower than expected yields can be explained by the nature of methane oxidation to methanol in these bacteria. This reaction, catalysed by methane mono-oxygenase, is a net consumer of reducing equivalents (NADH), which would otherwise be directed to ATP generation and biosynthesis. In simple terms the oxidation of methane to methanol consumes energy, lowering the yield.

Π From the above discussion, what do you think could be used instead of methane as a substrate?

Formaldehyde is highly toxic and reactive and not suitable as a growth substrate. Methanol is more soluble in water than methane and would seem a suitable substrate.

4.9.3 Stage one - methanol as a substrate

Methane can be chemically oxidised to methanol relatively cheaply. The heat liberated during such oxidation is higher grade than that liberated by biological conversion, and is thus easier to recover and use elsewhere. It was expected that using methanol as a fermentation substrate would, having by-passed the energy-inefficient methane oxidation step, lead to higher yield and lower the oxygen and cooling requirement. Methanol has the added advantage of being very much more soluble in water and easier to handle than methane.

methylotrophs

At that time few micro-organisms capable of using methanol as sole source of carbon and energy (methylotrophs) had been isolated, and so steps were taken to isolate such organisms from samples of soil, water and vegetation.

Π Protocols for isolating methylotrophic micro-organisms by enrichment from soil, with a view to SCP production, are given below. Choose the most appropriate one.

1) Add 10 g soil to 100ml liquid Medium A. Incubate with shaking in darkness for 24h at 22°C. Spread 10 x 0.1 ml volumes on agar plates of Medium A and incubate at 22°C until colonies appear.

2) Add 10 g soil to 100ml buffered saline. Shake vigorously for 10 minutes. Spread 10 x 0.1 ml volumes on agar plates of Medium A and incubate at 22°C until colonies appear.

3) Add 10 g soil to 100 ml liquid Medium B. Incubate with shaking in darkness for seven days at 22°C. Spread 10 x 0.1 ml volumes on agar plates of Medium B and incubate at 22°C until colonies appear.

4) Add 10 g soil to 100 ml liquid Medium C. Incubate with shaking in darkness for seven days at 22°C. Spread 10 x 0.1 ml volumes on agar plates of Medium C and incubate at 22°C until colonies appear.

Medium A: 0.5% w/v $(NH_4)_2SO_4$; 0.4% w/v NaH_2PO_4; 0.08% w/v $MgSO_4.7H_2O$; 0.01% w/v $FeSO_4$; 20.0% v/v methanol; pH 7.0

Medium B: 0.5% w/v $(NH_4)_2SO_4$; 0.4%W/V NaH_2PO_4; 0.08% w/v $MgSO_4.7H_2O$; 0.01 w/v $FeSO_4$; 1.0% methanol v/v; pH 7.0

Medium C: 0.5% w/v $(NH_4)_2SO_4$; 0.4% w/v NaH_2PO_4; 0.3% v/v methanol; 0.3%w/v yeast extract; pH 7.0

All media contain purified agar (solidifying agent) lacking organic nutrients

Response:

1) Not appropriate. Medium A contains 20% methanol which is likely to be inhibitory or lethal to most organisms. 24 hours is also a short time to allow for enrichment to occur.

2) Not appropriate. No enrichment occurs here, so the method is unlikely to pick up the very small numbers of organisms likely to be present in the soil. Medium A is also used, so response 1 also applies.

3) Appropriate. The medium contains a suitable methanol concentration and allows a suitable time for enrichment to occur. Medium B is elective for methylotrophs, that is, it selectively stimulates the growth of this type of organism (in dark conditions). This is the basis of the process of enrichment, in which the small numbers of a particular type of organism are stimulated to increase to a point where isolation is easier.

4) Not appropriate. Medium C would allow the growth and isolation of heterotrophs (due to the inclusion of yeast extract) and is thus not elective for methylotrophs. Also for an industrial process an organism that does not require growth factors (such as vitamins) is preferable. Medium C might well enrich for methylotrophs requring such expensive growth factors.

In practice, procedures such as described in 3) would be carried out over different periods of time, and at varying pH and temperature. Environments most expected to contain methylotrophs are swampy areas, where methane would be produced and then be oxidised to methanol. The organism finally chosen for the process we are considering here was isolated from a waterlogged soccer pitch!

4.9.4 Stage Two - Lab Scale

shake flask cultures

Once methanol-using organisms had been isolated they were screened in small-volume shake-flask cultures to determine their ability to grow in methanol-minimal-medium (such as Medium B described in the previous section) to produce high yields at high growth rates. Optimum growth temperatures and pHs were also determined.

Those organisms growing with high rates and yields and at relatively high temperatures (> 30°C) were selected for further study (growth at elevated temperatures would reduce cooling requirements). Studies on selected organisms were carried out in 5 litre continuous cultures to determine biomass yield coefficients, maximum growth rates (μ_{max}), affinities (Ks) for methanol and oxygen, and stability (of morphology and culture characteristics) on prolonged culture. Suitable cultures were identified.

Π Answer true or false to each of the following statements.

batch and continuous cultures

For SCP production continuous rather than batch cultures are preferred because:

1) continuous culture systems give higher outputs for a bioreactor of given size;

2) continuous culture systems are easier to operate;

3) continuous culture systems give more control over the culture environment and hence biomass quality;

4) continuous cultures are less susceptible to mutation;

5) continuous cultures convert higher proportions of substrate.

Our response:

1) True. Batch cultures give lower overall outputs than continuous cultures, as they suffer from non-productive down-time (the time taken to empty, clean, re-sterilise and re-fill the fermentor). After inoculation, considerable time can be taken for biomass to build up to a level where substrates are effectively utilised. Continuous cultures do not suffer such drawbacks once they are in operation.

2) False. Continuous culture systems are more difficult to operate than batch cultures. Medium is continuously added and withdrawn, making the process more prone to contamination. Maintenance cannot be carried out during lengthy culture runs, making the equipment more prone to breakdown. Batch cultures do not suffer such drawbacks.

3) True. In batch cultures, at the end of the growth phase when biomass concentration is high, the environment changes rapidly (as nutrients are becoming rapidly depleted) and growth rate decreases. Environment and growth rate can affect properties such as the protein content of cells, and thus biomass quality can vary from batch to batch. In continuous cultures the environment is constant. The specific growth rate (μ) is fixed by the dilution rate (D) so the product is less subject to variation. In fact at steady state $\mu = D$.

4) False. Continuous cultures operate for lengthy periods. Spontaneous mutations will arise and if they can compete successfully with the parent organism (by virtue of higher growth rate) they can predominate in the culture. Batch cultures have short growth times and so do not suffer such drawbacks.

5) False. Batch cultures can convert high proportions of substrates, as growth can be allowed to proceed until substrates are exhausted. In continuous cultures substrates are never fully converted, as medium is continuously removed. In fact, residual substrate concentration increases as the dilution rate increases, until virtually all of the medium remains unused. Continuous cultures usually recycle the medium after biomass removal to increase the efficiency of substrate conversion.

improved economics

Item 1 is the major factor in choosing continuous systems rather than batch systems for SCP production. Economics are improved by lower capital cost for the bioreactor (a major equipment cost, see Section 4.10) and by a higher output rate. Item 3 leads to greater control of product quality.

animal testing

Cultures with high affinities of O_2 and methanol were checked for amino acid profiles, protein and nucleic acid content, and lack of pathogenicity and toxicity in rats. Details of these toxicity tests are given in the resource material at the end of this chapter.

4.9.5 Stage three - 1 m^3 scale

obligate methylotroph

Methylophilus methylotrophus was the organism selected for further study. The organism is an obligate methylotroph, which means that it can use only C_1 organic molecules as a source of carbon. Suitable strains of organism were cultured in aseptic pilot-scale 1 m^3 continuous cultures to ensure stability at that scale in long-term cultures, and to produce material for extensive feeding and toxicity trials. It was important at this stage to ensure that there could be future flexibility in design and operation of a commercial-scale plant, so chemical composition, nutritional value and toxicological effects of the SCP product were examined with respect to variations in plant design and operation.

design flexibility

Π What aspects of plant design and operation would you have chosen to study at this stage in order to give maximum flexibility in design and operation of a future commercial-scale plant?

The factors chosen were variations in vessel size, the effect of non-aseptic operation and medium recycle, and variations in recovery and drying procedures.

In parallel with these studies, developments were underway to find the most economical method of large-scale culture and down-stream processing. The biochemistry of methanol and ammonia utilisation by *Methylophilus methylotrophus* was also studied to pinpoint possibilities for manipulation.

Π Why would you expect *Methylophilus methylotrophus* to be non-pathogenic?

As an obligate methylotroph, the organism would not be able to grow in animals, where substrate would not be available.

Preliminary trials showed that *Methylophilus methylotrophus* was a suitable source of feed. Changes in culture system, operation, mode of recovery and drying did not adversely affect this. The development team was confident that the pilot-scale and production-scale facilities could be developed effectively.

4.9.6 Stage four - 40 m^3 scale

mass transfer
heat transfer
mixing

It was expected that when scaling up the fermentation to about 40 m^3 the parameters most likely to be affected would be mass (oxygen) transfer, heat transfer and mixing. On the basis of the growth reaction, which can be written as

$$CH_3OH + NH_3 + O_2 \longrightarrow \text{biomass} + CO_2 + H_2O$$

and assuming conversion of methanol to either biomass or CO_2, the relationship between (i) yield and O_2 requirement, (ii) yield and minimum oxygen transfer rates (OTR), (iii) yield and heat production, and (iv) productivity and heat production (with various yields) can be estimated. You will find these data in the Resource Material at the end of this chapter.

Π The ideal requirements for the 40 m^3 system and projected larger production system are listed below. Study them all, then give reasons why you think each one was considered necessary.

1) Continuous operation.

2) Aseptic fermentation.

3) Optimum output.

4) An alternative to centrifugation for harvesting biomass.

5) Medium (water) recycle.

6) Aseptic harvesting procedure.

7) Appropriate drying procedure.

Our responses to these are:

1) If you cannot answer this question refer back to Section 4.9.4.

2) Medium recycle (item 5) means that organic materials released from *Methylophilus* can build up in the medium. These are available for the growth of contaminants. The process cannot be run at low pH to suppress the growth of contaminants as the organism itself requires a pH of about 7.0. Aseptic operating conditions are therefore required.

3) Continuous cultures can be run at different dilution rates. An optimum rate must be chosen, taking account of output, substrate utilisation and aeration and cooling requirement. We will work this out for ourselves later on.

4) Centrifugation is too expensive a method for the bulk recovery of bacterial cells, due to their low density.

5) Medium recycling is required to save water, save unused nutrients, and save on waste treatment costs (spent medium contains unused organic nutrients and so would otherwise have to be treated to reduce BOD).

6) Aseptic harvesting is necessary to overcome the need for medium re-sterilisation before recycling. Sterilisation costs are high. If biomass can be recovered by an aseptic process, the medium can be recycled without re-sterilisation. This excludes centrifugation, which cannot be operated under aseptic conditions.

7) Drying procedures need to be effective and economic.

The bioreactor

We will now deal with quite sophisticated issues concerning reactor design and performance. Although we deal with these in a sympathetic manner, you may find it useful to refresh your knowledge by reading more about bioreactors. The BIOTOL series offers opportunities to learn more about fermentors.

SAQ 4.9

With the aid of the data given in the Resource Material at the end of this chapter, answer the following. (Refer especially to Figures 4.10, 4.11, 4.12 and 4.13).

A 36 m^3 culture is grown at D = 0.3 h^{-1}, at a biomass concentration of 30 kg m^{-3} The biomass yield coefficient is 0.5 kg dry biomass per kg methanol. Estimate:

1) the output (kg biomass h^{-1});

2) the minimum OTR (kg O_2 required m^{-3} h^{-1});

3) the heat evolution rate (k Joules m^{-3} h^{-1});

4) the concentration of methanol required in the incoming medium to support a biomass concentration of 30 kg m^{-3} - assuming 90% utilisation at D = 0.3 h^{-1}.

air-lift pressure-cycle reactor

For a 40 m^3 pilot-scale fermentor an air-lift pressure-cycle bioreactor was chosen. The reactor (Figure 4.8) has a working volume of 36 m^3 with a 42 m high riser, fed with up to 80,000 m^3 air h^{-1} at pressures of about 3 x 10^5 Nm^{-2} (about 3 atmospheres). A maximum OTR of 10 kg O_2 m^{-3} h^{-1} is produced. The riser section is baffled at intervals to slow the circulation rate (and so enhance the OTR by increasing bubble residence time) and to promote bubble break up (again enhancing OTR). The rise of expanding bubbles is the sole source of energy input to the bioreactor, producing lower densities within the riser and thus causing liquid circulation down two downcomers. Bubbles coalesce in the top of the riser, their slip velocity (the rate at which they rise through the medium) thus increases and they disengage (break free of the liquid).

slip velocity

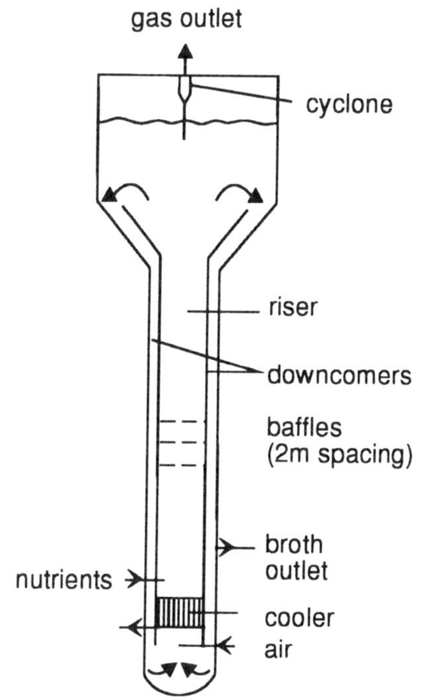

Figure 4.8 Pressure-cycle reactor.

Π Answer true or false to each of the following statements.

Compared to a conventional stirred bioreactor, an air-lift pressure-cycle bioreactor is more suitable for development of a SCP process from methanol because:

1) it is easier to construct and maintain at all scales;

2) the system gives higher OTRs and at greater energy efficiency;

3) it is easier to operate on a large scale because of defined and controllable flows;

4) it gives more even distribution of nutrients so that the culture is more likely to obey chemostat kinetics;

5) pilot and production-scale bioreactors of this type behave more like conventional stirred lab-scale bioreactors than equivalent-sized conventional bioreactors, thus making scale-up from lab-scale more reliable.

Our responses are:-

1) True. Stirred bioreactors require finely engineered aseptic seals on shaft bearings, which require maintenance. The larger the bioreactor the greater the problem. Air-lift bioreactors have no such moving parts.

2) True. Pressure-cycle systems need higher power input for air compression, but this is more than compensated for by saving on power for stirring. Air filtration costs are reduced because air is delivered at higher pressure, so actual volumes filtered are lower. Oxygen transfer is promoted by the high hydrostatic pressure generated at the base of the fermentor (at which point the air is introduced). Such systems give OTRs of up to 10 kg $O_2 m^{-3} h^{-1}$ for power inputs of 1-2 kW m^{-3}.

3) True. Pressure-cycle bioreactors have controllable and predictable flow patterns, which makes scale-up more predictable. Factors such as OTR and heat transfer are easier to arrange at large scales.

4) False. Pressure cycle bioreactors do not give even distribution of nutrients. If air and nutrients are introduced at a single point then O_2, CO_2 and nutrient concentration, as well as hydrostatic pressure, change in a cyclic manner as the medium flows around the reactor.

5) False. The behaviour of stirred bioreactors does not resemble closely pressure-cycle fermentors at any scale.

controlling bubble size

The characteristics of this pilot-scale system were extensively studied to discover how changes in medium viscosity, surface tension, flow rate and sparger and baffle design affected bubble size, slip velocity, disengagement and OTR. Bubble size and slip velocity proved to be crucial, as bubbles needed to be 1-10 mm in diameter for optimum gas exchange. Large bubbles (10 mm or more) produce inefficient gas exchange (with low surface:volume ratios and short residence times due to high slip velocities). Smaller bubbles (less than 1 mm diameter) fail to coalesce and disengage at the top of the riser.

baffles

Baffle design was adjusted to give appropriate bubble sizes with thorough lateral mixing, but avoiding high shear and high turbulence (which cause formation of bubbles of less than 1 mm). High rates of gas disengagement are achieved in an enlarged unbaffled section of the fermentor at the top of the riser.

Π How do you think that the enlarged unbaffled section at the top of the fermentor riser promotes bubble disengagement from the medium?

In the riser, baffles are placed at intervals to break up bubbles by increasing turbulence and shear. At the top of the riser the expanded section decreases the upward flow rate of the medium and this, together with the lack of baffles, decreases turbulence and shear, which in turn promotes coalescence of bubbles. Larger bubbles form which have increased slip velocity, so they more easily disengage from the medium.

Bubble behaviour was studied in the pilot-scale bioreactor so that a complete model of flow, OTR, mixing, cooling, energy requirement and disengagement could be developed for this system and larger production-scale vessels of similar type.

The yield problem

As well as understanding and optimising the working efficiency of the bioreactor it was also vital to optimise the efficiency of the fermentor - the micro-organism itself. The biochemistry of methanol utilisation is similar to that of methantrophs described in Section 4.9.2, and is shown in Figure 4.9.

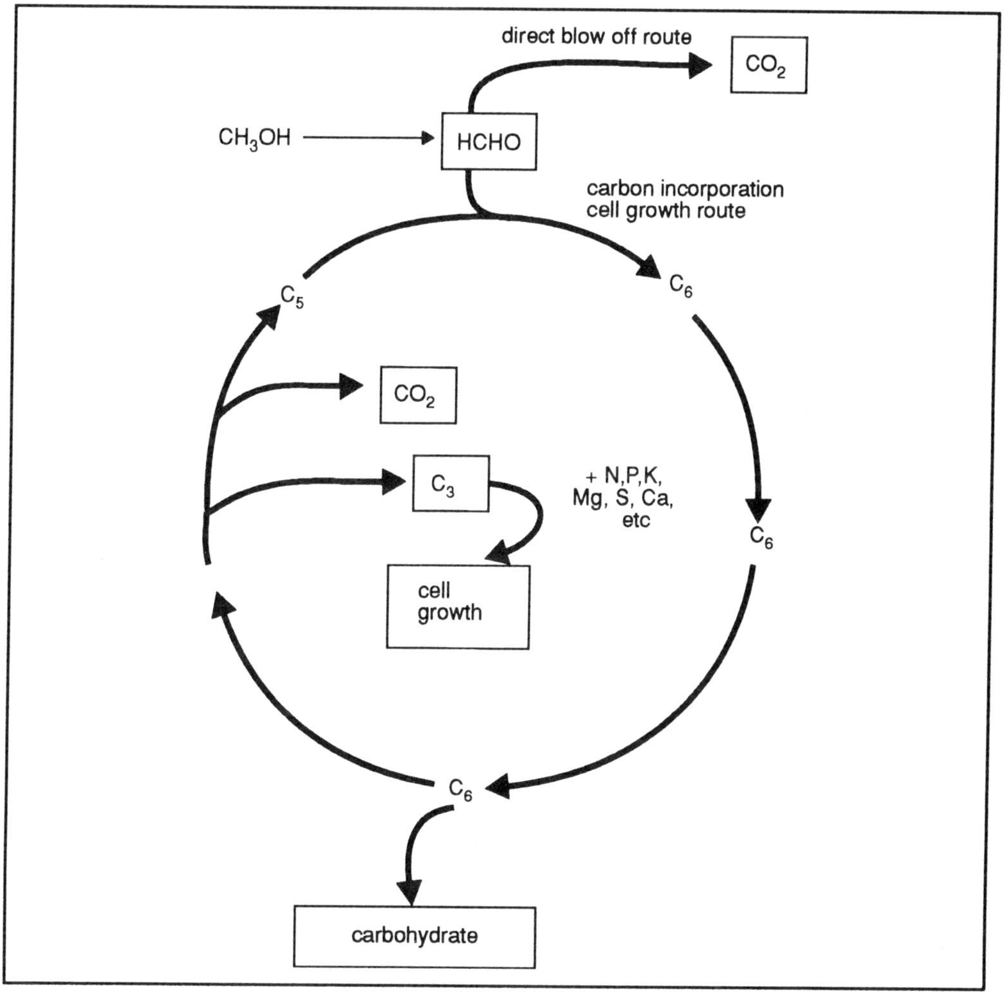

Figure 4.9 The utilisation of methanol as a carbon and energy substrate by methylotrophs.

Despite optimised culture conditions, yield coefficients for methanol were lower than the expected level of about 0.5. Yields were, in fact, lower than those achieved routinely with the $1m^3$ fermentation system. The problem was traced to the cyclical nature of the pressure-cycle system, which, due to introduction of air, ammonia, methanol and other nutrients at single points, leads to cyclical changes in nutrient concentrations as well as changes in pH, temperature and hydrostatic pressure. These parameters change in cycles as the medium flows around the reactor. Such changes do not occur in stirred vessels, accounting for differences in yield observed in comparison with those in the $1\ m^3$ system. Experiments showed that the cyclical changes in methanol concentration were the cause of reduced yields. It was assumed that this was due to an increase in maintenance energy. Maintenance energy is the energy needed to maintain cells. It is energy that is used for cell repair, maintaining osmatic balance, transport etc. It is not used for growth.

reduced yields

maintenance energy

Π How would you overcome the problem of low yield caused by cyclical changes in methanol concentration in the pressure-cycle bioreactor?

The solution was to introduce methanol into the medium at many points through nozzles in the bioreactor wall. In this way even methanol distribution was achieved, and yields of 0.5 kg biomass per kg methanol can be obtained.

genetic engineering

Genetic engineering was also carried out to improve yields even further. The production organism, strain AS1, assimilates ammonia according to the scheme:

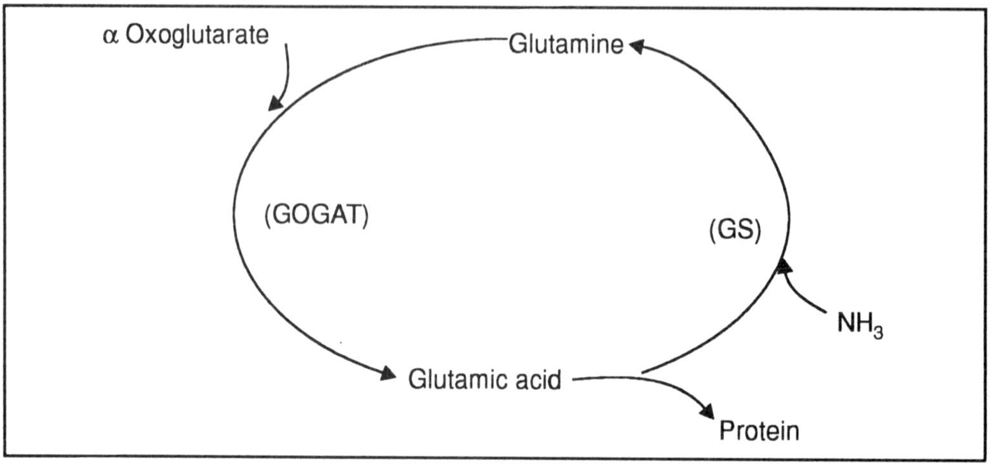

glutamate synthetase

Ammonia is accepted by glutamic acid in an energy (ATP consuming) step, and converted to glutamine. This reaction is catalysed by glutamate synthetase (GS) and can be written as:

$$\text{glutamic acid} + NH_3 + ATP \xrightarrow{GS} \text{glutamine} + ADP + Pi.$$

glutamate oxoglutamate amino-transferase

Glutamine then reacts with α-oxoglutarate to form 2 molecules of glutamic acid. This reaction is catalysed by glutamate oxoglutamate aminotransferase (GOGAT) and can be written as:

$$\text{glutamine} + \alpha\text{–oxoglutarate} + NAD(P)H \xrightarrow{GOGAT} 2\ \text{glutamic acid} + NAD(P)^+$$

One molecule of glutamic acid is recycled to accept ammonia and the other becomes available for protein synthesis.

glutamate dehydrogenase

Other bacteria such as *E. coli* assimilate ammonia by incorporating it directly into α-oxoglutarate in a reaction catalysed by glutamate dehydrogenase (GDH). This reaction can be written as

$$\alpha\text{–oxoglutarate} + NAD(P)H + NH_3 \xrightarrow{GDH} \text{glutamate} + NAD(P)^+$$

This by-passes the energy-consuming GS step of the GS/GOGAT pathway used by *Methylophilus methylotrophus* and is thus more efficient.

It was decided to use genetic engineering to put the more-efficient GDH system from *E. coli* into *Methylophilus methylotrophus* AS1. The strategy was as follows:

- select GOGAT-deficient mutants of AS1 after treatment with nitrosoguanidine; (these mutants would not be able to assimilate ammonia).
- isolate the GDH gene from *E. coli.*

strategy

- insert the GDH gene into a plasmid vector using the *Sal* I restriction enzyme;
- transfer the plasmid to the GOGAT deficient mutant of AS1;
- ensure the GDH gene was expressed in the new host.

The plasmid vector was chosen for its ability to be transferred into bacteria of different species. It was successfully transferred to the GOGAT-deficient mutant of AS1. These organisms then became able to assimilate ammonia and grow at normal growth rates. Yields of the engineered organism were about 5% greater than the parent organism in the pilot-scale culture system.

Π Select the correct ending to the following statement.

It is necessary to introduce the GDH recombinant plasmid into a GOGAT-deficient mutant strain of AS1 because:

1) GOGAT would otherwise compete with GDH for ammonia;

2) GOGAT would otherwise compete with GDH for α-oxoglutarate;

3) GOGAT would otherwise compete with GDH for glutamine.

The correct answer is 2). Both GDH and GOGAT use α-oxoglutarate as a substrate. GOGAT deletion ensures α-oxoglutarate is directed only through the GDH step.

legislation

Although the genetically engineered organism produced higher yields on methanol than the parent organism, it was not used for routine production. This was mainly because the legislation governing the large-scale production of genetically engineered organisms and their use as feed was, at the time, under review and was uncertain. It was therefore easier to establish a process based on a 'natural' organism.

Sterile Engineering

heat lysis and acidification

Sterility was essential in the bioreactor and in the biomass separation area, so that liquid could be recycled without resterilisation. This was achieved by careful design and operation. Cell recovery by centrifugation would have been expensive and difficult to achieve without risk of contamination, so a 2-stage process was developed. Stage 1, run under aseptic conditions, involved heat lysis of cells at 70°C followed by acidification. This caused flocculation of cells around gas bubbles released from the liquid, forming a froth of cell material which was skimmed off. This achieved a 10-fold thickening to 10-12% w/w. The froth is taken off through a seal to a non-aseptic, but clean area, to Stage 2, where it is centrifuged to a 20% w/w suspension for drying. An economic drying process was developed at this stage.

Pruteen

By this time, the product had proved, in continuing toxicity and feeding trials, to be safe and effective feed material. The process was patented and the trade name Pruteen given to the product. A licence to sell the product as feed was granted by the Ministry of Agriculture, Fisheries and Food.

4.9.7 Stage Five - The Pruteen Production Process

production scale bioreactor

The production-scale fermentation unit, with a projected annual capacity of over 50,000 tonnes was fully commissioned in 1980. The bioreactor (Figure 4.8) is 60 m high, with a 7 m base diameter and working volume 1,500 m^3. There are two downcomers and cooling bundles at the base. Initial sterilisation is with saturated steam at 140°C followed by displacement with heat sterilised water. Air and ammonia are filter sterilised as a mixture, methanol filter sterilised and other nutrients heat sterilised. Methanol is added through many nozzles, placed two per square metre. For start-up, 20 litres of inoculum is used and the system is operated as a batch culture for about 30 h. After this time the system is operated as a chemostat continuous culture, with methanol limitation, at 37°C and pH 6.7. Run lengths are normally 100 days, with contamination the usual cause of failure.

downstream processing

Harvesting is continuous and under aseptic conditions, with medium recycle without re-sterilisation. Residual nutrients are monitored to ensure there is no build up (for example of phosphate). The cream (biomass) from flocculation and centrifugation is adjusted to normal pH (remember it was acidified for flocculation), then dried in an air-lift flash drier. For this the cream is fed into the venturi of the drier tube and is carried upwards by the hot drying gas, after which it is recovered in high efficiency cyclones. The product is cooled and the very fine material (waste dust) is separated and remixed with cream from harvest for redrying. The granular product is lightly coated with vegetable oil to suppress dust and is stored pending quality control analysis. The product is sold in the granular form or is ground to a fine powder.

SAQ 4.10

Draw a flow diagram of the Pruteen process.

SAQ 4.11

The organism grows on methanol with a yield coefficient of 0.5. The maximum growth rate (μ_{max}) is 0.55 h^{-1}.

Theoretical considerations show that, with a methanol concentration of 70 kg m^{-3} in the incoming medium to the bioreactor, the biomass concentration and residual methanol concentration (in outgoing medium) would vary with dilution rate as outlined below.

Dilution rate (D) (h^{-1})	Biomass concentration (kg dry biomass m^{-3})	Residual methanol concentration (kg m^{-3})
0.1	34.3	1.4
0.2	33.1	3.7
0.3	31.1	7.8
0.4	26.3	17.3
0.5	2.5	65.0

For each dilution rate calculate:

1) the productivity (kg biomass produced m^{-3} h^{-1});

2) the percentage methanol utilisation;

3) the minimum OTR (kg oxygen required m^{-3} h^{-1}).

The bioreactor has a maximum OTR of about 11 kg O_2 m^{-3} h^{-1}.
From these data select the optimum working dilution rate for the bioreactor.

SAQ 4.12

For a 1500 m^3 system, working at D = 0.2 h^{-1} and biomass concentration 36 kg m^{-3}, with a yield coefficient of 0.5, calculate the annual:

1) output;

2) methanol requirement.

SAQ 4.13

Listed below are some of the steps in the development of an SCP process. List these steps in the sequence in which you think they would be performed, starting with the first and ending with the last. If you think that any steps would be performed concurrently (at the same time), list them side by side.

Marketing.
Measure protein content and amino acid profile.
Full-scale production.
Isolate organisms capable of using the substrate.
Apply for sales licence.
Measure affinity for substrate.
Perform feeding trials in animals.
Establish the size of the market.
Measure temperature optimum (for organisms' growth).

4.9.8 The Outcome

changing market conditions

The 1973 oil price rise caused an increase in the price of fuels, including the natural gas substrate for the process. However, this was offset by corresponding price increases in competing feedstuffs, mainly soya and fish meal. The high value of soya then stimulated increased production and in 1979 soya was over produced, leading to a collapse in the price of protein feeds. In 1983 the price of soya in Europe was below its 1973 level, whereas oil prices had again risen sharply in 1980. Pruteen was then selling at $600 per ton, more than twice the price of the conventional soya feed. In an attempt to maintain the economic viability of the process the product was promoted as a milk substitute (milk powder for animal weaning being more expensive than solid feeds like soya). However, the size of the market for milk substitutes was relatively small, and for a time the plant was operated at reduced capacity. By 1986 the price of methanol was $179 per ton, which at yield factors of 0.5, contributed $358 per ton to the cost of the product. This in itself exceeded the price of soya for feed, and the production plant was closed down. The cost of development of the Pruteen process has been estimated at $180 million (1986 prices).

success from the technological development

Despite the lack of economic success, the process has not been judged a failure. Extended feeding trials throughout the development and production stages showed that Pruteen was a satisfactory form of protein feedstuff, compared to conventional sources, and was without toxic side effects. The technology that had been developed for the Pruteen process has been sold under licence to other companies in the UK and elsewhere for SCP production (for example in Myco-protein production). A version of the pressure cycle bioreactor has also been developed for effluent treatment. ICI have also applied the technology to other biotechnological products, such as biological feedstocks for plastic production.

4.10 Economics of SCP production

confidential data

Detailed economics of individual industrial processes, including SCP processes, are usually regarded as confidential, out of fear that publication may lend advantage to competitors. In addition, 'economy of scale' rule generally applies (that is as the production capacity increases, the cost of the product decreases), so that direct comparisons can only be made between systems of similar capacity. Some economic data on SCP processes have been published and are presented in the Resource Material at the end of this chapter. You should appreciate that the data are outdated by more than a decade, during which time substrate costs will have varied relative to each other, and technology will have improved. This means that the comparative costs presented in Table 4.13, for example, may not be now as presented there. Nevertheless the data presented do provide an outline of the economics of SCP production. The processes referred to in the Resource Material are not necessarily those mentioned in the text and so you may find some differences in detail.

SAQ 4.14

Use the data given in the Resource Material to complete the following summary of economics of SCP production. Delete inappropriate choices and replace Xs with figures. (You will find Tables 4.9 - 4.15 most useful in making your choices).

1) The most significant production cost in SCP production is the cost of raw materials/labour/running costs, ranging from X to X% of total production cost.

2) The substrate contributes least to production costs when it is a waste/non-waste.

3) For liquid substrates the most significant equipment cost is medium preparation/fermentation/harvesting/drying, ranging from X to X% of the total running costs.

4) The most significant running cost (for liquid processes) is inoculum preparation/medium preparation/fermentation/harvesting/drying, ranging from X to X% of the total running costs.

5) The most significant running cost of fermentation is sterilisation/aeration/cooling, ranging from X to X%.

6) The most significant cost of fermentation equipment is the air compression system/the bioreactor/the sterilising system/the cooling system.

7) The most significant cost of harvesting and drying equipment is drying/filtration/centrifugation.

8) The most significant running cost of harvesting and drying is drying/filtration/centrifugation.

SAQ 4.15

Use the data in the Resource Material to answer the following question.

It is 1977. The bacterial SCP from methanol plant referred to in Table 4.9 does not produce protein at a price that competes with soya protein. By how much would the cost of methanol have to fall in order that the protein from such a plant can be produced competitively with soya protein?

You can assume i) that the SCP processes referred to in Tables 4.7 and 4.9 to 4.15 are of 2×10^5 tons annual capacity, ii) that yield on methanol is 0.5kg biomass per kg methanol, iii) bacterial SCP contains 60% protein.

SAQ 4.16

For the yeast process from *n*-alkanes, which of the following contributes most to the cost of yeast production, and which contributes least? (Tables 4.7; 4.11 and 4.12 are helpful).

1) Aeration (during fermentation).

2) Cooling (during fermentation).

3) Ammonia.

4) Labour.

5) Drying.

SAQ 4.17

You are developing a process to produce SCP from a carbohydrate (sugar) substrate. The price of this substrate is $0.2 per kg sugar. You have a yeast, *Candida ifulika*, which is suitable for food or feed and which grows on the sugar substrate with a yield factor 0.48. You have also a filamentous fungus, *Fusarium inolika*, which is also suitable for food or feed and which grows on the sugar substrate with a yield factor of 0.45.

Both fungi will grow at pH 2.5, at which non-aseptic processes can be operated (that is without sterilisation). However, the SCP grown in non-aseptic systems is suitable only as feed. The SCP from both organisms can be used as a high-protein food additive, but *Fusarium* sp. must be ground up (powdered) for this. In addition, the filamentous fungus can be used to make meat substitutes. For this the SCP must be prepared deep-frozen and not dried.

The selling price of the SCP must be the same as or less than competing food and feedstuffs The price of conventional competing protein feeds is $0.80 per kg protein. The price of conventional competing high-protein food additives is $1.55 per kg protein. For a meat substitute, the SCP can be priced at $1.05 per kg biomass.

An existing fermentation unit has been made available for use (it would be the same for both organisms). Thus there are the following fixed costs for production of either organism.

		Contribution to production cost ($ per kg biomass)
Substrates other than the carbohydrate substrate		0.1
Fixed running costs (including fermentation)		0.055
Labour		0.05
	Total:	**0.205**

SAQ 4.17 contd

Thus, for the production of either organism you have, apart from the cost of the carbohydrate substrate costs of $0.205 per kg biomass. Assume both organisms give the same output.

You have the choice of whether or not to operate an aseptic or non-aseptic system, and which down-stream processing operations to use. The costs of these operations are given below.

	Contribution to production cost ($ per kg biomass)
Sterilisation	0.04
Centrifugation (with washing)	0.055
Dewatering (filtration with washing)	0.001
Drying	0.02
Deep-freezing	0.04
Milling (grinding)	0.01

Which of the following products could be most profitable?

1) *Candida* sp. as feed.

2) *Candida* sp. as high-protein food additive.

3) *Fusarium* sp. as feed.

4) *Fusarium* sp. as high-protein food additive.

5) *Fusarium* sp. as meat substitute.

Summary and objectives

A variety of SCP processes have been developed with a view to producing food and feed from alternative or waste carbon sources. Processes have been based on CO_2, carbohydrates, hydrocarbons and their derivatives. Technology has ranged from relatively simple open lagoons or solid-substrate fermentations to large-scale aseptic continuous cultures in fermentors. Feed processes have been developed mainly in Europe, Japan and the former USSR, where feedstuffs are in short supply. Processes have not generally been successful due to unfavourable economics (rising substrate costs and decreasing cost of soya), or political and social pressures (particularly against SCP produced on oil-based substrates). In the former USSR, where different economic and political systems applied, feed processes were operated on a large scale.

For food, SCP has been produced on a smaller scale and is mainly limited to yeasts, the use of which in food has been traditional. The exception is Myco-protein, which is being promoted as a health food, rich in protein and lacking animal fat, and which can be used as meat substitutes in high-value vegetarian convenience foods.

Industrialised countries that have developed the technologies are not the countries suffering population explosions and food shortages. The so-called developing countries that have the population and food problems, do not generally have the industrial base and technological expertise necessary to operate large-scale processes. These factors limit the application of SCP processes.

Perhaps SCP will become a thing of the future, when conventional proteins might be in short supply in industrialised countries. Unfortunately we might not have to wait long for this situation to arise. 1988 and 1989 saw drought conditions over significant parts of the USA - a reminder that climate is not fixed, but undergoes changes. Fish stocks in many oceans are not managed or conserved. At least we can draw comfort from the fact that SCP technology has been tried and proven, and is available if need be.

After studying the material in this chapter, you should be able to:

- describe the advantages of using micro-organisms as food and feed, compared to conventional protein sources;
- describe the different physiological types of micro-organism used in SCP production;
- use knowledge of the organisms and technologies involved in the development of SCP processes to make valued judgements on alternative strategies for process development;
- compare the economics of SCP processes with those of conventional protein production.

Resource Material

Toxicological tests on SCP in animals

To test for possible toxic effects, rats, mice, dogs, chickens pigs, and fish were fed Pruteen at various levels in their diet, ranging from 30% to 60%. Control animals were fed casein (milk protein) instead of Pruteen. Feeding periods ranged from 28 days to 3 years. The parameters measured are outlined below.

Feed intake rate.

Growth rate.

Haematological characteristics (haemoglobin level, red blood and white blood cell counts, platelet count, blood clotting time).

Biochemical characteristics (plasma levels of alanine and aspartate transminases, alkaline phosphatase, triglycerides, cholesterol, urea, uric acid, allantoin, glucose, protein, albumin, sodium, potassium, calcium, magnesium, phosphorus; urine levels of protein and glucose).

Histopathological examination (up to 40 different tissues were post-mortem and examined for abnormality).

Effects on reproduction [size of litters/broods; evidence of teratogenicity (physical defects) in foetuses].

Oxygen requirements and heat outputs of organisms grown on methanol

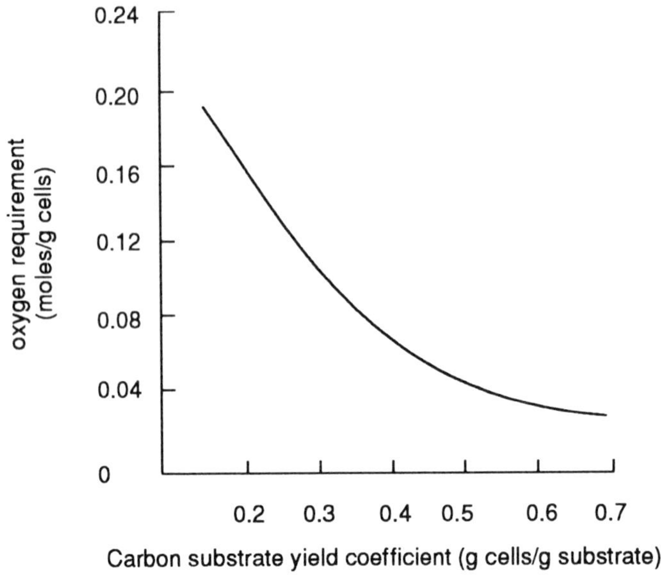

Figure 4.10 Typical effects of yield coefficient on oxygen requirement when only biomass and CO_2 are produced (methanol as substrate)

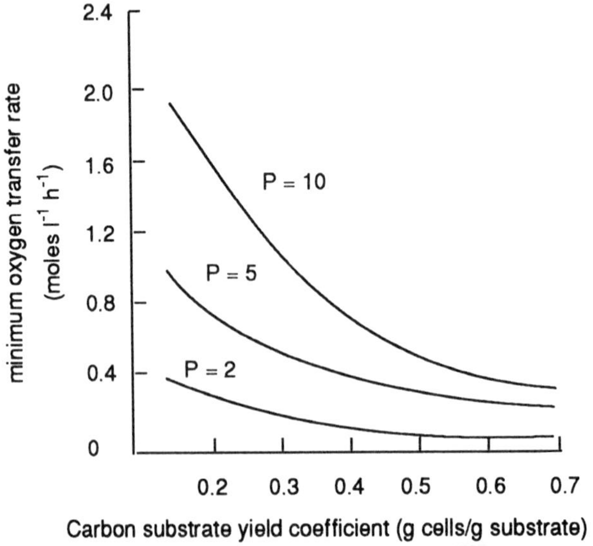

Figure 4.11 Typical relationship between minimum oxygen-transfer rates and yield coefficients at various productivities (methanol as substrate).

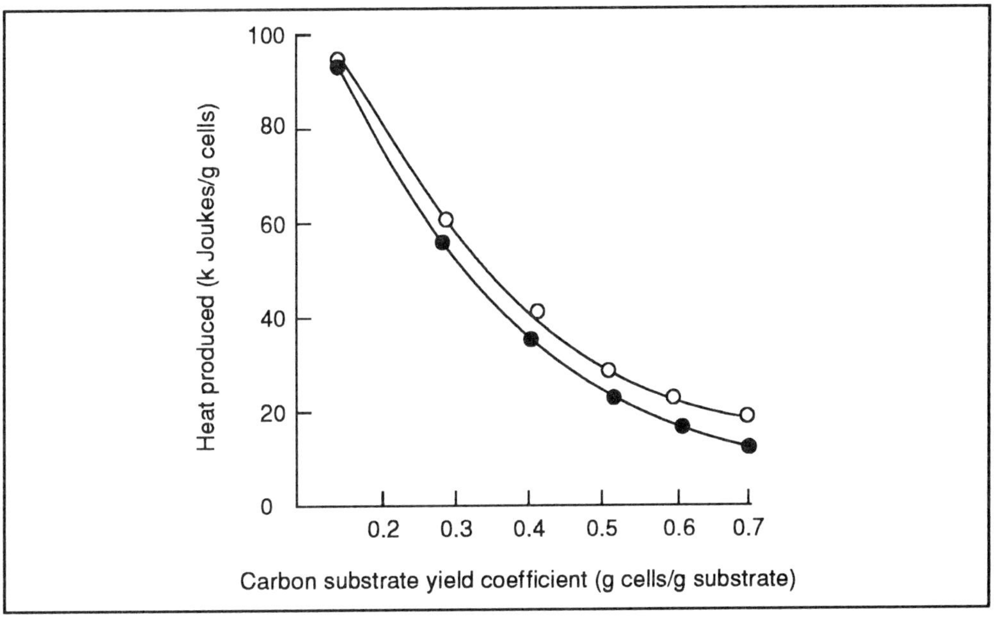

Figure 4.12 Estimated relationships between heat production, calculated from heats of combustion and yield coefficients (●)or calculated from experimentation (O). Methanol as substrate.

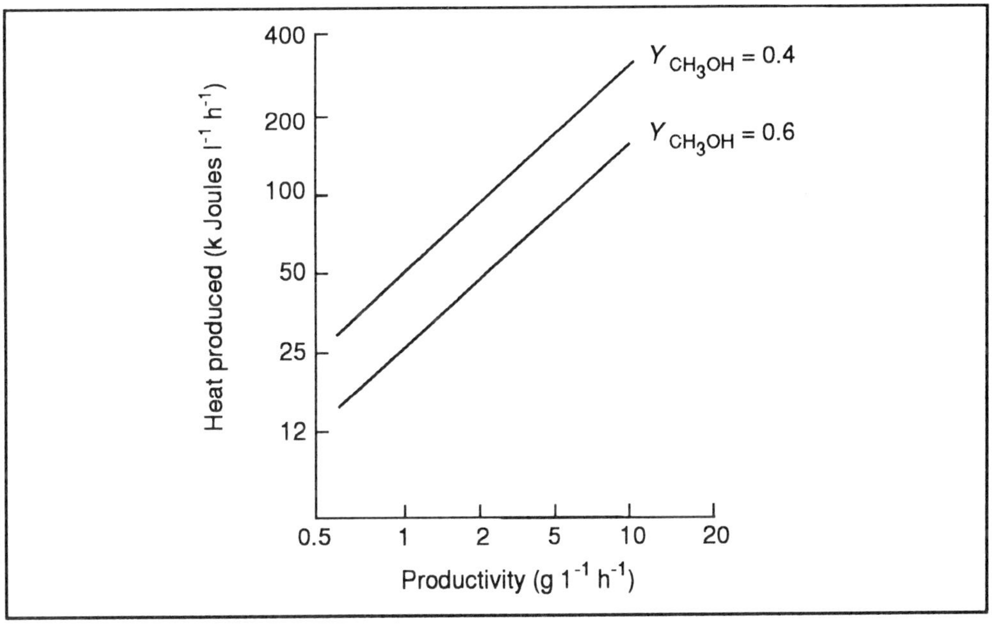

Figure 4.13 Estimated relationships between heat production and productivity at various yield coefficients

Economics of SCP Production

	$ kg-1	% Protein	$ per kg protein
Beef	1.54	15	20.3
Pork	1.10	12	19.1
Poultry	0.66	20	6.6
Cheese	0.78	24	6.5
Milk Powder	0.46	36	2.5
Soya Flour	0.15	52	0.6
Peanut Flour	0.15	59	0.5
Yeast (from *n*-alkanes)	0.42	53	1.4
Yeast (from molasses)	0.33	53	1.3
Fungi (from cellulosis)	0.15	43	0.7
Algae	0.66	46	2.8

Table 4.7 The production cost of various proteins (1980 figures, USA, prices in $ US).

10^4 tonnes per year capacity, 1975, Japan, producing yeast				
	***n*-Alkanes**	**Methanol**	**Ethanol**	**Molasses**
Investment cost	1	1.2	0.85	1.1
Production cost	1	1.4	1.6	1.3
10^4 tonnes per year capacity, 1975, Israel, producing yeast				
	***n*-alkanes**	**Methanol**	**Molasses**	
Investment cost	1	0.68	0.77	
Production cost	1	0.75	1.24	
2×10^5 tonnes per year capacity, 1977, USA				
	Yeasts, *n*-Alkanes	**Bacteria, Methanol**		
Investment cost	1	0.73		
Production cost	1	0.98		

Table 4.8 Cost of SCP grown on various substrates in comparison to the *n*-alkane process.

The tables that follow give the costs of various SCP production processes in comparative rather than in actual form. To see what this means examine Table 4.9. The production cost of raw materials for yeasts grown on *n*-alkanes is given as 58.5. This means that the cost of raw materials accounts for 58.5% of the total production costs of this process. The same cost for bacteria grown on methanol is 73.8. This means that in this case 73.8% of the total production cost is accounted for by raw materials. This does not mean that the actual cost of raw materials for the methanol process is more than that for the *n*-alkanes process, as the total costs of the two processes are not necessarily similar.

	Yeast, *n*-alkanes	Bacteria, Methanol	Yeast, Ethanol	Fungus, Sulphite waste liquor
Depreciation (of value of production plant)	9.3	5.8	5.8	9.1
Raw matrials (total)	58.5	73.8	77.1	55.1
a) Carbon substrate	29.4	47.4	63.9	17.0
b) Ammonia	11.1	11.8	3.2	16.2
c) Phosphoric acid	9.9	12	4.8	13.3
d) Mineral salts	2.9	2.6	1.9	4.2
e) Other	5.2	-	3.3	4.4
Labour	8.4	6.2	5.1	11.0

Table 4.9 Relative equipment costs of various SCP processes.

	Yeast, *n*-alkanes	Bacteria, Methanol	Yeast, Ethanol	Fungus, Sulphite waste liquor
Supplies Storage	3.3	4.9	4.0	4.9
Medium Preparation	1.1	1.5	1.2	1.6
Inoculum Preparation	0.9	1.1	0.9	1.3
Fermentation	51.2	43.4	50.0	50.6
Harvesting	14.2	11.0	13.1	7.6
Drying	17.8	23.1	18.7	17.0
Product storage	11.5	15.0	12.1	17.0

Table 4.10 Relative production costs of various SCP processes.

	Yeast, *n*-Alkanes	Bacteria, Methanol	Yeast, Ethanol	Fungus, Sulphite waste liquor
Supplies Storage	1.2	0.2	0.2	0.2
Medium Preparation	1.6	2.4	1.0	1.0
Inoculum Preparation	1.8	2.5	2.3	2.6
Fermentation	67.9	61.7	53.5	76.8
Harvesting	5.9	3.8	16.0	3.6
Drying	21.0	28.8	26.4	15.1
Product storage	0.6	0.6	0.6	0.7

Table 4.11 Relative running costs of various SCP processes (The cost of the electricity, fuels and water used in various processes).

	Yeast, *n*-Alkanes	Bacteria, Methanol	Yeast, Ethanol	Fungus, Sulphite waste liquor
Sterilisation	-	21.4	-	-
Aeration	88.9	70.9	86.3	92.3
Cooling	7.9	5.1	12.5	6.7
Others	3.2	2.6	1.2	1.0

Table 4.12 Relative running costs of fermentation of various SCP processes.

	Yeast, *n*-Alkanes	Bacteria, Methanol	Yeast, Ethanol	Fungus, Sulphite waste liquor
fermentor	40.1	34.5	42.1	34.4
Air supply system	20.0	21.2	16.4	24.4
Cooling system	37.4	32.0	40.6	39.1
Sterilisation system	-	10.6	-	-
Others	2.5	1.7	0.9	2.1

Table 4.13 Relative costs of fermentation equipment for various SCP processes.

	Yeast, *n*-Alkanes	Bacteria, Methanol	Yeast, Ethanol
Filtration	11.9	-	-
Centrifugation	26.8	31.1	27.7
Drying	55.6	68.9	57.4
Others	5.7	-	14.9

Table 4.14 Relative costs of harvesting and drying equipment

	Yeast, *n*-Alkanes	Bacteria, Methanol	Yeast, Ethanol
Filtration	7.4	-	-
Centrifugation	10.0	11.7	8.5
Drying	77.7	88.3	61.8
Others	4.9	-	29.7

Table 4.15 Relative running costs of harvesting and drying.

The large scale production of organic acids by micro-organisms

The large scale production of organic acids by micro-organisms

Overview

In this Chapter we shall look at the use of micro-organisms to produce organic acids of commercial importance. Although all of the examples to be mentioned are relatively simple chemically, they are interesting in that they are metabolically diverse. Some are genuine end products of metabolism, while others are compounds considered to be central metabolites in all living cells. These central metabolites are normally present in relatively small, constant amounts. However, some micro-organisms can be "persuaded" to produce enormous yields of these metabolites.

The first part of this Chapter, the Introduction, will identify some of the organic acids produced by micro-organisms and highlight those which are of commercial interest. The products considered in this chapter are metabolites of the tricarboxylic acid (TCA) cycle or oxidative derivatives of glucose. Since most of the biological commerical processes involve interference with the metabolism of micro-organisms, we present a section discussing relevant pathways together with the control mechanisms involved. The industrial production of the most commercially important organic acid, citric acid, is then considered in depth. Finally, we outline the biochemistry, formation and downstream processing of other TCA cycle intermediates (malic acid, fumaric acid) and oxidation products of glucose (itaconic acid, gluconic acid and its derivatives).

5.1 Introduction

organic acid definition

We must first indicate exactly what is meant by the term 'organic acid' in the context of this Chapter. Inevitably it will be far more restricted in scope than the literal definition, which essentially means 'any organic compound which is acidic'.

Π Can you give one or more examples in each of the categories below of organic acids which are produced by living cells?

1) Acids of non-carbohydrate origin which are produced by all living systems.

2) Acids continuously produced by all living systems.

3) Acids of carbohydrate origin which are constantly produced by living systems and are not considered as waste products.

4) Acidic examples of continuously produced waste products.

Acceptable answers to part 1) include amino acids and fatty acids or specific examples of each, such as glycine or stearic acid respectively. The obvious answer for part 2) is the central metabolite pyruvate, though all of the acids of the TCA cycle would be appropriate. Answers to part 3) include the principal acid of the hexose monophosphate pathway, 6-phosphogluconate, and the acid intermediate of glycolysis, 1,3-diphosphoglycerate. Answers to part 4) include an enormous number of acids since all living systems produce acid end products. However, in the context of this chapter the waste products produced by bacteria growing anaerobically are particularly relevant. These include: lactate, produced in great quantity by the lactic acid bacteria; lactate, acetate, formate and succinate produced by the *Enterobacteriaceae*; butyrate and acetate produced by *Clostridium* species.

There are many more correct answers to each part of the question. To simplify matters, all of the answers are compounds which are acidic because they contain the carboxyl group (-COOH). This chapter does not consider any organic acids which are organic compounds made acidic by the presence of, for example, phosphate or sulphate groups. Further, to warrant discussion in this Chapter, an organic acid has to satisfy the following criteria:

- there has to be a micro-organisms which will produce it in commercially significant quantities;
- there has to be a demand for the compound industrially;
- the overall costs of producing and extracting the acid have to be economic.

end products and intermediates

In metabolic terms there are three clearly distinguishable types of compound to deal with. Firstly, compounds which are obviously waste products - end products of one or more pathways which would normally be excreted from the cell (for example lactic acid). Secondly, compounds which are end products of pathways but which are not waste products and whose synthesis is normally very carefully controlled (for example amino acids). Thirdly, compounds which are intermediates of pathways and hence not normally considered as end products or wastes at all (for example citric acid).

manipulation of metabolic pathways

Production of large quantities of organic acids by micro-organisms would, on the face of it, seem easier if we are dealing with acids which are genuine waste products rather than non-waste compounds such as central metabolites. It is merely a technical problem to encourage certain bacteria to produce a waste product such as lactate as this compound is normally excreted into the surrounding medium. The removal of spent medium regularly and harvesting of lactate could allow continuous production of lactate. However, production of metabolites such as amino acids is more complex and to obtain sufficient quantities of intermediate compounds such as citric acid is even more of a problem. The key problem is how to encourage a micro-organism to produce a vast excess of an organic acid whose synthesis is normally controlled very efficiently at relatively low concentration. A detailed discussion of this problem will occur later, but we can generalise here and identify the four main ways in which metabolic pathways can be manipulated:

- by altering the environmental conditions, eg temperature, pH, medium composition (especially the elimination of ions and cofactors considered essential for particular enzymes);
- by disrupting a pathway using substrate analogues;
- by mutation - giving rise to mutant organisms which may only use part of a metabolic pathway or regulatory mutants;
- by genetic engineering.

It should be noted that apart from a passing reference to natural selection of wild-types with enhanced specific properties, the genetics of organic acid producing micro-organisms is beyond the scope of this chapter.

5.1.1 Generalised scheme for fermentation

There are several stages common to most fermentation processes but before identifying these it is appropriate to define some terms which will appear during this Chapter. Let us first distinguish between primary and secondary metabolites.

primary and secondary metabolites

Primary metabolites are compounds which are essential to the growth and well being of the cell and, during the growth phase, are produced continuously. *Secondary metabolites* are those compounds not essential to the life of the cell and not produced continuously; often but not always, they are produced during non-growth phases of the cell. The growth phase where primary metabolites are produced is sometimes referred to as the *trophophase*, whereas the phase during which secondary metabolites are formed (usually the stationary phase) is termed the *idiophase*.

trophophase idiophase

All of the compounds we shall study in this Chapter are primary metabolites though both phases of growth will be studied. For example, as we shall see, citric acid is produced continuously at low levels during trophophase but only accumulates at high concentration during idiophase.

Finally the word 'fermentation' will be used in its industrial sense, that is a commercially viable process in which a micro-organism produces a required product or change.

Let us now consider the essentials of the fermentation process, largely as a revisionary exercise, before looking at the individual examples in detail.

SAQ 5.1

Study the unlabelled block diagram, and then replace the question marks with the words and phrases to give a generalised scheme of an industrial fermentation. Assume in this example that the product is excreted from the microbial cells.

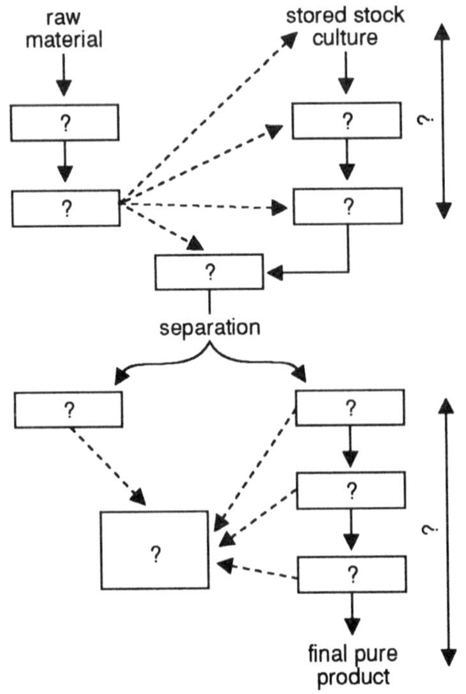

Words/Phrases

Product extraction
Effluent and waste disposal
Medium preparation
Seed vessel
Purification
Cell free supernatant
Cell biomass
Production bioreactor
Downstream processing
Medium sterilisation
Primary culture
Upstream processing

You should note that the figure in SAQ 5.1 is a simple outline as fermentations generally have more steps than indicated; for example many have a multiple purification step. If the product were the whole cell (for example in single cell protein processes) then purification of the cell biomass would be necessary. If the required product were an intracellular compound then some stage of cell breakage would be essential.

5.1.2 Organic acids relevant to this Chapter

Table 5.1 shows the organic acids relevant to this Chapter together with the usual substrate, the micro-organisms employed to produce them and finally the potential end uses. Such acids are often grouped into two broad categories, those which are members of or related to the TCA cycle and secondly, those which are oxidation products of glucose. The carbon source for the latter group is usually quite specific, either glucose itself or a polysaccharide which yields glucose easily. The carbon source for the former group can be much more diverse and/or complex, for example: glucose or its polymers (molasses, starches); byproducts of industry (methanol, methane); waste products of industry (sulphite waste liquor from paper manufacture) or plant waste (lignins, cellulose derivatives). Acetate, lactate and some amino acids however do not readily fit into either of the groups.

organic acid	substrate	producer micro-organism	end use(s)
* citric acid	sugar(s)	*Aspergillus niger*	Flavouring for beverages and confectionery. Pharmaceutical food syrups, resins, dye mordants, antifoaming agents, sequestering agents.
* malic acid	glucose	*Lactobacillus brevis*	Food and drink manufacture
* fumaric acid	glucose	*Rhizopus delemar*	Used in the plastics industry and, to a lesser extent, in the food industry
* itaconic acid	glucose	*Aspergillus terreus* other *Aspergillus* spp	Intermediate for organic syntheses, eg acrylic resins
* gluconic acid	glucose	*Aspergillus niger* *Gluconobacter suboxidans*	Pharmaceutical industry and as a washing and softening agent preventing a build up of scale. Retards setting of building materials.
acetic acid	ethanol	*Acetobacter aceti*	Vinegar, food industry
lactic acid	lactose	Lactic acid bacteria	Dairy industry as a preservative/flavour enhancer

Table 5.1 Example of organic acids produced commercially by micro-organisms; organic acids considered in this chapter are labelled with *. A related acid, α-oxoglutaric acid, is easy to produce microbiologically but has no current end use; succinic acid is produced chemically. Amino acids are beyond the scope of this chapter.

A detailed study of the amino acids is beyond the scope of this Chapter. However, industrial production of amino acids is considered in Chapter 8 of this text.

5.2 Metabolic pathways and metabolic control mechanisms

5.2.1 Revision of the reactions of the tricarboxylic acid (TCA) cycle

A detailed revision of the TCA cycle is necessary to ensure an understanding of the mechanisms and reasons governing the choice of process conditions for encouraging production of any selected TCA cycle intermediate or related compound. Descriptions of the cycle can be found in many text books, for example in the open learning BIOTOL text entitled 'Principles of Cell Energetics'. Chapter 7 of that text describe in great detail the TCA and glyoxylate cycles, both of which are relevant to this Chapter. The most relevant parts of the cycle are the control mechanisms and processes involved in the intermediary metabolism; these are influenced and exploited in efforts to upset the balance of normal metabolism leading to overproduction of the desired organic acid.

Let us first examine Figure 5.1. It is worth spending some time on this in order to understand the rationale of the remainder of this Chapter.

glycolysis

Glycolysis (the Embden Meyerhof pathway) is a ten enzyme pathway which is summarised in Figure 5.1. During the course of this pathway, glucose is cleaved to two pyruvate molecules - a process involving the utilisation of two ATP (generating two ADP) but later there is formation of four ATP from four ADP. Thus a net yield of two ATP is achieved. Another consequence of glycolysis is the reduction of two molecules of nicotinamide adenine dinucleotide (NAD^+) to NADH + H^+. Although the exact mechanism of the individual reactions of glycolysis is not really necessary for this Chapter, we must bear in mind the fact that during the process energy and reducing power are formed. The two reactions indicated by * are two further enzymatic steps which, along with three reactions of the TCA cycle, constitute the glyoxylate cycle.

In the reaction that bridges glycolysis and the TCA cycle, for each pyruvate degraded to acetyl CoA, one CO_2 is released and a further NAD^+ is reduced to NADH + H^+.

A variety of starting materials other than glucose or its derivatives is possible for use by some micro-organisms; the four shown in Figure 5.1 are all initially converted to acetyl CoA for entry into the central metabolic pathways.

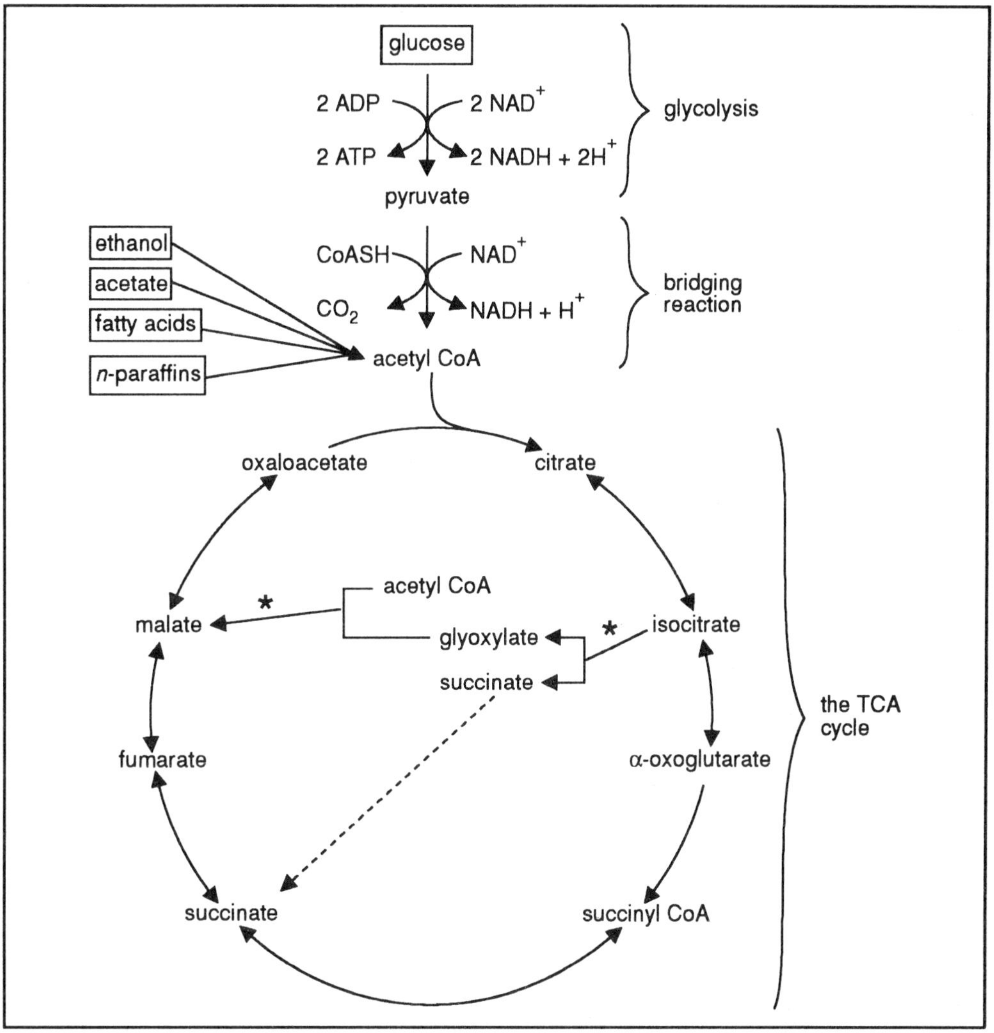

Figure 5.1 A simplified diagram of glycolysis and the tricarboxylic acid (TCA) cycle showing the entry points for various substrates. * indicates the two reactions specific to the glyoxylate cycle. Compounds in boxes are potential substrates for entry into the TCA cycle, via acetyl CoA.

5.2.2 The relationship between anabolism and catabolism

At this point we need to consider the two halves of metabolism -anabolism and catabolism - and in particular the metabolic control involved.

catabolism

Catabolism is the process by which intra- or extracellular molecules are degraded to yield smaller ones which are either waste products or building blocks for biosynthesis. During catabolism reducing power in the form of NADH + H^+, $FADH_2$ or NADPH + H^+ is generated. Subsequent reoxidation of these cofactors (particularly NADH) by aerobic cells releases energy which is converted to ATP, the major short term energy storage currency. The mechanisms involved in this ATP formation are the electron transport chain and oxidative phosphorylation. These processes are intimately linked - a bit like two parts of a zip fastener - in that when oxidation of NADH takes place, the formation of 3 ATP from 3 ADP usually takes place.

anabolism

Anabolism is the building up or biosynthesis, of complex molecules such as protein, nucleic acids and polysaccharides, from raw materials originating from intra- or extracellular sources. The biosyntheses are energy (ATP) requiring processes.

Catabolism and anabolism have to be carefully regulated and are inevitably intimately linked.

Π What are the three areas where the processes of catabolism and anabolism are linked?

Catabolism produces ATP, reducing power and intermediates. Anabolism requires all three, thus these are the three main links.

No living cells can store large amounts of ATP. There is a finite amount of 'adenine' distributed between AMP, ADP and ATP. Thus if the cell has a relatively high concentration of ATP, the concentrations of AMP and/or ADP must be lowered. The balance alters like a "see-saw", as one goes up the other must come down. In addition the total amount of NAD^+/NADH and $NADP^+$/NADPH in the cell is constant.

Π What is the advantage of the see-saw type of change to the ratio of the concentrations?

The answer is that such a system is far more sensitive to small changes in concentration of the respective compounds. The cell is recognising a change of the ratio of compounds, rather than the rise or fall of a single compound.

Although cells cannot store ATP, they must always have a minimum amount available to keep them alive. Thus a constant level of ATP must be maintained indicating that catabolism and anabolism occur constantly under normal conditions.

Π One compound controls the overall metabolism of the cell regulating and balancing anabolism and catabolism. Can you name it?

The answer in practice is ATP though you would have been theoretically correct if you had said ADP and AMP. Indirectly, NAD^+ or NADH are also compounds which regulate the anabolic/catabolic balance.

allosteric enzymes

The metabolic control is exercised on certain key regulatory enzymes of a pathway called allosteric enzymes. These are enzymes whose catalytic activity is modulated through non-covalent binding of a specific metabolite at a site on the protein other than the catalytic site. Such enzymes may be allosterically inhibited by ATP or allosterically activated by ATP (some by ADP and/or AMP).

energy charge

Thus ATP is the effective controller of metabolism but because AMP + ADP + ATP is constant, it is really the ratio of adenine nucleotides which is important. This ratio is termed the adenylate charge or energy charge and is expressed as:

$$\text{Energy charge} = \frac{0.5\ [\text{ADP}] + [\text{ATP}]}{[\text{AMP}] + [\text{ADP}] + [\text{ATP}]}$$

The theoretical limits are 1.0 (all ATP) and 0 (all AMP) with a normal working range of 0.75 to 0.9. The involvement of energy charge in the integration and regulation of metabolism is considered further in the BIOTOL text entitled 'Biosynthesis and the Integration of Cell Metabolism'.

After revising the TCA cycle reactions in more detail we shall return to the subject of metabolic control by ATP.

Figure 5.2 shows a detailed version of the TCA cycle indicating cofactor changes and the individual intermediates.

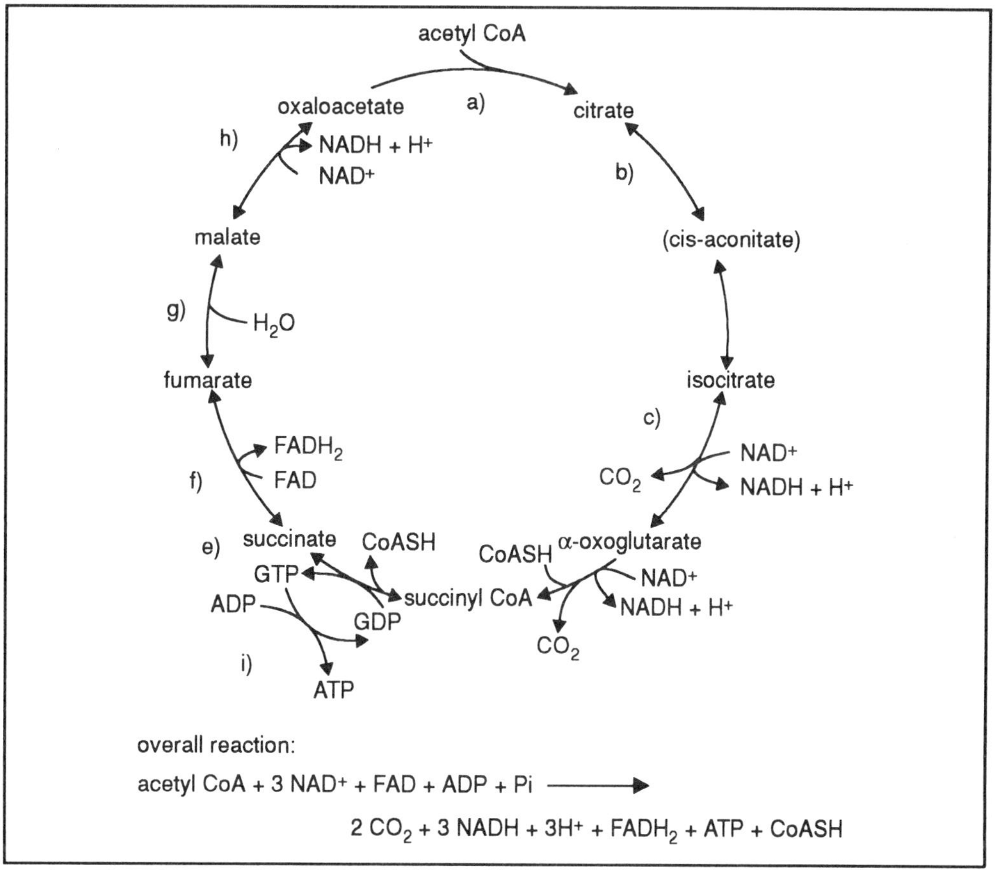

Figure 5.2 The tricarboxylic acid cycle.
Enzymes: a) citrate synthase; b) aconitase; c) isocitrate dehydrogenase; d) α-oxoglutarate dehydrogenase; e) succinyl CoA synthetase; f) succinate dehydrogenase; g) fumarase; h) malate dehydrogenase; i) nucleoside diphosphokinase.

5.2.3 The control of metabolism

In section 5.2.2 we considered a simple equation expressing the energy charge of the cell in terms of the ratio of adenine nucleotides. Figure 5.3 summarises the principal allosteric enzymes of glycolysis and the TCA cycle and indicates how the individual adenine nucleotides influence the activity of a variety of enzymes. The enzymes to the right of the glucose to pyruvate pathway are those involved in glycolysis; those to the left are involved in gluconeogenesis, ie the synthesis of glucose from pyruvate.

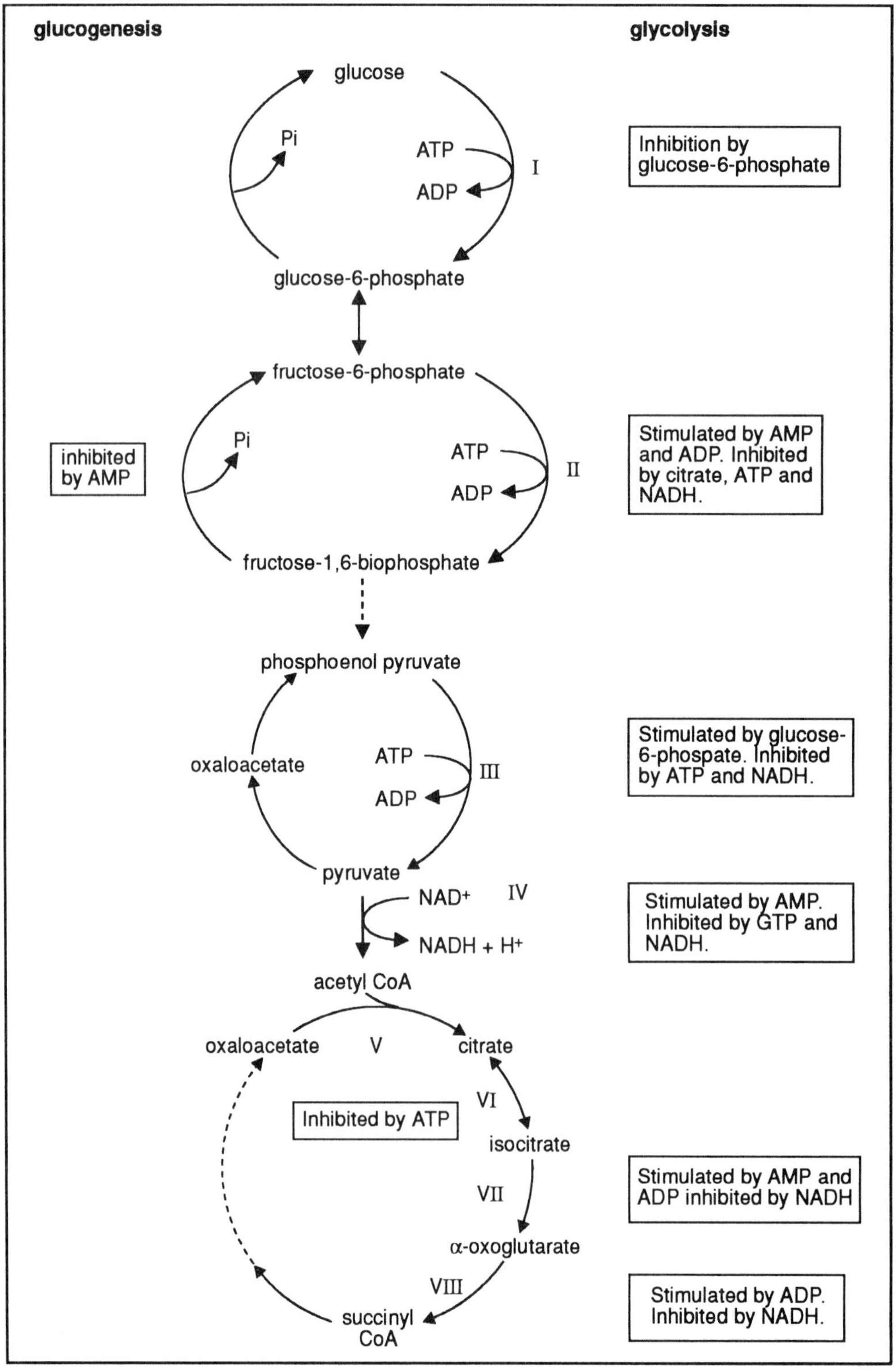

Figure 5.3 Major control points of glycolysis and the TCA cycle. Enzymes: I, hexokinase; II, phosphofructokinase; III, pyruvate kinase; IV, pyruvate dehydrogenase; V, citrate synthase; VI, aconitase; VII, isocitrate dehydrogenase; VIII, α-oxoglutarate dehydrogenase.

Let us now consider how the control system operates. Overall the TCA cycle produces large amounts of reducing power in the form of NADH + H^+. As we noted earlier, subsequent reoxidation of NADH + H^+ must be accompanied by oxidative phosphorylation, the process by which ATP is produced from ADP. When the cell has a high energy charge (high ATP concentration and low [ADP + AMP] concentrations) then temporarily no more ATP is required by the cell. Thus the logical thing to do is to stop or 'switch off' the TCA cycle. The first reaction of the cycle is the most appropriate one to inhibit and we know that citrate synthase is inhibited by high ATP concentration. As the ATP is used by the cell, its level will fall with a concomitant increase in the level of ADP (and AMP). Thus the inhibition of citrate synthase is gradually removed. In a more positive sense, the TCA cycle is stimulated at this stage because the next two enzymes of the cycle, aconitase and isocitrate dehydrogenase are stimulated by increased concentrations of both ADP and AMP.

citrate synthase

phospho-fructokinase

Within glycolysis, the main allosteric control is exercised by phosphofructokinase, a complicated enzyme unusual in that its activity is stimulated by one of its products (ADP) and inhibited by one of its substrates (ATP). One further point about this enzyme which will be important to us later: in *Aspergillus* spp., elevated levels of ammonium ions relieve phosphofructokinase of inhibition by citrate.

aconitase

A further way in which metabolic control may be exercised is the artificial deprivation of required ions and cofactors, for example aconitase must have ferrous ions for activity. Conversely, addition of toxic ions is possible, for example aconitase is inhibited by cupric ions. Finally the use of metabolic analogues is possible. If monofluoroacetate is added to cells then monofluorocitrate is produced by citrate synthase and this compound inhibits the activity of aconitase. Great care has to be taken when using metabolic analogues, however, they are often less than 100% specific and may have unexpected and unwanted serious side effects.

5.3 The industrial production of citric acid

5.3.1 Historical introduction

Historically the production of citrate has been an important development in the pioneering of fermenter technology. It was shown back in 1893 by Wehmer that a fungus, *Citromyces* (now reclassified as a *Penicillium* spp.) would accumulate citric acid in liquid culture. Wehmer in fact tried to scale up the process to an industrial level but there were two main problems. Firstly, the duration of the process under his conditions took far too long: of the order of several weeks. Secondly, a problem was caused by Wehmer's incorrect belief that citric acid only accumulated around neutral pH and lengthy incubation at this pH inevitably leads to contamination.

world demand

The world demand for citric acid around 1900 amounted to some 10,000 tonnes per annum. This was realised by pressing citrus fruits and precipitation of the citric acid as calcium citrate. An Italian, government-led cartel had virtual monopoly of this process and as such the price of citric acid was very high.

Aspergillus niger

A major breakthrough in the fermentation process came in 1916 - 1920 when it was found that *Aspergillus niger* grew well at pH values below 3.5, producing citric acid in days rather than weeks. The faster incubation and highly acid conditions (often below pH 2.0) also served to minimise potential problems caused by contamination.

Industrial production began in Belgium in 1919 followed by America in 1923 and England in 1927. In these early processes, high sugar concentrations were employed using pure ingredients. Newer materials were continually being tried, for example sugar beet molasses which was used commercially for the first time in 1928.

By the mid 1930's over 80% of the world's citric acid was produced by fermentation. At present virtually all of the world production comes from this process. By 1981 over 200,000 tonnes were produced annually (possibly as high as 300,000 tonnes); the industry in the United Kingdom at that time being worth some £20 million per annum, one tenth of the world's turnover.

One of the more recent innovative approaches was to look for new micro-organisms and novel carbohydrate substrates. The early fermentations used sugar beet or cane molasses, various syrups, sweet potato starch or glucose itself and the micro-organism was always an *Aspergillus* spp. In the early 1930's it was found that yeasts would produce citric acid from acetate. Since then a variety of yeasts, principally *Candida* spp., has been shown to convert glucose, *n*-alkanes or ethanol to citric acid with great efficiency.

yeasts

n-alkanes

The realisation that yeasts would produce citric acid from *n*-paraffins was very attractive in the late 1960's. Petroleum byproducts were plentiful and very cheap and there was detailed knowledge available on these processes because the use of hydrocarbon-utilising yeasts for single cell protein was well developed. The strategy was to use *n*-alkane to produce high yields of citric acid-producing *Candida* spp. and to harvest two useful end products rather than just one. The process has not been commercially successful however. *Candida* spp. produce mixtures of citric acid and isocitric acid and the latter is not a useful product. In addition, since 1973 when petroleum prices rose sharply and have in fact continued to rise, the *n*-paraffins are no longer a cheap substrate.

In summary the majority of the world citric acid production is still via microbial fermentation of carbohydrate substrates (derived from plants) using *Aspergillus niger*.

5.3.2 Current uses of citric acid

food industry

Citric acid, being an intermediate of the TCA cycle, is considered to be non-toxic and very safe for human consumption. As such it has long since had unlimited approval by the World Health Organisation Expert Committee as a food additive. Over 60% of the production is used by the food industry, particularly in soft drinks, jams, jellies, sweets and wines. Around 10% of the production is used in the pharmaceutical industry and in cosmetics. Increasingly, less pure grades of citric acid are being used to produce citric acid esters which are used as plasticisers in the plastics industry. Citric acid is used freely as a builder in detergents and is being used in laundries because it has the advantage of being totally biodegradable.

5.3.3 The biochemistry of citric acid production

Broadly we can say that molasses are converted to glucose which is converted via glycolysis to pyruvate. Citric acid is then produced via acetyl CoA. Although this statement is largely correct, we must develop the details because there are two main problems which have to be overcome.

Π What, briefly, are the problems which have to be overcome?

If you are in doubt as to one or both of them, consider the following question - how is citrate *normally* produced and what would normally happen to it?

We know that citric acid is formed from acetyl CoA and oxaloacetate by the reaction:

acetyl CoA + oxaloacetate → citrate

biochemical problems

The first problem is that if citric acid is removed, there is apparently no way of regenerating oxaloacetate. The second problem is that to accumulate citric acid, aconitase has to be blocked to avoid citric acid being converted to aconitate.

Examine Figure 5.1 again. There is an apparent conflict; if citric acid accumulates there would appear to be no way in which the TCA cycle could continue to regenerate oxaloacetate. On the other hand continuation of the TCA cycle would regenerate oxaloacetate but there will be correspondingly less citric acid accumulated.

ferrous ions

In fact the aconitase enzyme in *A. niger* is active even when citric acid is accumulating. This aconitase, if allowed to come to equilibrium, yields 90% citrate, 3% cis-aconitate and 7% isocitrate. To lower the activity of the enzyme, ferrous ions (essential for activity) are not added to the medium and may have to be removed from complex sources of carbon (such as molasses) before they are used.

isocitrate dehydrogenases

α-oxogluterate dehydrogenase

The fact that isocitrate is being produced means that some way of blocking the next enzyme, isocitrate dehydrogenase would be desirable. This enzyme or rather these enzymes, because *A. niger* has four of them (they differ in cofactor requirements), are largely inhibited by higher-than-average concentrations of citric acid. Thus there is little, if any, transfer of isocitrate to α-oxoglutarate. Finally, a comment about α-oxoglutarate dehydrogenase. Synthesis of the enzyme in *A. niger* is repressed by glucose and by increased ammonium ion concentration. In citric acid fermentation the latter is above physiological levels; thus α-oxoglutarate dehydrogenase is inactive.

SAQ 5.2

Draw your own version of the reactions from pyruvate onwards to incorporate your understanding of the TCA cycle in *A. niger* during citric acid formation.

The diagram looks very promising in terms of citric acid formation in that α-oxoglutarate dehydrogenase is inactive, isocitrate dehydrogenase has very low activity and aconitase equilibrates 90% towards citric acid.

Π What problem would you expect the organism to have if the scheme shown in the diagram (response to SAQ 5.2) was operating?

The answer is that not only is there no apparent regeneration of oxaloacetate; there is seemingly no way of producing any of the TCA cycle intermediates from succinyl CoA through to oxaloacetate.

pyruvate carboxylase

Originally it was thought that the glyoxylate cycle was important in regenerating at least some oxaloacetate in *A. niger*. However it is now known that there is a particularly active pyruvate carboxylase which operates in glucose grown *A. niger*. This enzyme carboxylates pyruvate to oxaloacetate and is not subject to metabolic regulation (it is a constitutive enzyme), therefore:

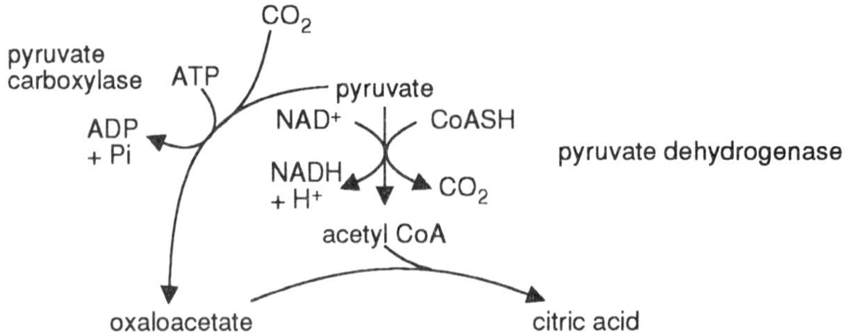

The oxaloacetate and acetyl CoA can both be generated if necessary in a 1 to 1 ratio to produce citrate without loss of carbon dioxide.

Π Look at the answer to SAQ 5.2. Can you see another advantage of carboxylating pyruvate to produce oxaloacetate?

The answer is that there is now a mechanism by which all of the other TCA cycle intermediates from oxaloacetate to succinyl CoA can be produced (all of these reactions are reversible).

There are several apparent problems which we still need to resolve in this section before we have a better understanding of citric acid production.

citrate synthase

pyruvate kinase

Let us consider Figure 5.3 again. Both pyruvate kinase and citrate synthase (enzymes III and V) are inhibited by elevated ATP concentrations. During citric acid production ATP concentrations are likely to arise (ATP produced in glycolysis) and either of these enzymes could, if inhibited, slow down the process. In fact all of the evidence suggests that both enzymes are modified or controlled in some way such that they are insensitive to other cellular metabolites during citric acid production.

phosphofructokinase

We have indicated earlier that phosphofructokinase is the major regulator of glycolysis. It is inhibited by high citrate and ATP concentrations and stimulated by high AMP and ADP concentrations. During citric acid production the cell would normally have low AMP + ADP and high citrate plus ATP levels. Phosphofructokinase should, therefore, under 'normal' conditions be inhibited during citric acid production. However, research has shown that this enzyme becomes insensitive to these cellular metabolites when elevated levels of ammonium ions are present. Such conditions are brought about during manganese deficiency.

manganese deficiency

We have not mentioned manganese (Mn^{2+}) before in this Chapter but its presence, or rather its absence, is crucial to successful citric acid production. Manganese deficiency in *A. niger* causes:

- lowering of the TCA cycle enzymes (synthesis of enzymes repressed);
- a severe limitation of protein turnover;
- elevated ammonium ion levels;
- changes to the cell surface - altering permeability and also the shape of the hyphae.

Figure 5.4 summarises the changes occuring in *A. niger* in citric acid production mode when compared to conventional metabolism. It is worth studying the Figure for some time because it explains some of the features necessary for a successful fermentation process.

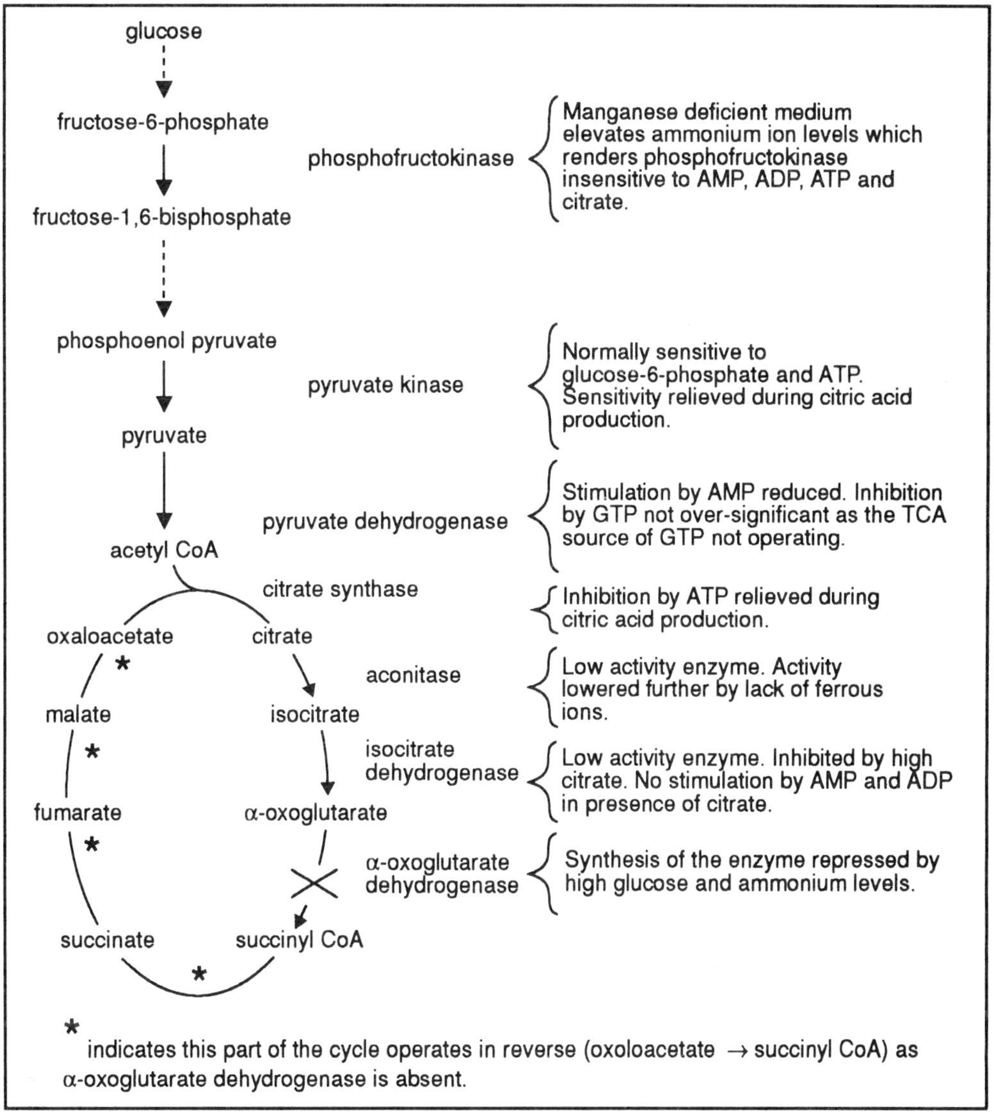

Figure 5.4 Changes in certain allosteric enzyme sensitivities of *A. niger* in citric acid producing mode compared to general metabolism. Conditions required for citric acid accumulation: manganese deficiency; ferrous ion deficiency; low pH; high concentration of glucose.

There is one, final, major problem to be addressed. Compare Figures 5.3 and 5.4.

Π Can you identify exactly what still has to be resolved?

The answer is that many of the enzymes are sensitive to elevated NADH concentrations. Glycolysis and the bridging reaction will produce NADH in a relatively high quantity.

Π If we suggested that a 'simple' answer is - "that all of the relevant enzymes are insensitive to NADH levels during citric acid production" - would this solve the problem?

Looking at Figure 5.3 in isolation it would appear to be a satisfactory solution. However in practice it is not sufficient. Remember that there is only a finite amount of NAD^+ and NADH in cells. Therefore the NADH has to be continually reoxidised to NAD^+ making the latter available for continued operation of glycolysis.

Π Can you remember the name of the two processes involved in the oxidation of NADH? What is the relationship between these two processes?

The two processes are electron transport and oxidative phosphorylation. NADH is reoxidised by the process of electron transport using the electron transport chain and the energy released from this process is harnessed by oxidative phosphorylation to generate ATP. We noted earlier that the two processes are intimately linked or coupled. Normally one cannot proceed without the other.

During citric acid production there is massive generation of NADH but little demand for ATP. Thus the situation could quickly arise where there is no further ADP available for oxidative phosphorylation within the cells. This means that the electron transport chain cannot operate and no further oxidation of NADH can occur.

two electron transport chains

It has been shown however that *A. niger* whilst producing citric acid is unusual in that it operates two separate electron transport chains (Figure 5.5). One is very conventional, similar to that in other organisms; it is tightly coupled to oxidative phosphorylation and inhibited by antimycin and cyanide. The other, still not completely understood, is not linked to oxidative phosphorylation and is inhibited by salicyl-hydroxamic acid (SHAM). Thus oxidation of NADH can occur independently of ATP synthesis; the energy generated is not conserved but is lost as heat energy.

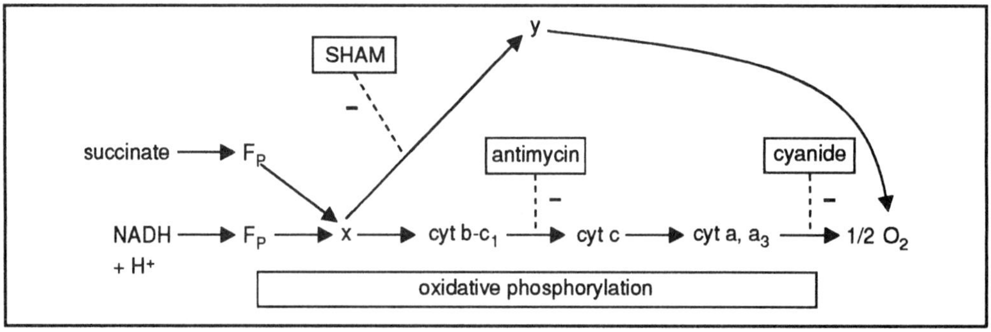

Figure 5.5 Respiratory chains in *A. niger*. SHAM = salicyl-hydroxamic acid; F_P = flavoprotein; - = inhibition; x and y are unidentified components.

The overall reaction for the metabolic conversion of glucose to citric acid is:

glucose + 0.5 O_2 → citric acid + $2H_2O$

effect of O_2 starvation

Thus, in theory at least, there are two substrates for citric acid production: glucose and oxygen. The latter has to be supplied at all times because if oxygen supply is interrupted, even for a few minutes, then citric acid production drops dramatically and does not recover, even after a rapid resumption of the oxygen supply. Interestingly, growth is unaffected by a transient cessation of oxygen supply. It has been shown that the conventional electron transport chain is associated with normal growth mechanisms but the second, special electron transport chain, is specifically involved with citric acid production.

SAQ 5.3

Insert the missing words from the list below into the following paragraph, which describes the mechanism of citric acid accumulation in *A. niger*.

A large intracellular pool of __a)__ ions (caused by a negative effect of severe limitation of __b)__ on protein turnover) and an increased respiratory activity, which in part is not coupled to __c)__ synthesis, stimulates metabolic flux through glycolysis without significant metabolic control. This, together with __d)__ pyruvate carboxylase and the peculiarities in the operation of the TCA-cycle, results in elevated cellular concentrations of __e)__. This in turn enhances citric acid accumulation by inhibiting __f)__ dehydrogenase.

Word list: isocitrate, constitutive, ammonium, ATP, manganese, citrate.

SAQ 5.4

For each of the following enzymes in *A. niger*, select metabolites from the list provided that either inhibit or stimulate activity during a) balanced growth and b) during citric acid accumulation (caused by Mn^{2+} deficiency).

enzyme	balanced growth		citric acid accumulation	
	inhibits	stimulates	inhibits	stimulates
phosphofructokinase				
pyruvate kinase				
pyruvate carboxylase				
pyruvate decarboxylase				
citrate synthase				
isocitrate dehydrogenase				

metabolites:

ATP
ADP
AMP
GTP
citrate
glucose-6-phosphate

In summary, and regarded at its simplest, *A. niger* degrades suitable carbon sources by glycolysis which is occurring at an uncontrolled rate. There is inevitably an overflow at the pyruvate level which is converted to citric acid.

5.3.4 Medium requirements and environmental factors involved in citric acid production

An understanding and awareness of the biochemistry of citric acid production has given us the opportunity to predict suitable medium and environmental conditions.

glucose

The substrate must obviously be either glucose, sucrose or a mixture of simple compounds such as molasses. It is important that the *A. niger* can obtain glucose readily and rapidly. Many polysaccharides are not suitable here because the rate at which they are degraded becomes limiting. The optimal concentration of sugars in the medium seems to be of the order of 15 - 25%.

unwanted metal ions

One of the possible problems of using the complex mixtures is the presence of unwanted materials such as certain metal ions, particularly manganese and iron. In addition, inhibitory compounds such as acetic acid may be present. All need to be removed before fermentation can proceed.

nitrogen

Nitrogen is normally supplied as an ammonium compound in citric acid fermentations and sufficient has to be supplied to enable the effect of manganese deficiency (increased levels of ammonium in the metabolic pool) to occur. Remember that increased metabolic pool ammonium has the effect of releasing the allosteric controls exerted on phosphofructokinase.

phosphate

In the presence of sufficient metal ions such as zinc, phosphate deficiency is known to inhibit growth and increase yields of citric acid. However, phosphate is added not only as a source of phosphorus but also as phosphoric acid to acidify the medium. Restricted growth but good citric acid yield is also achieved by maintaining iron and zinc deficiency; hence low phosphate levels are not necessary.

potassium hexacyano-ferrate

Trace metals have to be removed, notably manganese, ferrous ions and zinc. This is often accomplished using the compound potassium hexacyanoferrate which precipitates or complexes the metals and, in excess, acts to inhibit growth and indirectly promotes citric acid production. The amount of potassium hexacyanoferrate required is variable depending on the nature of the ion content of the carbon source.

Finally, aeration levels have to be carefully monitored and maintained continuously. Remember that it is particularly important that oxygen supply is not interrupted, not even temporarily.

SAQ 5.5

List the main ingredients and properties of a possible growth medium together with any comments regarding optimal environmental conditions. Try to give brief reasons for your choices.

5.3.5 The fermentation process

Currently citric acid is still manufactured almost entirely using *A. niger* growing on carbohydrate substrates. Three types of fermentation process have been used:

- the Japanese Koji process;
- the surface culture process;
- the submerged culture process.

The Koji process

Koji process

The use of the Koji process is very restricted and apart from this paragraph, it will not be discussed further. The substrates are either starch or the residue of sweet potatoes which are placed in shallow trays. Water is added to a 70% weight ratio. Steaming of the mixture yields a pasteurised paste which is inoculated with *A. niger*. The initial pH is 4.5 and an incubation temperature of 30°C is used. The pH drops to 2.0 during citric acid production and the process is completed, optimally, in four days. Citric acid is obtained as an aqueous solution from crushed mycelial/substrate residue and purified by conventional means, as will be discussed shortly. The advantage is that the process time is short and local resources can be used. However, the yield is low and it has proved difficult to remove effectively the trace metals from the substrate economically. Reliance therefore has been placed on obtaining *A. niger* strains which accumulate citric acid in the presence of such elements - so far with limited success.

The surface culture process

surface culture process

A flow diagram of the process is presented in Figure 5.6. The surface culture process is so called because the fermentation takes place in shallow trays. The trays are conventionally stacked up to ten high and several stacks may be placed together in the same room. The requirements for the plant are simple: limited cleaning of the room to produce aseptic conditions at the beginning and a facility to pass warm, moist air over the trays.

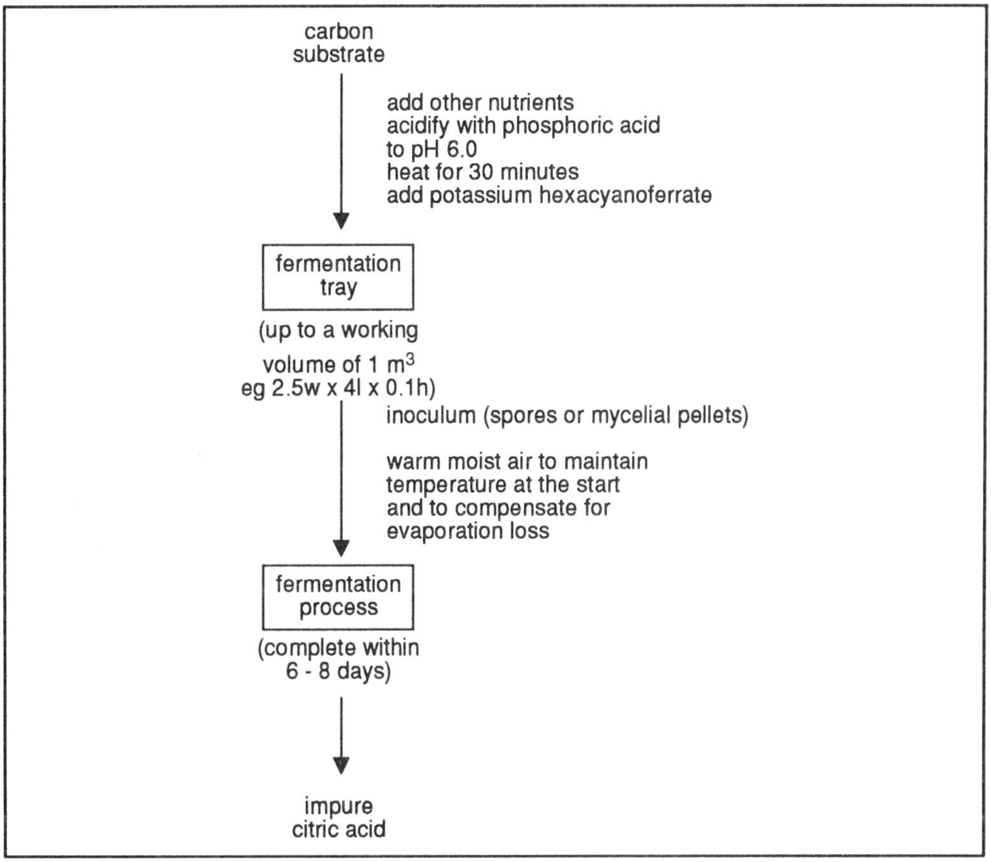

Figure 5.6 The surface culture process.

The raw substrate, usually molasses at 20 to 25%, together with added nutrients is acidified and heated to reduce the level of contamination in the medium. Sufficient amounts of potassium hexacyanoferrate are added to precipitate or chelate the trace metals, together with a controlled excess to function as a metabolic inhibitor. Care is required since a vast excess will cause unwanted effects in the fungus and dramatically lower the yields. A prior knowledge of the trace metal content of the starting material is required because the amount of potassium hexacyanoferrate to be added will depend upon it.

The trays are charged with the hot medium and, on cooling to 30°C, the inoculum is added. Two days are normally required for germination and growth and towards the end of this period the pH drops. These first two days are crucial in a microbiological sense because it is here that problems of contamination are most likely to occur.

The process is usually completed within eight days after which a yield of 210 to 250 kg m^{-3} of citric acid may be obtained, assuming a conversion ratio of 100 g glucose to 75 g citric acid.

The submerged culture process

submerged culture process

The process (outlined in Figure 5.7) is more sophisticated in terms of initial plant requirements and is also more difficult and costly to operate. However, the volumes which can be processed exceed by far those used in the surface culture method. In addition, the relative space required is much less per cubic metre of medium.

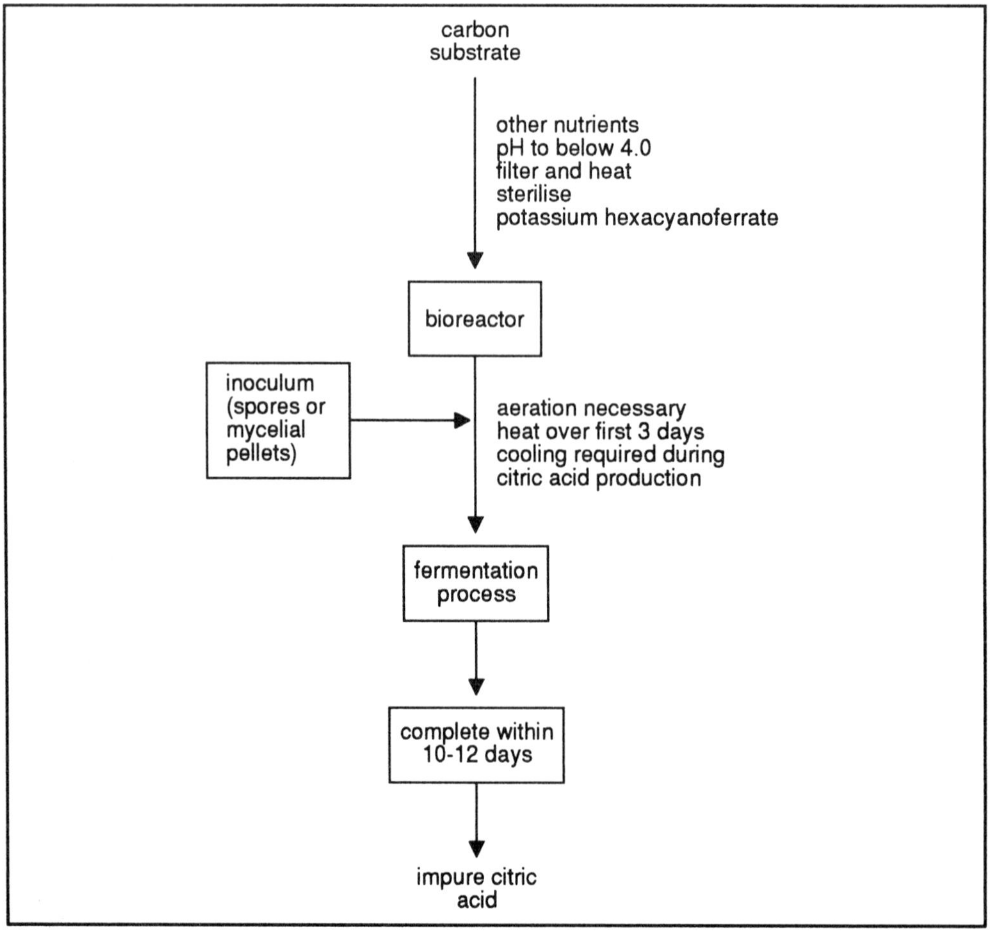

Figure 5.7 The submerged culture process.

The carbohydrate (again often molasses, 15 - 25%) and added nutrients are pH-adjusted to below 4.0 and, for this process, have to be sterilised. It is necessary to add potassium hexacyanoferrate but greater care is required in this process compared to surface culture. The *A. niger* seems to be more sensitive to and more easily inhibited by hexacyanoferrate in submerged culture. It is essential however to lower the ferrous and manganese concentrations, probably below 200 and 5 $\mu g\ l^{-1}$ respectively, to optimise the performance of *A. niger*.

working volumes

Conventional stirred reactors with working volumes of 50 to 150 m^3 have been used routinely for citric acid production whereas tower bioreactors, currently 200 m^3 and larger (greater than 600 m^3) are envisaged.

Inoculation is by conidia of *A. niger* or alternatively using pre-cultured mycelial pellets. Broken mycelial masses are slow to grow initially and are unsuitable here. Two to three days are required for germination during which heat input to maintain 30°C is required. During citric acid formation, cooling is necessary.

The process is complete after 10 - 12 days with yields of up to 125 kg m^{-3}.

Π Can you think of a reason why heat generation during citric acid accumulation is relatively high?

The reason for this is that reoxidation of NADH via the alternative electron transport chain (not coupled to oxidative phosphorylation) liberates heat.

In summary, the Koji process serves as a small scale, relatively localised and specialised process designed largely to take advantage of the available carbohydrate source.

The surface culture technique is undoubtedly easier to perform and cheaper to install. The restricted volumes of the system are, however, a disadvantage.

comparison of processes

The submerged culture process continues to increase in terms of the percentage of citric acid produced compared to that produced by the surface culture method. Tower bioreactors are preferred over stirred reactors because they cost less, there is less risk of contamination and they are less limited by size.

Continuous culture is not considered suitable for citric acid production; the requirement for a multi-tank system to separate growth and product formation would make the process uneconomic.

5.3.6 Downstream processing

After fermentation, large volumes of spent medium containing citric acid and mycelium remain.

Π Where do you think the citric acid will be located in the liquid culture?

The correct answer here is in both the medium and the mycelium. It is known that some 15% may remain in the mycelium immediately after harvesting.

rotary filtration

Usually the citric acid outweighs the biomass by a ratio of 5 : 1. The initial task is to remove the mycelium from the medium, a process usually carried out by rotary filtration. The wet biomass is crushed and rewashed to obtain most of the 15% citric acid contained within and the washings are added to the spent medium.

calcium hydroxide

The next stage is to precipitate the soluble citric acid as insoluble calcium citrate using calcium hydroxide.

Π Care has to be taken to ensure that the calcium hydroxide is very low in magnesium salts. Can you think of a reason for this?

The answer is that the objective is to precipitate out all of the citric acid as insoluble calcium citrate. Magnesium citrate is very soluble and would, therefore, be lost in the aqueous phase during the next separation.

The insoluble calcium citrate is dissolved in sulphuric acid yielding soluble citric acid and insoluble calcium sulphate and other calcium salts. The solution is then evaporated, crystallised as necessary to purify it, centrifuged, washed and dried; this leaves purified citric acid.

percentage recovery

Downstream processing recovers up to 95% of citric acid in the culture. The process is summarised in Figure 5.8.

The final product is marketed as an anhydrous crystalline or monohydrate crystalline compound available as a powder or in granular form.

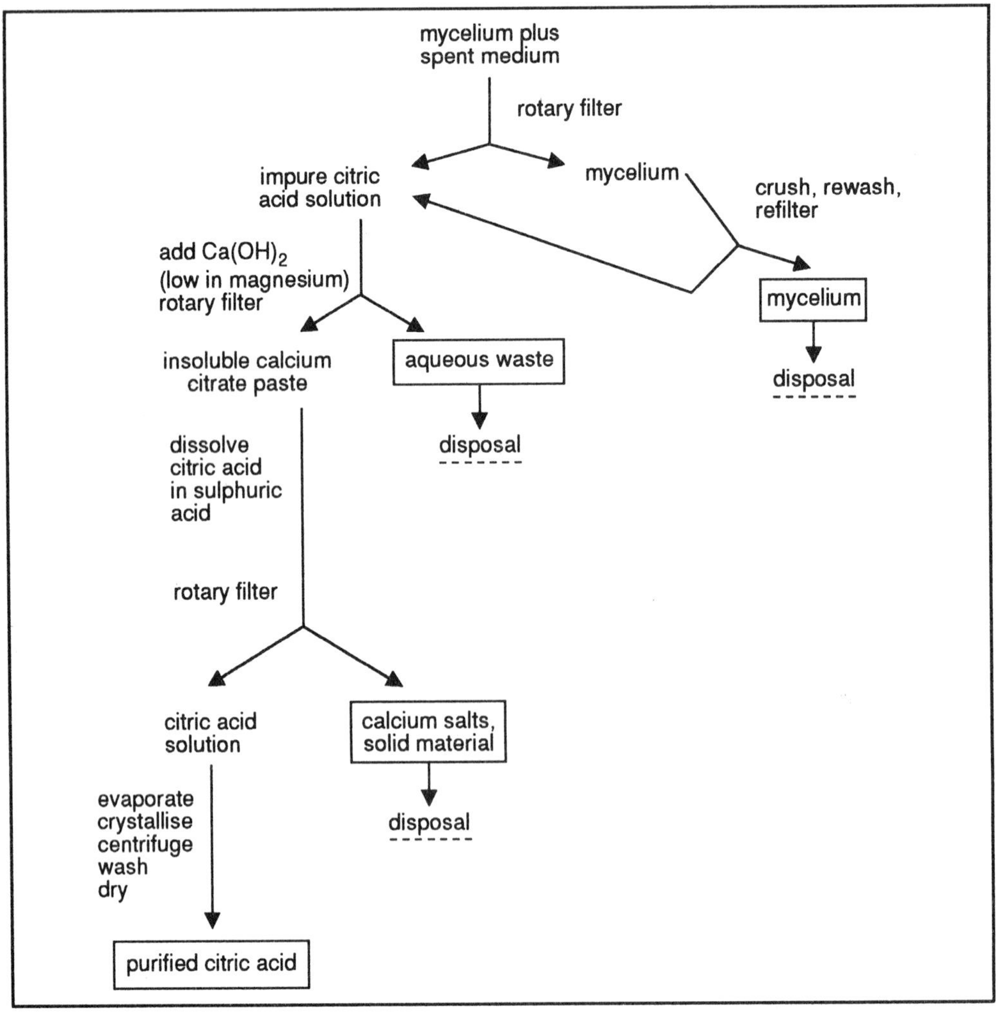

Figure 5.8 Downstream processing in citric acid production.

SAQ 5.6

Match each of the following statements with a process for citric acid formation using *A. niger*.

Statements

1) The process produces the highest yields of citric acid.
2) The process is complete within four days.
3) The starting pH of the medium is less than 4.0.
4) Nutrients are sterilised prior to inoculation.
5) The substrate is starch or the residue of sweet potatoes.
6) The medium is acidified with phosphoric acid to pH 6.0.
7) Cooling required during citric acid accumulation.

Processes

The surface culture process.
The Koji process.
The submerged culture process.

5.4 The production of other TCA cycle intermediates

In addition to citrate, there are other acidic intermediates of the TCA cycle and several of these are used in industry. Even though some have been produced by fermentation, production is currently from other, cheaper sources.

5.4.1 Fumaric acid

fumaric acid

Fumaric acid is used in the plastics industry, in the food industry and as a source of malic acid. Although demand has increased rapidly over the last 30 years its production from fermentation has been totally replaced by a chemical method. It is now produced far more cheaply by the catalytic oxidation of hydrocarbons, particularly benzene. With the continuing uncertainties concerning the availability and cost of petroleum, however, fermentation may yet be a viable alternative.

Fumaric acid can be produced in high yield by several genera of fungi, notably *Aspergillus* and *Rhizopus*, using glucose, corn steep liquor and other substrates. Large amounts of calcium carbonate have to be added to neutralise the fumaric acid.

Several *Candida* spp. will metabolise *n*-hydrocarbons to produce fumaric acid; the exact process is not fully worked out although glycolysis and reverse TCA are central features of the biochemistry of the process.

5.4.2 Malic acid

malic acid

Malic acid has a limited use in the food industry as an acidifying agent where it is an alternative to citric acid. In nature, only L(-) malic acid is found whereas the relatively cheap, chemical synthetic methods yield D/L mixtures. The favoured industrial way to produce the L(-) acid is by enzymic transformation from fumaric acid. Either whole cells or isolated and immobilised enzymes can be used, with high conversion efficiencies.

In the late 1970's attention again turned to microbial fermentation processes. Several *Candida* spp. were shown to produce high yields of malic acid from *n*-paraffins. The problem is that large amounts of both malic acid and succinic acid are produced, and only malic acid is desired as succinate can be produced very cheaply by chemical means. Careful adjustment of media conditions has enabled *Candida brumptii* to produce 0.88 g malic acid per g of *n*-paraffin, yielding 25 g per litre. Growth occurs over two days followed by a four day product formation phase. Biochemically, the glyoxylate cycle is very important during malic acid production, channelling isocitrate and acetyl CoA to two malate molecules in the presence of a much reduced complete TCA cycle.

glyoxylate cycle

SAQ 5.7

Draw a simple sketch diagram (refer to Figure 5.2 if you need help) and then, by using equations, show that the following statement is correct: 'the glyoxylate cycle and part of the TCA cycle are important in malic acid production'.

5.5 The industrial production of itaconic acid

5.5.1 Introduction

Itaconic acid was discovered in 1929 as a metabolite of an *Aspergillus* species which was subsequently named *Aspergillus itaconicus*. A short time later *A. terreus* was shown to accumulate the acid and one particular strain of this, superior to all others, is still the current industrial producer of itaconic acid.

methylene succinate

The compound itaconic acid - or methylene succinate - is a substituted methacrylic acid having the structural formula:

$$\begin{array}{l} CH_2 = C - COOH \\ \qquad\quad | \\ \qquad\;\; CH_2 \\ \qquad\quad | \\ \qquad\; COOH \end{array}$$

It is used largely in the plastics and paints industries and, when used at the 5% level in acrylic resins, it imparts the ability to hold printing inks. It is sold as the free acid in two grades: industrial grade or the purer, more expensive, refined grade.

5.5.2 The biochemistry of the process

Originally itaconic acid was produced chemically by the pyrolysis of citric acid. The treatment results in water loss converting citric acid to aconitate. Subsequent decarboxylation of the latter gives two isomers, itaconic acid and citraconic acid. The process was not very successful, partly because succinate and itatartaric acids were also produced.

two pathways

It was originally thought that microbial fermentation by *Aspergillus terreus* followed a similar pattern in that citric acid - produced by glycolysis plus the bridging reaction and the first reaction of the TCA cycle - was converted to aconitic acid. This compound was then decarboxylated to itaconic acid. This pathway is the one to the right of Figure 5.9 (alternative B). However, further studies of the process showed that the left hand scheme (alternative A) was also a possible metabolic route to itaconic acid. In the latter the presence of a condensing enzyme which combines pyruvate and acetyl CoA to yield citramalic acid has been confirmed. Subsequent conversion to itaconic acid via citraconic acid and itatartaric acid has been demonstrated.

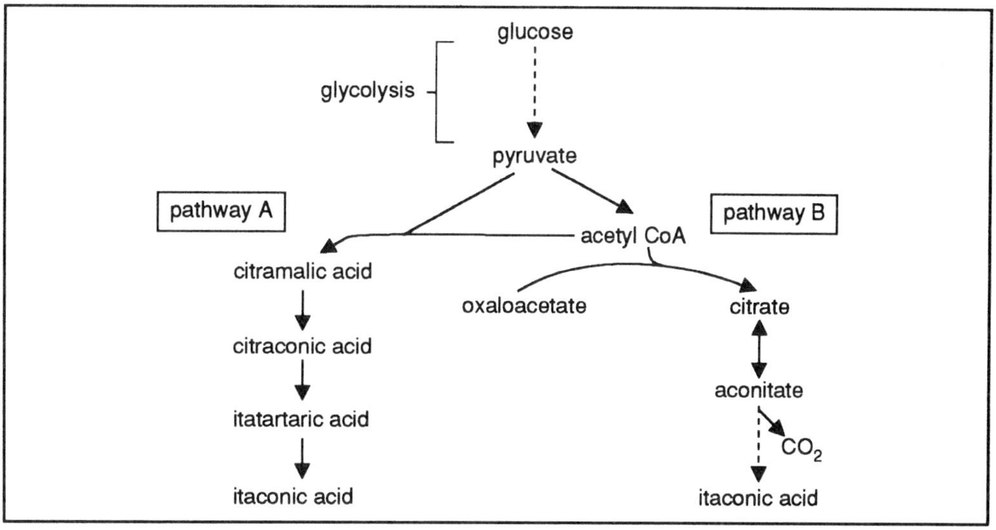

Figure 5.9 Two routes of formation of itaconic acid.

copper ions

The effects of copper ions on the process were elucidated as follows. It was known that copper ions inhibit aconitase activity.

Π What would be the effect on itaconic acid yield of adding excess copper ions if 1) scheme B was the major pathway and 2) scheme A was the major pathway (see Figure 5.9)?

1) With B as the major pathway the yield would fall dramatically if aconitase was totally inactivated.

2) With A as the major pathway, it is reasonable to assume that there would be no effect on itaconic acid production since aconitase is not involved. In fact it was shown that the presence of copper ions increases the yield of itaconic acid by a factor of up to 3.

These facts indicate that the major or only pathway involved in itaconic acid production is pathway A.

The empirical formula for glucose is $C_6H_{12}O_6$ and for itaconic acid it is $C_5H_6O_4$, so itaconic acid production is another 'aerobic fermentation' and requires aeration.

$$C_6H_{12}O_6 + 1.5O_2 \rightarrow C_5H_6O_4 + CO_2 + 3H_2O$$

Like in citric acid production relatively modest aeration rates are required. Stopping aeration for literally a minute or so will permanently arrest product formation.

5.5.3 The fermentation process

sugar beet molasses

3 day fermentation

Much of the details of the process remains in the hands of the manufacturers but an overall description will be presented here. Conventional stock cultures of *A. terreus* are maintained and, when required, are germinated in large volumes of sugar beet molasses (15% sugar) medium. Continuous aeration for 18 hours and incubation at 33 - 37°C ensures germination into mycelia of a suitable biochemical type. By the time of inoculation of the fermenting vessel the pH will have dropped from 7.5 to around 4.0. Inoculation using one fifth of the volume together with maintenance of a higher temperature, 45°C, reveals a pH change of 4.1 down to 3.1 over the next 24 hours. Fermentation is allowed to continue for a further 2 days during which the pH will be brought back to pH 3.8 using either calcium hydroxide or ammonia. If ammonia is used, ammonium itaconate is found in the purified free acid product; this is an irritation rather than a problem and is preferable to addition of calcium hydroxide. The reason for this is that the resulting calcium itaconate, which is relatively insoluble, would stick to the mycelium and increase aeration problems. Throughout the fermentation vigorous aeration with agitation is required.

5.5.4 Downstream processing

filtration and acidification

A simplified scheme is shown in Figure 5.10. The fermentation mixture is filtered to remove mycelium and suspended solids. The solution is treated with hot carbon and then filtered. During the process acidification is necessary to reverse the neutralisation by calcium hydroxide or ammonium hydroxide employed during fermentation.

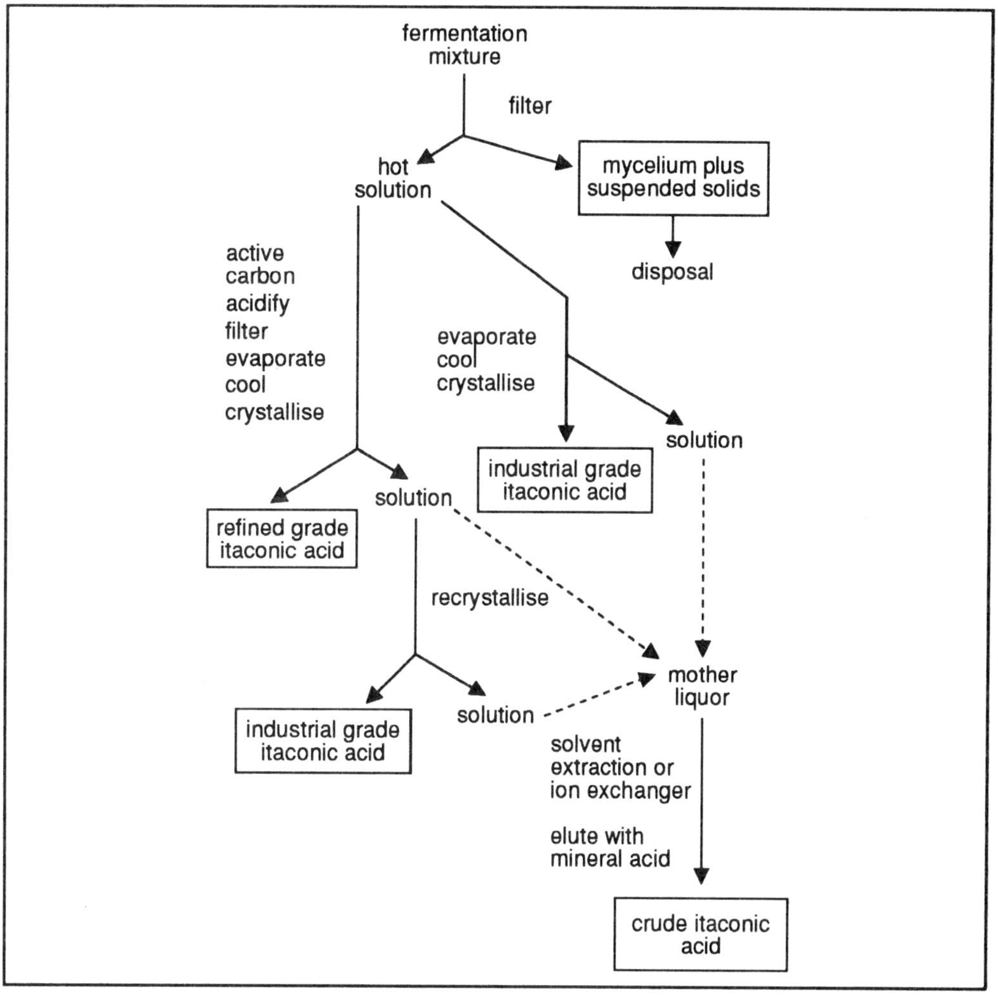

Figure 5.10 Downstream processing in itaconic acid production. Any of the four products can be esterified as required.

The filtrate following carbon treatment is evaporated, allowed to cool and crystallised. This yields refined grade itaconic acid. The solution may be further evaporated to yield a second sample - this time designated industrial grade. Industrial grade samples can also be made directly by evaporating the fermentative filtrate and thus avoiding the activated carbon step.

Finally the mother liquor can be extracted by solvents (such as *n*-amyl alcohol) or passed through a suitable ion exchange resin. Elution is by mineral salts to yield a crude grade mixture.

If itaconic acid esters are required, esterification of any of the grades can be carried out yielding esters of varying purity.

SAQ 5.8

Give three reasons why it is believed that the formation of itaconic acid in *A. terreus* is via a route which does not involve TCA cycle intermediates.

5.6 The industrial production of gluconolactone and gluconic acid

5.6.1 Introduction

one-step and two-step reactions

The production of gluconolactone and gluconic acid are, in a biochemical sense, the simplest of the processes studied in this Chapter. From glucose they are only one-step and two-step reactions respectively. Their production by bacteria was first detected in 1880 and gluconic acid was then demonstrated in fungal culture filtrates in 1920. *Aspergillus niger* was found to produce large quantities of gluconic acid, particularly when the acid was neutralised by the addition of calcium carbonate. Investigation and use of surface cultures and various types of submerged cultures occurred. Finally, a submerged method using sodium hydroxide to neutralise the acid produced was developed.

submerged processes

Currently only submerged processes are used commercially. It is interesting that even though the conversion is very simple, a chemical process is not favoured. In fact several different oxidising agents are available but the processes have proved to be too costly and less efficient.

The ability of bacteria - particularly *Pseudomonas* spp. and *Gluconobacter* spp. - to produce gluconolactone and gluconic acid has been exploited and the process is used commercially, mainly in the production of the lactone.

sodium gluconate sequestering agent

Whilst world wide demand continues to increase, the requirement is far lower than that for citrate. Gluconic acid is commercially available as a 50% technical grade solution. In addition technical grade calcium and sodium salts of gluconic acid together with gluconolactone are produced. Quantitatively the greatest demand is for sodium gluconate which is a very efficient sequestering agent in neutral or alkaline environments. The hydroxyl groups of the gluconate bind di- and tri-valent metals (for example calcium, magnesium and iron) in soluble form, preventing precipitation. There is an industrial demand for sodium gluconate in the cleaning of glassware, for example bottles, particularly when automated washing equipment is used. In addition it is used for washing walls, metals and other surfaces to remove insoluble metal carbonates.

calcium and iron deficiencies

Calcium gluconate is one of the relatively few soluble calcium salts and is used in the pharmaceutical industry as a source of calcium for patients with calcium deficiency. Many drugs are supplied as the gluconate derivatives. Other gluconates such as iron gluconate can be used, in this case to treat iron deficiency.

The free acid is a mild acidulant and is used in a variety of foods. It is also used as a cleansing agent, for example in the dairy industry to prevent the build up of 'milk stone'.

All of the gluconates can be used as a cement or plaster additive where they retard setting times whilst increasing the strength and water resistance of the materials.

Finally gluconolactone - largely produced from *Gluconobacter suboxydans* is used in baking powder and bread mixes and other areas where its effervescent properties may be exploited.

5.6.2 The biochemistry of the production of gluconic acid and derivatives

hexose monophosphate pathway

Virtually all living systems have a degradative pathway called the hexose monophosphate pathway which has the initial reactions:

glucose → glucose – 6 – phosphate → 6 – phosphogluconolactone → 6 – phosphogluconate

The dehydrogenase catalysing the glucose-6-phosphate to 6-phosphogluconate step is specific for the phosphorylated derivative of gluconate.

glucose oxidase

A similar but independent process involving non-phosphorylated derivatives is found in several fungi, notably *Aspergillus* and *Penicillium* spp. and also bacteria notably *Pseudomonas* and *Gluconobacter* spp. In this pathway free glucose is converted to gluconolactone directly by the enzyme β-D-glucose: O_2 oxidoreductase (glucose oxidase).

$$\text{D-glucose} + O_2 \xrightarrow[\substack{\beta-\text{D}-\text{glucose}:O_2\\ \text{oxidoreductase}}]{} \text{D-gluconolactone} + H_2O_2$$

Glucose oxidase is specific for β-D-glucose and where α-D-glucose is the available substrate, prior conversion from the α to the β form is required.

continuous aeration

In this reaction, hydrogen peroxide is produced which is toxic to cells and has to be removed quickly and efficiently. This is carried out by the enzyme catalase. The equation also indicates the need for molecular oxygen and the fermentation process needs a continuous supply of air.

Conversion of the gluconolactone to gluconic acid occurs under certain conditions by spontaneous hydrolysis, though a rather specific lactonising enzyme is present in cells.

The control of glucose oxidase production is still not fully elucidated, nor is the role of gluconic acid in cells. However, elevated levels of glucose oxidase are found in conditions of high glucose concentrations and above normal oxygen tension.

5.6.3 The fermentation process

The first culture technique, reported in 1927, to be attempted commercially was a surface-culture, shallow-pan technique, though this method has not been used for many years. Relatively soon after this, in 1933, production using a submerged culture technique was reported and this method has been in use continuously since then. Various significant developments have been made, notably the addition of calcium carbonate to neutralise the acids produced in order to increase yields (1937) and the use of sodium hydroxide for neutralisation (1952).

Π Write down the fermentation conditions which you think may encourage optimal gluconic acid production.

There are several conditions that you could have listed which have been mentioned so far in Section 5.6. For example:

1) Perhaps the most important of all is that there should be a high concentration of glucose, between 15 and 35% depending on the type of fermentation.

2) Secondly the concentrations of the nitrogen source should be carefully calculated and kept relatively low. Ideally the process requires minimal growth and maximum product accumulation. High concentrations of nitrogen source would probably encourage far too much growth thus lowering the yield efficiency.

3) Sufficient quantities of other essentials such as minerals must be provided.

4) The pH has to be controlled - the acid which is produced has to be neutralised maintaining a pH in excess of 6.0. Below pH 3.0 the glucose oxidase is inactivated and in fungal systems low pH encourages citric acid production.

5) High aeration rates are required to stimulate glucose oxidase production and activation.

Depending on which of the products, calcium gluconate, sodium gluconate or gluconic acid (free acid) is required, the fermentations have some basic differences.

Calcium gluconate

low solubility

Either a spore suspension or mycelium can be used to inoculate the production vessel. The medium contains a maximum glucose concentration of 15%. The upper limit reflects the low solubility of calcium gluconate which is normally about 4% at 30°C, but it can form supersaturated solutions up to about 15% without risk of precipitation.

Sterilised calcium carbonate slurry is added in increments to the sterilised medium to maintain pH. Addition of all the necessary calcium carbonate at the beginning of the process would irreversibly and adversely affect the inoculum.

Fermentation usually occurs in a conventional stirred vessel at 30°C (with cooling) and vigorous aeration. The process from start to finish can take as little as 24 hours thus absolute sterilisation is not crucial. However, several processes reuse the mycelium many times and in these circumstances clean conditions are a minimum requirement.

Sodium gluconate

The two main differences between calcium gluconate and sodium gluconate production are that, in the latter, pH control is performed by addition of sodium hydroxide and the initial glucose concentration is different.

high glucose concentrations

pH control by sodium hydroxide is much easier and more precise than pH control using calcium carbonate slurry. Because the sodium gluconate is far more soluble (40% at 30°C) than calcium gluconate, higher glucose concentrations can be used to give higher production yields.

Medium containing up to 35% glucose is steam sterilised by a steam jacket around a conventional stirred tank reactor. pH adjustment starts as the medium cools and is maintained at 6.5. Vigorous aeration is again required. Inoculation is usually by using a mycelial suspension. Under these conditions a 30% glucose solution can be almost quantitatively converted to sodium gluconate within 36 hours.

5.6.4 Downstream processing

Downstream processing to obtain calcium gluconate is different to that employed to obtain either sodium gluconate or the free gluconic acid.

calcium gluconate

In calcium gluconate extraction, the mycelium and suspended solids are removed by filtration and the resulting filtrate is then usually decolourised by active carbon treatment. After further filtration to remove the carbon the solution is heated and evaporated to a 20% calcium gluconate concentration. On cooling, crystallisation - which may need encouraging by addition of seeding crystals - begins. After a short time the process is complete and the crystals are removed by centrifugation and washed with cold water. The pure crystals are then dried at 80°C or converted to other end products.

sodium gluconate

Sodium gluconate is easier to extract than the calcium salt. To obtain commercial grades of sodium gluconate the culture filtrate, after filtration to remove mycelium and suspended solids, is simply concentrated to 45% solids, pH-adjusted to 7.5 with sodium hydroxide and drum dried. Purer products may be obtained by active carbon treatment of the hot solution before drying or by carbon treatment and recrystallisation.

free gluconic acid and gluconolactone

Free gluconic acid and the gluconolactone may be obtained from a calcium gluconate fermentation or, most often, from a sodium gluconate fermentation. Calcium gluconate is treated with sulphuric acid and the calcium sulphate is subsequently removed by centrifugation to leave the free acid. Sodium gluconate is converted to the free acid by ion exchange.

gamma and *delta* lactones

Aqueous solutions of gluconic acid are equilibrium mixtures of gluconic acid itself, together with the *gamma* and *delta* lactones. The relative concentrations of each of these can be profoundly influenced by the temperature at which crystals are allowed to separate out, a situation aided by seeding with the appropriate required crystal. At temperatures below 30°C, and particularly between 0 to 4°C, free gluconic acid predominates. Between 30 and 75°C the predominant crystals are mainly *delta* lactone and above 70°C the principal crystals are the *gamma* lactone.

SAQ 5.9

Identify each of the following statements as true or false; if false give a reason for your response.

1) Free gluconic acid can be obtained from calcium gluconate by treatment with sulphuric acid.

2) During sodium gluconate formation, pH control is by addition of sodium hydroxide.

3) Starting glucose concentrations in calcium gluconate fermentations are higher than those for sodium gluconate fermentations.

4) D-gluconolactone is an intermediate in the hexose monophophate pathway.

5) Production of D-gluconolactone from glucose in several fungi is a one-step process catalysed by glucose oxidase.

6) Glucose oxidase is specific for α-D-glucose.

SAQ 5.10

Draw a flow diagram of downstream processing for gluconic acid and its derivatives. Commence with "fermentation mixture" and end with the following five products: calcium gluconate, sodium gluconate, free gluconic acid, *gamma* lactone and *delta* lactone. (Use the style of the flow diagram shown in Figure 5.10).

Summary and objectives

This chapter has described the industrial production of organic acids that are intermediate in the TCA cycle (citric acid, malic acid, fumaric acid) or oxidative derivatives of glucose (itaconic acid, gluconic acid and its derivatives). We have seen how an understanding of metabolic pathway regulation and the physiology of micro-organisms can be exploited to 'overproduce' central metabolites. The approach involves careful consideration of medium design and culture conditions, rather than genetic manipulation. Subsequent downstream processing depends on the type of organic acid product and may involve precipitation, evaporation and crystallisation stages. Further purification stages give rise to one or more commercial grades of the desired organic acid or its derivatives.

Now that you have completed this Chapter you should be able to:

- list organic acids from central metabolic pools that are in demand industrially and are produced commercially by micro-organisms;
- list the reactions of the TCA cycle leading to the formation of organic acids of commercial significance;
- discuss the mechanisms controlling glycolysis and the TCA cycle;
- describe in depth the metabolic pressures imposed on selected micro-organisms in order to obtain high yields of citric acid;
- compare and contrast types of citric acid fermentations;
- broadly describe the commercial production by fermentation of fumaric acid, malic acid, itaconic acid and various gluconic acid and gluconolactone derivatives.
- outline schemes for downstream processing of various organic acids.

Production and diversification of antibiotics

Production and diversification of antibiotics

6.1 An introduction to antibiotics

The discovery and production of antibiotics has been of tremendous importance to human and animal health care. Prior to their discovery about half a century ago, many bacterial infections caused debilitating diseases and fatalities were high. The discovery of antibiotics was a major step in the treatment of infectious diseases, especially those caused by bacteria. Today about 50,000 tonnes of antibiotics are produced annually. About a third of this consists of penicillins, whilst tetracyclines make up about a quarter of the market.

The first of the antibiotics that found practical use as a therapeutic was penicillin. The success of penicillin initiated a vast screening process all over the world, which resulted in the isolation of a large number of antibiotic substances from various natural sources. Many of these compounds were produced by micro-organisms and prove to be lethal for other micro-organisms. Many of these compounds were also very toxic to humans and could not be used therapeutically. Nevertheless a large number of classes of useful compounds were produced. The chemical structures of members of some of the most important classes are shown in Figure 6.1.

Π Examine Figure 6.1 and see if you can identify the β-lactam structure in the first four structures shown.

β-lactam ring

The structure you are looking for is a four membered ring containing three carbon atoms and a nitrogen atom in the ring. The structure you should have identified is

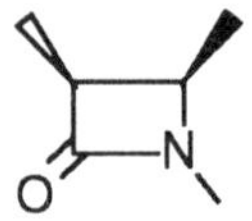

You will see that this structure contains a tertiary amide.

You should examine the other structures shown, but we would not expect you to remember the details of these structures. You should, however, be aware of the general forms of:

major classes of antibiotics

- quinolones;
- tetracyclines;
- glycopeptides;
- sulphonamides;
- aminoglycosides;
- macrolides;
- streptomycin.

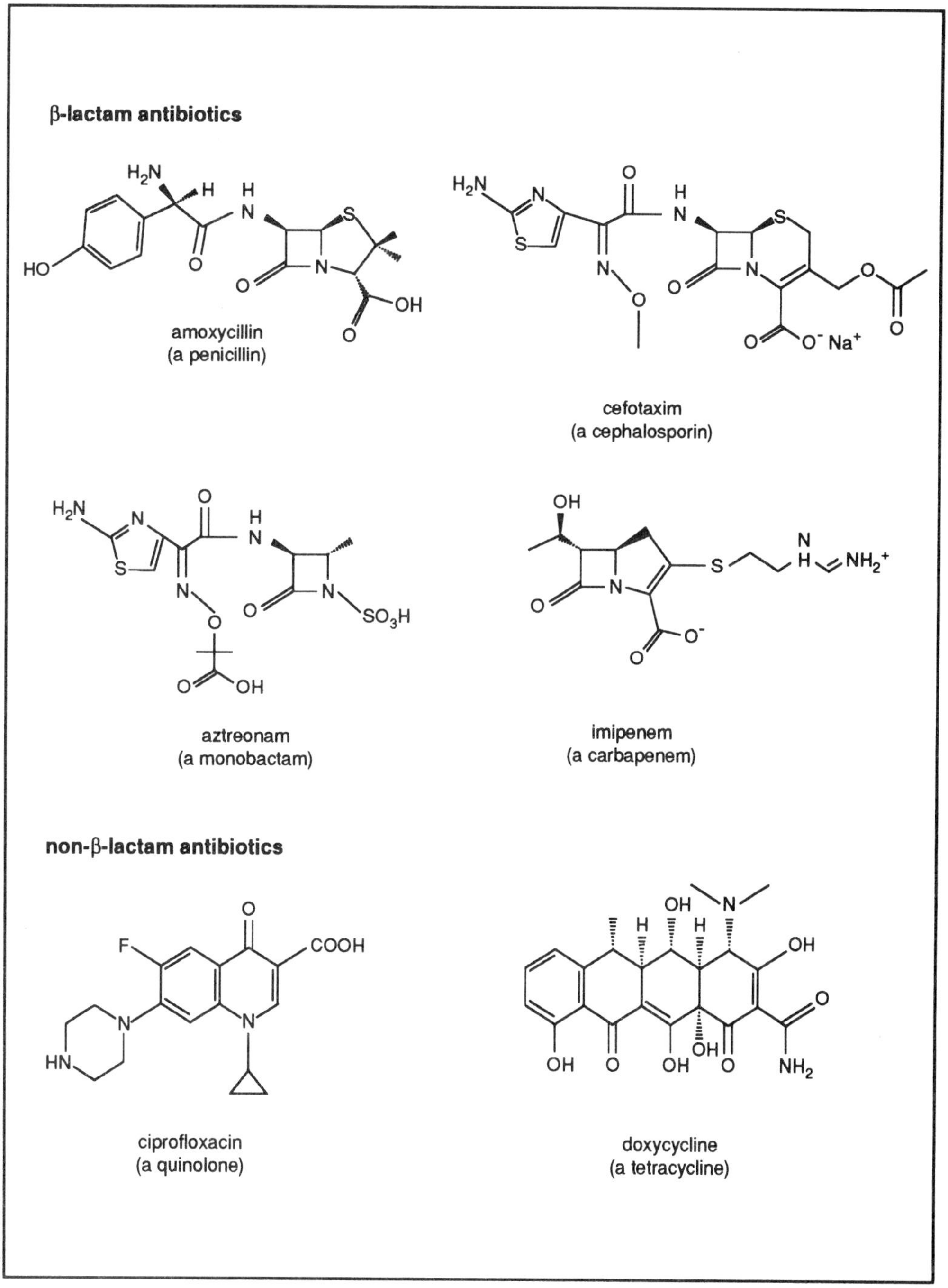

Figure 6.1 The chemical structure of some members of the important classes of antibiotics.

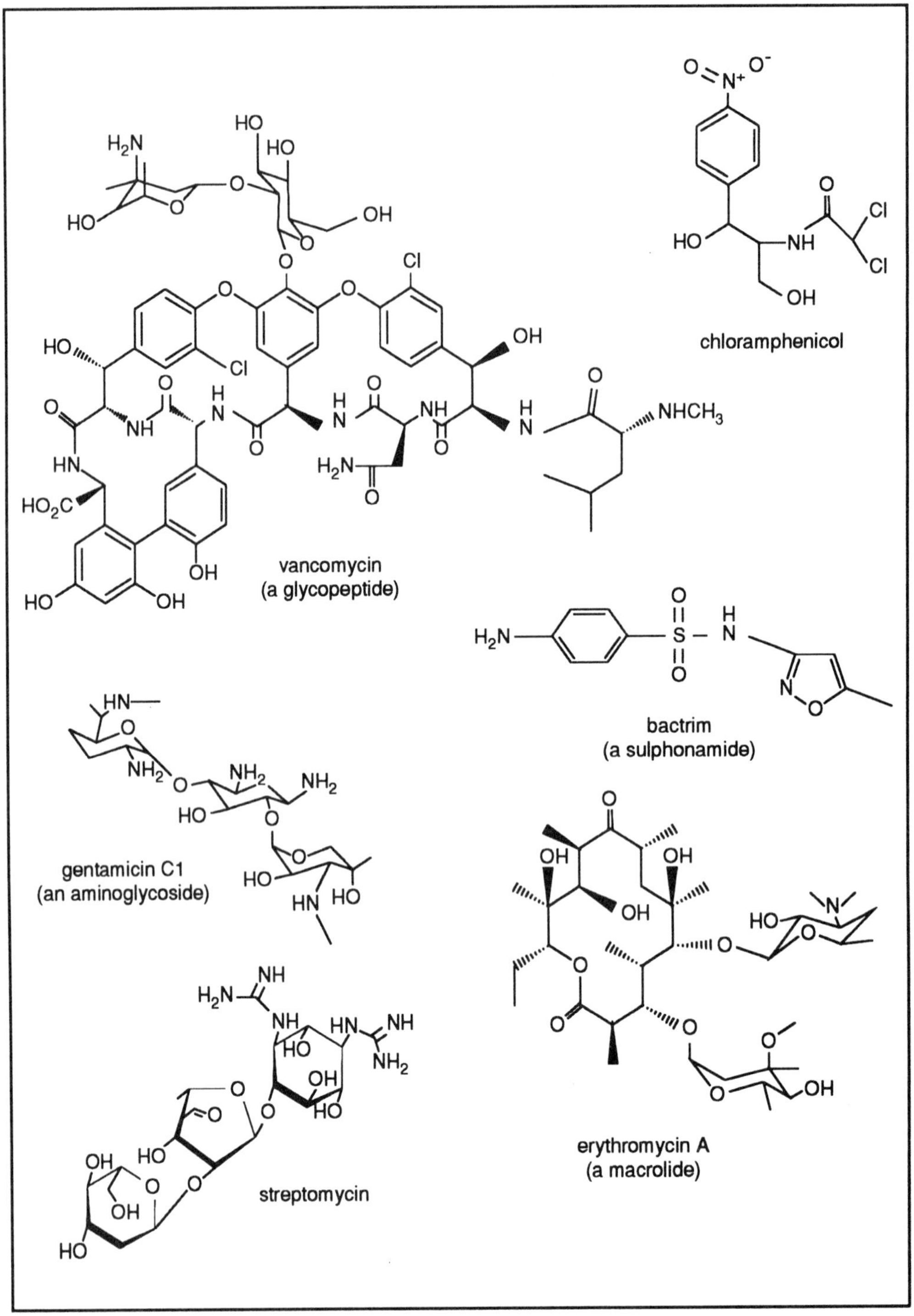

Figure 6.1 .... Continued.

Π From these structures, would you expect each of these groups of antibiotics to act in the same way in target organisms?

modes of action

You should have concluded that because these structures are very diverse, it is unlikely that they will act in the same way. This is, in fact, true; the mode of action differs from one class of antibiotics to another. We have listed some modes of action of antibiotics in Table 6.1.

Antibiotic	Mode of action
β-Lactams	Inhibition of synthesis of, or damage to, cell wall
Penicillins	
Cephalosporins	
Monobactams	
Carbapenems	
Vancomycin	
Bacitracin	
Cycloserin	
Fosfomycin	
Quinolones	Inhibition of synthesis or metabolism of nucleic acids
Rifampin	
Nitrofurantoins	
Nitroimidazoles	
Polymyxins	Inhibition of synthesis or damage to cytoplasmic membrane
Polyene antifungals	
Sulfonamides	Modification of energy metabolism
Trimethoprim	
Dapsone	
Isoniazid	
Aminoglycosides	Inhibition of protein biosynthesis
Tetracyclines	
Chloramphenicol	
Erythromycin	
Clindamycin	
Spectinomycin	
Mupirocin	
Fusidic acid	

Table 6.1 Modes of action of antibiotics.

We have this far established that the antibiotics are a diverse group of compounds that are produced industrially in large amounts which are of great value in health care. We need to establish one further point.

Π Write down two reasons why society needs so many different antibiotics and explain why we do need to continue to find ways of producing new antibiotics?

There are many reasons why we need to produce a large variety of antibiotics. Different disease causing micro-organisms have different structures and different metabolisms. Thus you should have anticipated that a particular antibiotic may be effective against a particular type of micro-organism but not against others. For example, traditional penicillins were more effective against Gram-positive bacteria than against Gram-negative bacteria. Furthermore, because of the large numbers of cells involved, their rapid rates of growth and the ability to transfer genetic material between often quite unrelated organisms, new varieties of disease-causing micro-organisms arise quite frequently. Amongst the changes that are detected amongst disease-causing micro-organisms is the development of resistance to antibiotics. This resistance may, for example, depend upon the production of enzymes that destroy the antibiotic or on changes to structural components of cells which result in the antibiotic not being taken up by cells, or on its failure to interact with the target component. The widespread use of antibiotics will itself act as a selection mechanism leading to the proliferation of antibiotic resistant strains. Thus, when penicillin was first introduced, most disease-causing Gram-positive bacteria were sensitive to this antibiotic. Now many such organisms are resistant to this antibiotic. In most cases, this resistance is based upon the production of an enzyme which hydrolyses either the β-lactam ring (β-lactamase) or the secondary amine linking the lactam ring to another moiety (penicillinase). In many instances the genes coding for the resistance factor are encoded in plasmids and are, therefore, readily transmitted from organism to organism. Many strains of bacteria now carry multiple antibiotic resistances.

resistance to antibiotics

It is for these reasons that a search for new antibiotics must continue.

In this chapter, we will examine strategies for producing antibiotics. We have had to be selective and have chosen to confine discussion largely to the β-lactams, with particular emphasis on the diversification of the primary antibiotics using biotransformation. We have adopted this strategy in order to produce a manageable study, while enabling us to explain the main principles involved.

chapter overview

We will begin by giving a brief overview of the strategies that may be employed to produce desirable antibiotics. Then we will give a brief review of the history of the production of penicillin. We will then examine the mode of action of β-lactam antibiotics and briefly describe the biosynthetic pathways of β-lactam antibiotic production. Subsequently we will examine, in greater depth, the biotransformation of penicillins. A consideration of cephalsporin production will follow and will be compared with the production and diversification of penicillins. In the final part of this chapter we will briefly describe the new β-lactams.

6.2 General strategies for the production of antibiotics

secondary metabolites

screening techniques

The major classes of antibiotics are secondary metabolic products of micro-organisms. Many were discovered by empirically screening culture filtrates or cell extracts for antimicrobial activity. A range of techniques (examples are methods using, impregnated discs, porous cylinders, cut wells, see Figure 6.2) have been used to carry out such screening.

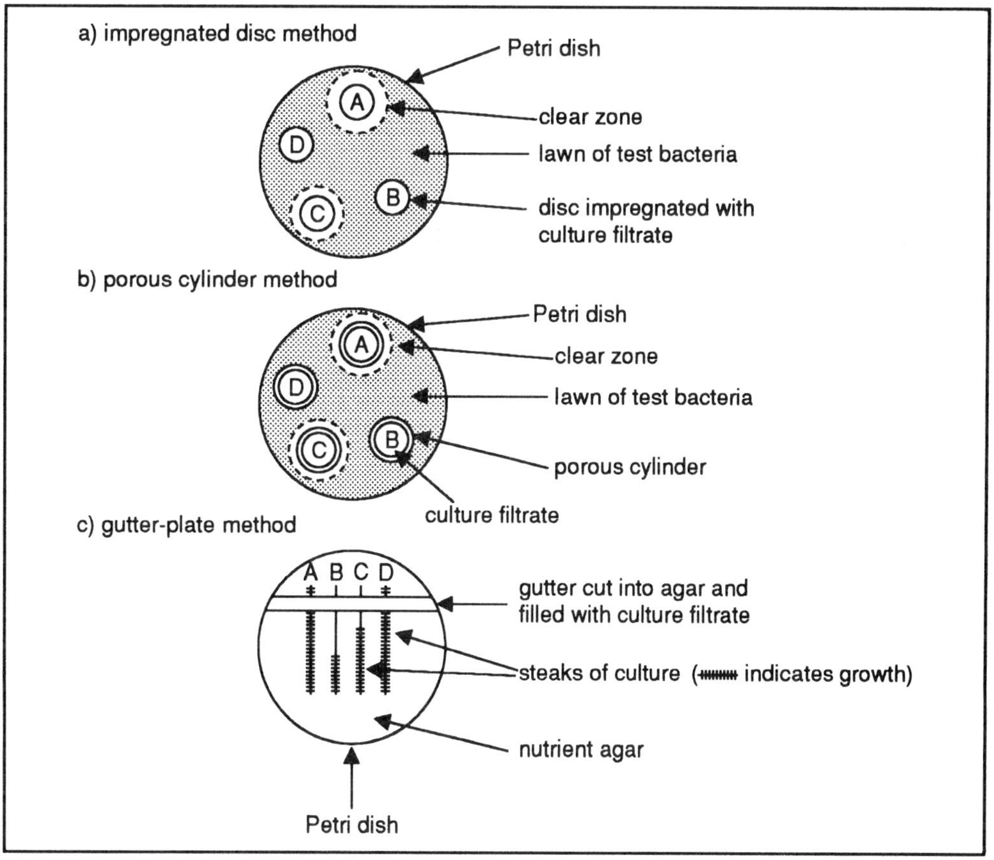

Figure 6.2 Examples of techniques used to screen microbial cultures for antibiotic activity. In a), filter discs are impregnated with the test sample and placed on the surface of a nutrient agar filled Petri dish, which had been seeded with bacteria. Anti-microbial activity is detected by the inhibition of growth around the impregnated disc. b) Illustrates a similar approach except in this case the test substances is placed in the centre of a porous cylinder. Antimicrobial substances diffuse out of the cylinder and inhibit growth around the cylinder. c) Illustrates another procedure that depends upon the diffusion of antibiotics through agar. In this case, a single sample is tested for antimicrobial activity against a range of organisms.

Π Examine Figure 6.2 carefully. Which test sample in a) appears to have the greatest antimicrobial activity?

The most obvious answer is A because the zone of inhibition is greatest around the disc impregnated with A. However, there are other possibilities. The size of the zone of inhibition depends on:

- the amount of antimicrobial agent present (ie its concentration);
- the rate at which it diffuses (this depends on its molecular mass);
- the sensitivity of the organism that has been used to seed the plate to the antimicrobial component.

Thus, it could be for example that B produces an anti-microbial agent but this is ineffective against the test organism. The gutter plate method (Figure 6.2c) provides a method for testing samples for antimicrobial activity against a range of organisms.

Π Which organism(s) seem to be:

a) most sensitive to the antimicrobial activity of the sample illustrated in Figure 6.2c?

b) insensitive to the sample being tested in Figure 6.2c?

The answer to a) is B. The growth of this organism is prevented at quite some distance from the gutter. The answer to b) is A and D, neither of these organisms appear to be inhibited by the test substances.

Once an antibiotic producer has been identified, the next stage is to produce sufficient of the antibiotic to evaluate its potential for therapeutic use. Questions, such as, is it toxic to humans?, is it effective against disease organisms?, does it possess suitable characteristics (for example solubility, chemical stability) for use as a medicine?, need answering. Let us assume that a new, potentially useful antibiotic has been discovered. The key questions then become, how can the desired material be produced in the most cost effective way? is it possible to produce variants of the antibiotic which have desirable properties, such as greater effectivity against infection, cheaper ways to produce it or increased stability?

Π See if you can list some approaches that may be used to reduce costs/maximise yields of antibiotic production.

new strains, changed by culture conditions

There are many approaches that may be used here. One approach is to screen related organisms to see if a higher yielding strain may be obtained. Alternatively, the culture conditions used to cultivate the antibiotic-producing strain may be modified with the objective of increasing antibiotic production. This may include manipulation of physical conditions such as pH and temperature, addition of precursors of the antibiotic or specific inhibition of particular metabolic activities. We might also use genetic manipulation (for example mutation or genetic engineering) to enhance product yield. You will meet specific examples of these strategies in our discussions of penicillin production.

Π Assume you have an antibiotic-producing organism. See if you can list some approaches that may enable you to use the same organism to produce a range of different, but related antibiotics.

There are several possibilities you may have suggested, for example, using slightly different precursors which may lead to the production of slightly different end products. Alternatively, metabolic inhibition might be used or you may have considered using mutants. Another approach would be to isolate the antibiotic first and to modify it *in vitro* using chemical or biotransformation (enzymatic) methods. All of these approaches have found practical applications. We will again use the β-lactams to illustrate these strategies.

strategies for improving yields and diversifying products

In Figure 6.3, we have provided a summary of the possible strategies for improving yields of antibiotics produced by microbial cultures and for diversifying the nature of the products that are manufactured.

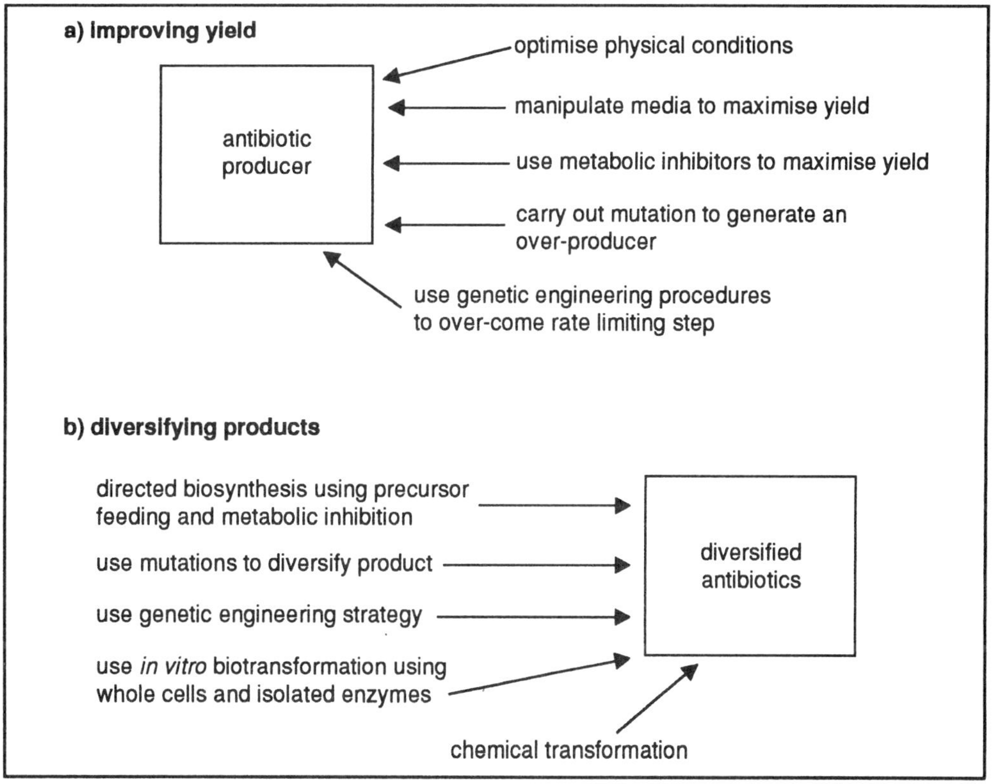

Figure 6.3 Summary of the strategies available for improving yields and for diversifying the products made by antibiotic-producing micro-organisms.

Consideration of pencillin production serves to illustrate the success of these strategies. Penicillin was produced at a concentration of about 1 ppm by the first penicillin producers that were isolated. By manipulation of culture conditions together with genetic manipulation, yields in excess of 10 g l^{-1} (excess of 10,000 ppm) are routinely achieved. This development has also been paralleled by the diversification of the product and a wide variety of penicillins are now available. For these reasons, together with the fact that the history of penicillin production includes most of the important innovations now taken for granted in newer fermentation, it is worthwhile briefly reviewing the history of penicillin production.

SAQ 6.1

The structure of penicillin G is drawn below in a different form from that illustrated for β-lactams in Figure 6.1.

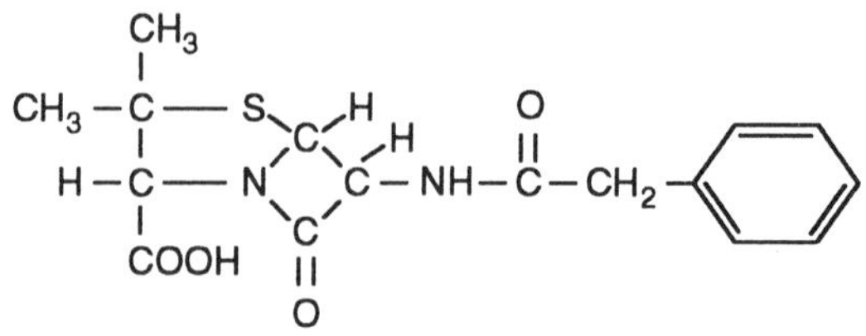

Make a comparison of the structure of penicillin G and amoxycillin and briefly explain a strategy that might be used to diversify penicillins.

6.3 A brief history of penicillin production

6.3.1 Surface cultures and product diversification

floating mycelium

Penicillins, like most antibiotics, are secondary products whose synthesis is not directly linked to growth. The enzymes that produce secondary products are normally repressed or inhibited under conditions which favour rapid growth. In the early work on penicillin, *Penicillium notatum* was grown as a floating mycelium on about 2 cm depth of liquid medium. The mycelium absorbed nutrients from the medium and penicillin was excreted into the medium. The mycelium and spent medium are readily separated.

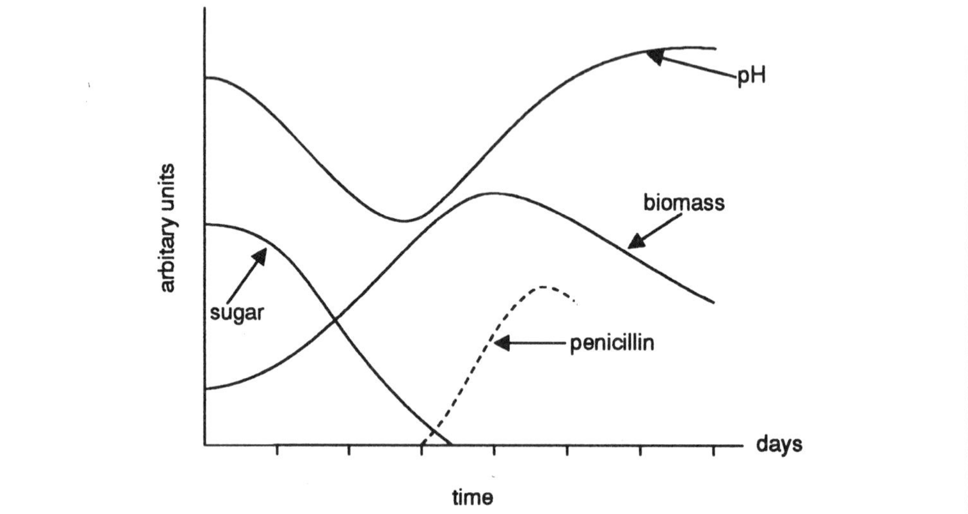

Figure 6.4 Stylised representation of changing parameters and penicillin production in cultures of *Penicillium notatum*, grown as a surface culture on Czapek-Dox medium (adapted from Hockenhull DJ-D "Production of Antibiotics by Fermentation" in Essays in Applied Microbiology edited by Norris J R & Richmond M H 1981. John Wiley & Sons Ltd Chichester).

Czapek-Dox medium

Initially, Czapek-Dox medium was used. This medium, containing mineral salts, sodium nitrate and sugar (usually sucrose or glucose), allowed rapid growth but only very small amounts (around 1μg ml^{-1}) of penicillin were produced. Effectively what happened in these cultures was rapid growth with no penicillin production until virtually all the sugar had been used. At the same time, the pH dropped dramatically as the sugars were being metabolised, but rose sharply when the mycelium began to lyse. The culture was, therefore, only in the optimum pH range (pH 6.5-7.0) for penicillin production for a very short time. These observations are illustrated in Figure 6.4.

Π The important question was, how could the period of penicillin production be extended? See if you can list two or three ways in which this might have been achieved.

slow release of assimilable carbohydrate

There are several possibilities. One would be to ensure that carbohydrates were available at such a rate that they did not cause excessively rapid growth. This was achieved by using lactose in place of the more readily assimilated glucose or sucrose. The fungus hydrolyses this substrate only slowly, thereby releasing the more readily assimilated glucose and galactose at a slow rate. Thus the mycelium behaved as though it was semi-starved and the biomass produced penicillin over a much longer time period (day 2 to 7). Also under these conditions the pH was maintained nearer to the pH optimum of 6.6-7.0 for penicillin production. The result was an improvement in penicillin yield of about 5-10 fold.

complex nitrogen source

Further improvements were achieved by using ammonium acetate or ammonium lactate as nitrogen source in place of nitrate. This reduced the rise in pH observed at the end of the fermentation. Replacement of these nitrogen sources by complex nitrogen sources, such as casein hydrolysate, improved the long term availability of nitrogen and further stabilised the pH with concomitant improvements in yields. Thus by changing the energy, carbon and nitrogen sources significant advances in product yield were achieved as a consequence of slower growth and pH stabilisation.

corn steep liquor

The subsequent advance was rather fortuitous and rested more with serendipity than with scientific logic. A search was made for cheaper more effective replacements for casein hydrolysate. Amongst the tested materials was corn steep liquor (CSL). CSL is a by-product of the manufacture of starch from maize kernals. Whole maize is incubated in warm water, at 50°C acidified with SO_2. Thermophilic bacteria hydrolyse proteins and other components of the kernals, thereby loosening the starch granules. These are removed, leaving behind the steep liquor which is used to treat further maize kernals. Ultimately, the liquor is too viscous to re-use and the liquor is concentrated and used as cattle feed. It was this material that was used for penicillin fermentation. Surprisingly, the yield of penicillin increased by a further 5-10 fold giving yields of 50-100 μg ml^{-1}.

In part, the increase in yield could be attributed to the capacity of the CSL to buffer the pH and to facilitate the slow release of carbohydrate. However, by far the most important factor was that CSL provided a precursor that led to the production of a more stable and easier to isolate form of penicillin, what we now call penicillin G.

CSL contained β-phenyl ethylamine

In essence, what was happening was that the CSL supplied β phenylethylamine. This had been produced from the amino acid phenylalanine by the action of the microflora in the CSL. Thus:

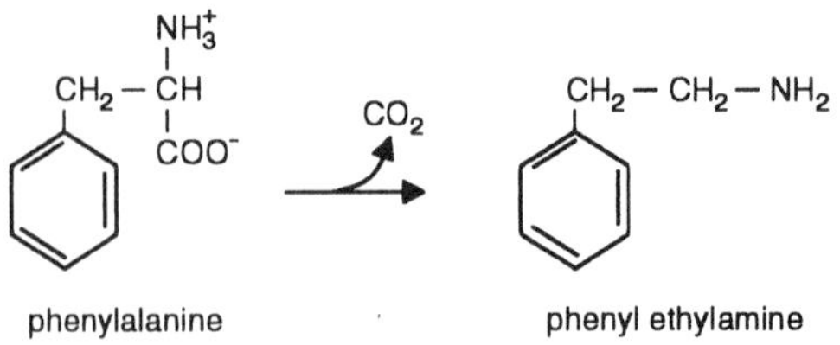

penicillin G

The *P. notatum* took up the β-phenyl ethylamine, converted it to β-phenylacetate, which was subsequently attached to the 6-amino group of penicillanic acid to give benzyl penicillin (penicillin G). We can represent this process by:

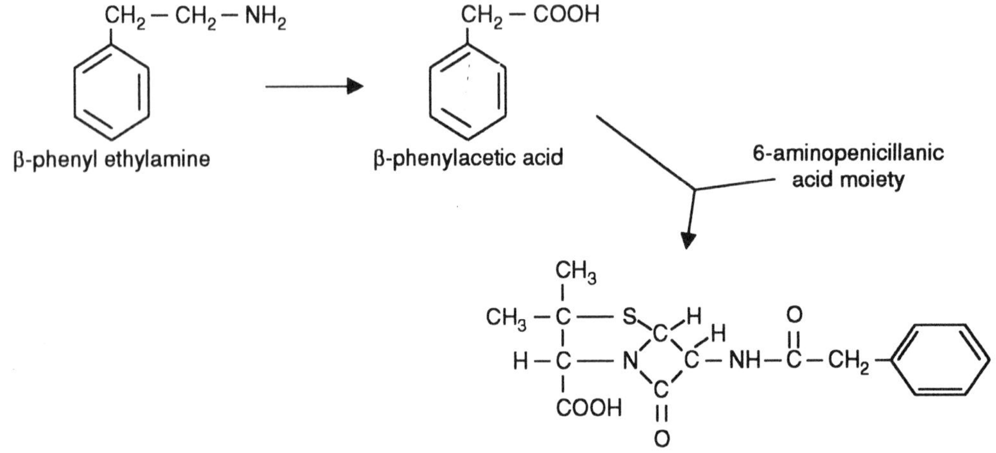

penicillin G

easily crystallised and improved stability

Previous penicillins had aliphatic groups attached to the 6-amino penicillanic acid moiety. Penicillin G has many advantages over the aliphatic derivatives, eg it is more easily crystallised and it is more stable.

Π Suggest an alternative way of producing penicillin G, without using CSL.

The obvious way is to include β-phenylacetic acid or β-phenylethylamine in cultures. Indeed, when β-phenylacetic acid was added to cultures grown in CSL, the yields of penicillin were enhanced further. Typical yields were 100-150 $\mu g\ ml^{-1}$.

Π The results obtained by the addition of β-phenylacetic acid to cultures suggest a method of producing a wide variety of penicillins. See if you can explain what this is.

In principle, by adding derivatives of acetic acid to culture media, we might be able to produce a wide range of penicillins. This strategy was adopted, eg inclusion of phenoxyacetic acid led to the production of penicillin V:

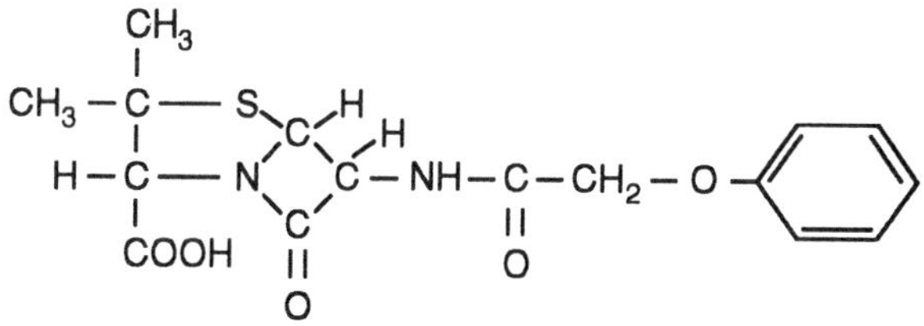

penicillin V

oral administration

Penicillin V has advantages over some other penicillins as it is stable at low pH and can be administered orally.

We may describe the production of diverse penicillins using this strategy as directed biosynthesis using precursor feeding. We have listed some examples of penicillins in Table 6.2.

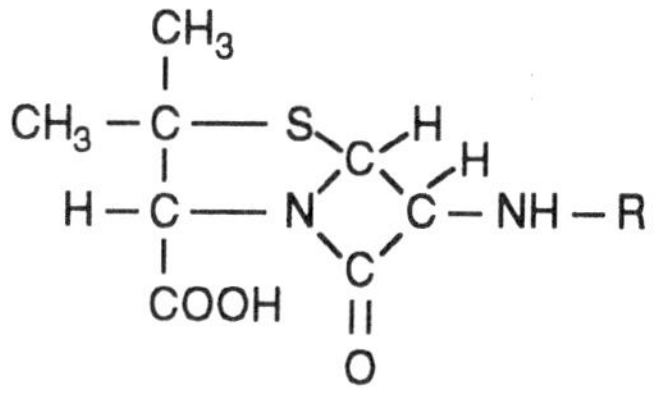

6-aminopenicillanic acid (6-APA) [if R is H]

R =	name of penicillin
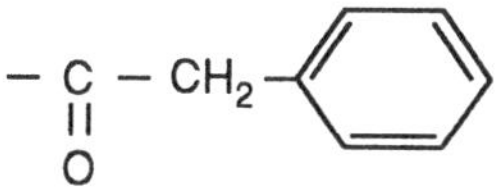	G
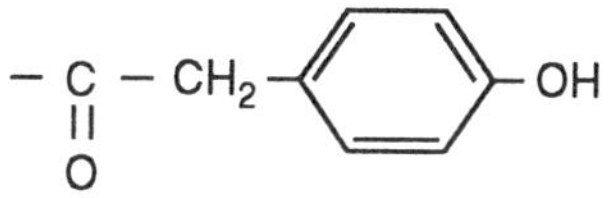	X
$-C(=O)-CH_2-CH=CH-CH_2-CH_3$	F
$-C(=O)-CH_2-(CH_2)_5\ CH_3$	K
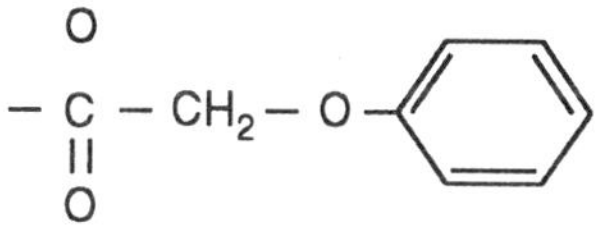	V
$-C(=O)-CH_2-(CH_2)_3-CH(COOH)-NH_2$	N

Table 6.2 Examples of penicillins

SAQ 6.2

Suggest what precursors should be fed to cultures to produce each of the penicillins shown in Table 6.2. Read our response carefully as it contains some additional information.

6.3.2 Use of deep cultures in the production of penicillin

So far we have shown how, by manipulating the formulation of media, improvements in product yield and product diversification were achieved in the early years of penicillin production. We have deliberately selected the high points of these development activities. We will now turn our attention to another aspect of the development of penicillin production: the switch from surface to deep culture.

surface culture

Initially, penicillin was produced in shallow earthenware "penicillin pots" that resembled bedpans used in hospitals. Milk bottles were then used. The problems with these approaches stemmed from the costs of the multiple inoculations that were needed and the costs of harvesting from multitudinous small cultures. Replacement of these small vessels by larger tray-like vessels, however, was not entirely successful: the trays often warped during sterilisation.

Π See if you can list some advantages of using surface cultures.

The main advantages of using shallow surface cultures are that there are few problems ensuring that the cultures remain aerated and, because of the large surface area and thin layer of medium, there are few problems with localised overheating.

commercial consideration

Despite these advantages, deep (submerged) cultures were still deemed to be the most viable route to satisfying the market demand for penicillin. It was estimated that a surface culture equivalent to 2 hectares would be required to produce the same amount of penicillin as a deep culture equivalent to 5×10^4 litres. The desire to switch to deep cultures was thus driven by commercial consideration.

Penicillium chrysogenum

In practice, *P. notatum* was found to be unsatisfactory for deep culture. It grew in large tightly packed pellets. This lead to oxygen starvation in the centre of the pellets. Alternative organisms were sought which would combine good growth characteristics (loose, confluent growth) with high penicillin yield. A strain of *Penicillium chrysogenum* was selected. This strain, NRRL 1951, produced only 50 µg penicillin ml^{-1}. Nevertheless, its growth characteristics made it desirable. Subsequently, variants of this strain (for example B25, X1612), produced by mutagenesis, gave higher yields. Strain X1612 produces above 400-500 µg penicillin ml^{-1}.

trophophase and idiophase

two phase processes

The development of the deep culture approach followed conventional routes, including optimising inoculum density, oxygen dispersal and control of temperature, pH and foam. These aspects of process technology are dealt with elsewhere in the BIOTOL series (for example "Operational Modes of Bioreactors" and "Bioreactor Design and Product Yield") so we will not deal with them in any detail here. You should appreciate that the development of the technology recognised the apparent distinction between the growth of the producing organism and the biosynthesis of the desired product. The concept arose that two phases in penicillin production could be distinguished. In the first phase, rapid growth of the organism took place, while in the second phase little growth occurred, but this phase was marked by penicillin production. So there was separation into a growth phase and a production phase which were later called the "trophophase" and "idiophase" by Bu'llock. Although the separation of these two phases may not be quite as distinct as may be implied by the use of these two terms, it provides the basis of contemporary penicillin production processes. In these processes, the culture is first cultivated under conditions which favour growth. Once the culture has fully grown, culture conditions are manipulated to favour penicillin production. Although these manipulations may be carried out in a single vessel, it is more usual to physically separate them into two vessels. Such two phase processes have become the norm for the production of secondary products. A stylised system is shown in Figure 6.5.

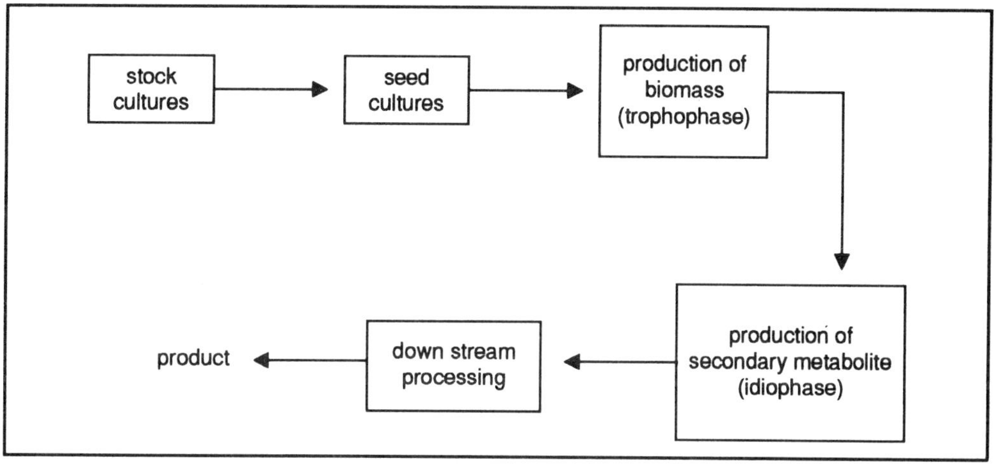

Figure 6.5 A stylised system for producing secondary metabolites such as penicillin.

Π Although processes differ in detail, this generalised scheme is adopted for the production of secondary metabolites. Usually, the vessel used for biomass production is smaller than that used to produce the desired product. Explain why this is so.

batch or continuous transfers

The aim in most processes is to produce biomass as quickly as possible (that is use conditions which allow rapid growth), but to maintain cells in production as long as possible. If for example it took the cultures 3 days to grow, but they could be maintained in a productive state (that is in the idiophase) for 10 days, then in principle the ratio of vessel volumes could be 3:10, trophophase:idiophase. Again this is a simplification, since in some processes the cultures leaving the growth phase may be concentrated before being transferred to the production phase, in order to establish very high biomass concentrations in the idiophase tank. Such processes may be operated in a batch-wise manner, the biomass produced in the trophophase being transferred *en bloc* into the idiophase. Alternatively, biomass may be transferred continually from the trophophase vessel into the idiophase tank.

In this description we have made a clear distinction between growth and secondary product synthesis. You should, however, realise that the distinction is not quite so sharp in practice. Thus we might expect some, albeit a small amount, of secondary product formation in the trophophase and some growth of new cells replacing dead ones in the idiophase. Nevertheless, the separation of the process into two phases enables the optimisation of conditions for growth in one phase and the imposition of conditions which maximise production of antibiotic in the other.

Before we leave this description of the production of penicillin, we should point out that it is not essential that growth and production phases are physically separated. It is possible by using a pre-set feed pattern to carry out both processes in the same vessel.

single vessel

In Figure 6.6 we have illustrated the production of penicillin in a single vessel. You will notice that sugars are fed into the vessel, first slowly because there is little biomass to support, but as this increases the rate of sugar input is also increased. Once growth is complete, the sugar feed rate is again reduced to a level which maintains the grown culture and allows penicillin production. Notice also the change in pH during growth and its maintenance at an optimum for penicillin biosynthesis in the production phase.

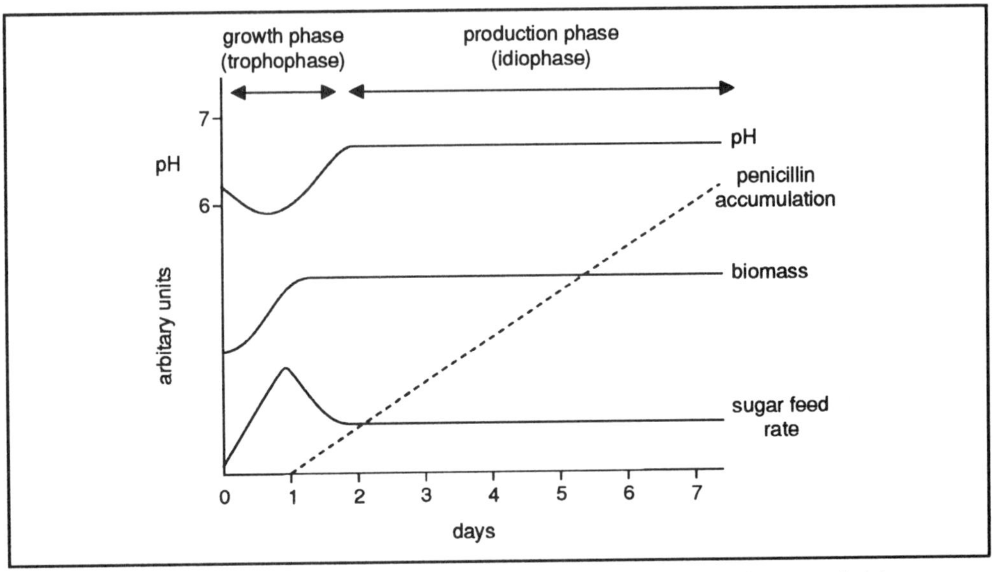

Figure 6.6 Production of penicillin using pre-set feed patterns. Note that increase in biomass occurs over the first short phase, while the penicillin production phase is maintained for a much longer period.

sugar feed rate pattern

In Figure 6.6 we have used sugar feed rate as an example of a pre-set feed pattern. In practice we can adjust a wide variety of parameters during the incubation.

Π Suggest some parameters that might be adjusted to enhance penicillin production.

You may for example have suggested that the supply of other nutrients, such as ammonia and O_2, may be adjusted to respond to the different needs of growth and penicillin production. You may also have suggested that precursors of specific penicillins, such as β-phenylacetic acid, may be added after the growth phase is complete. You may have also considered altering physical parameters such as pH and temperature.

oxygen supply and heat removal

The development of deep cultures for the production of penicillin posed a number of important technological questions. A typical large (50 m^3) bioreactor uses 0.5 tonne of sugar per day. The heat generated by the metabolism would, if not removed, cause the temperature to rise by almost 2°C per hour. Similarly, such a vessel would consume almost 0.5 tonne of oxygen per day. Thus important questions of how to remove the excess heat and supply the necessary oxygen without causing foaming had to be solved.

This is as far as we want to take our discussion of the development of deep cultures for the production of penicillin in this chapter. It is an aspect of the production of secondary metabolites, such as penicillin, more appropriately dealt with in the technology texts in the BIOTOL series.

'take home' messages

The lessons we hope you will take from this brief description of the history of penicillin production are that:

- yields of secondary metabolites may be greatly influenced by the composition of the culture media and by physical parameters such as pH;
- growth of biomass and accumulation of secondary metabolite frequently occur in two separate phases (trophophase and idiophase) and the optimum conditions for each may be quite different;
- growth and production phases may be carried out in separate vessels or, by using pre-set feed rates and parameter control, in a single vessel;
- the exact form of the product may be influenced by components in the culture media and precursor feeding may be used to direct the biosynthesis of a specific product.

SAQ 6.3

Two figures showing changing parameters in surface cultures of *Penicillium notatum* any given below. The medium used in each case was based on that of Czapek-Dox. In the upper figure, the substrate was glucose, while in the lower figure the substrate was an unspecified sugar X. Explain why the yield of penicillin was greater for X than it was with glucose.

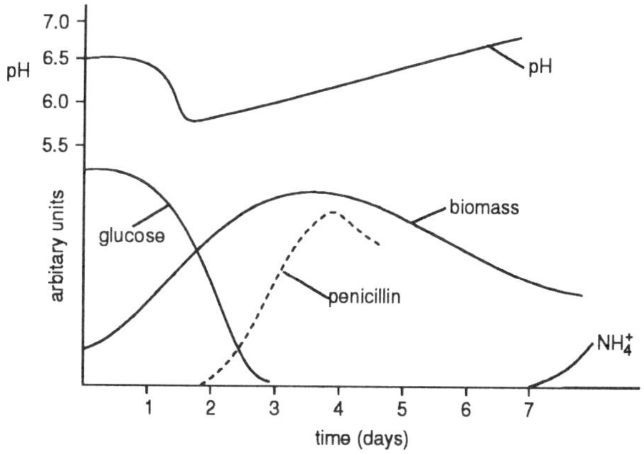

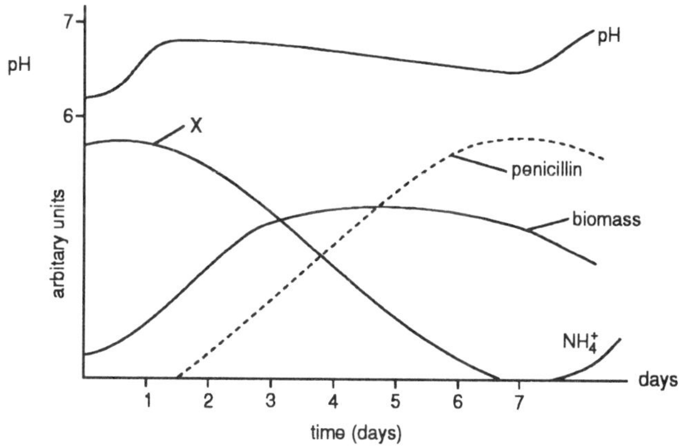

Now that we have provided you with an overview of the history of penicillin production, we will examine some more details of the biotransformation of β-lactams. We will briefly outline the normal biosynthesis pathways that lead to their production and then consider how these products may be diversified *in vitro* to give a wider range of valuable compounds. We begin by briefly explaining how the β-lactam antibiotics are effective as therapeutic agents.

6.4 Antibacterial mode of action of β-lactam antibiotics

selective toxicity

The origin of the success of β-lactam antibiotics mainly results from the extreme low toxicity of these compounds with regard to human beings. In other words, β-lactam antibiotics have a highly selective toxicity for bacteria since they do not interfere with human metabolism, but inhibit the formation of the cell wall of growing bacteria.

peptidoglycan layer

In bacteria the cytoplasmic membrane is covered with a peptidoglycan layer, which determines cell shape and imparts the rigidity necessary to protect the bacterium from osmotic rupture. The peptidoglycan structure consists of alternating units of the amino sugars N-acetylglucosamine and N-acetylmuramic acid. The N-acetylmuramic acid units are linked to peptide chains. Many of these peptide chains are also cross-linked to each other. In *Staphylococcus aureus*, for example, the cross-linking is achieved when the amino group of a terminal glycine unit is inserted in the bond that links an alanylalanine unit of another chain (Figure 6.7).

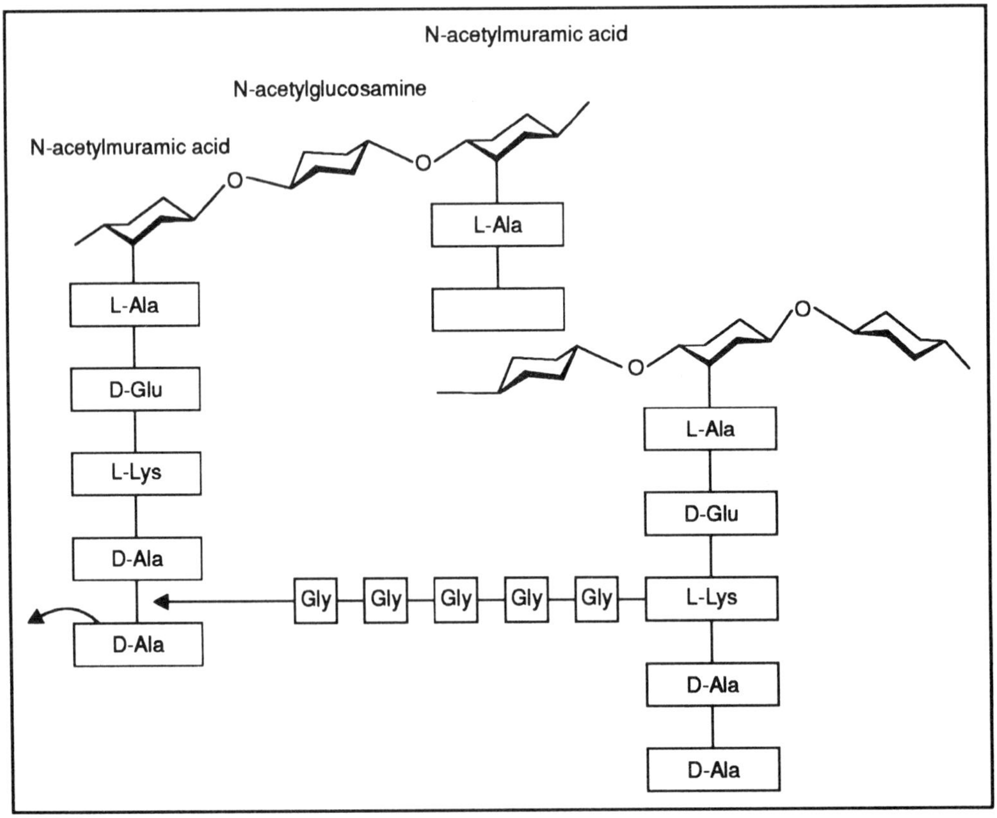

Figure 6.7 Formation of cross-linkage between individual peptide chains in the peptidoglycan layer of *S. aureus*.

transpeptidase

This insertion is accomplished by an enzyme called transpeptidase. β-Lactam antibiotics function as substrates for the transpeptidase, thereby establishing selective inhibition of bacterial cell wall synthesis. The structural similarity between β-lactam antibiotics and the alanylalanine unit is remarkable as can be seen in Figure 6.8.

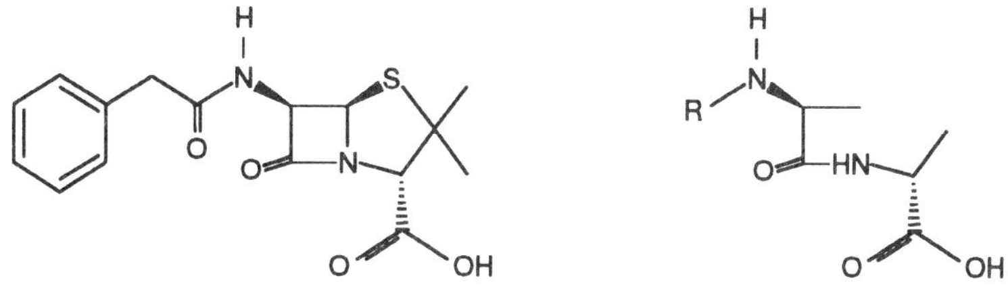

Figure 6.8 Pencillins are similar to the bacterial peptidoglycan terminal alanylalanine moiety. Because of this similarity, the enzyme transpeptidase recognizes β-lactam antibiotics as substrate. As a result of this the β-lactam is incorporated in the peptide chain thereby making peptide-peptide cross-linking impossible. The occurrence of this phenomenon stops the construction of the bacterial cell wall.

6.5 Biosynthesis of penicillins and cephalosporins

Use Figure 6.9 to help you follow the description given below.

many identical enzymatic steps

Penicillins and cephalosporins are products of biosynthetic pathways that have many identical enzymatic steps. It is generally accepted that the tripeptide, δ-(L-α-aminoadipyl)-L-cysteinyl-D-valine (LLD-ACV), is the direct precursor to both penicillin and cephalosporin C.

The first biosynthetic steps are two reactions that generate ACV from its constituent amino acids L-α-aminoadipic acid, L-cysteine and L-valine. L-α-aminoadipic acid and L-cysteine are condensed by the enzyme 'AC synthetase' and, in the next step, the resultant δ-(L-α-aminoadipyl)-L-cysteine is coupled with L-valine. In this step the configuration of L-valine is inverted to D-valine.

The tripeptide LLD-ACV is then cyclised to isopenicillin N by an oxidative reaction involving the removal of four protons. The enzyme that catalyses this reaction is isopenicillin N synthetase or synthase (IPNS).

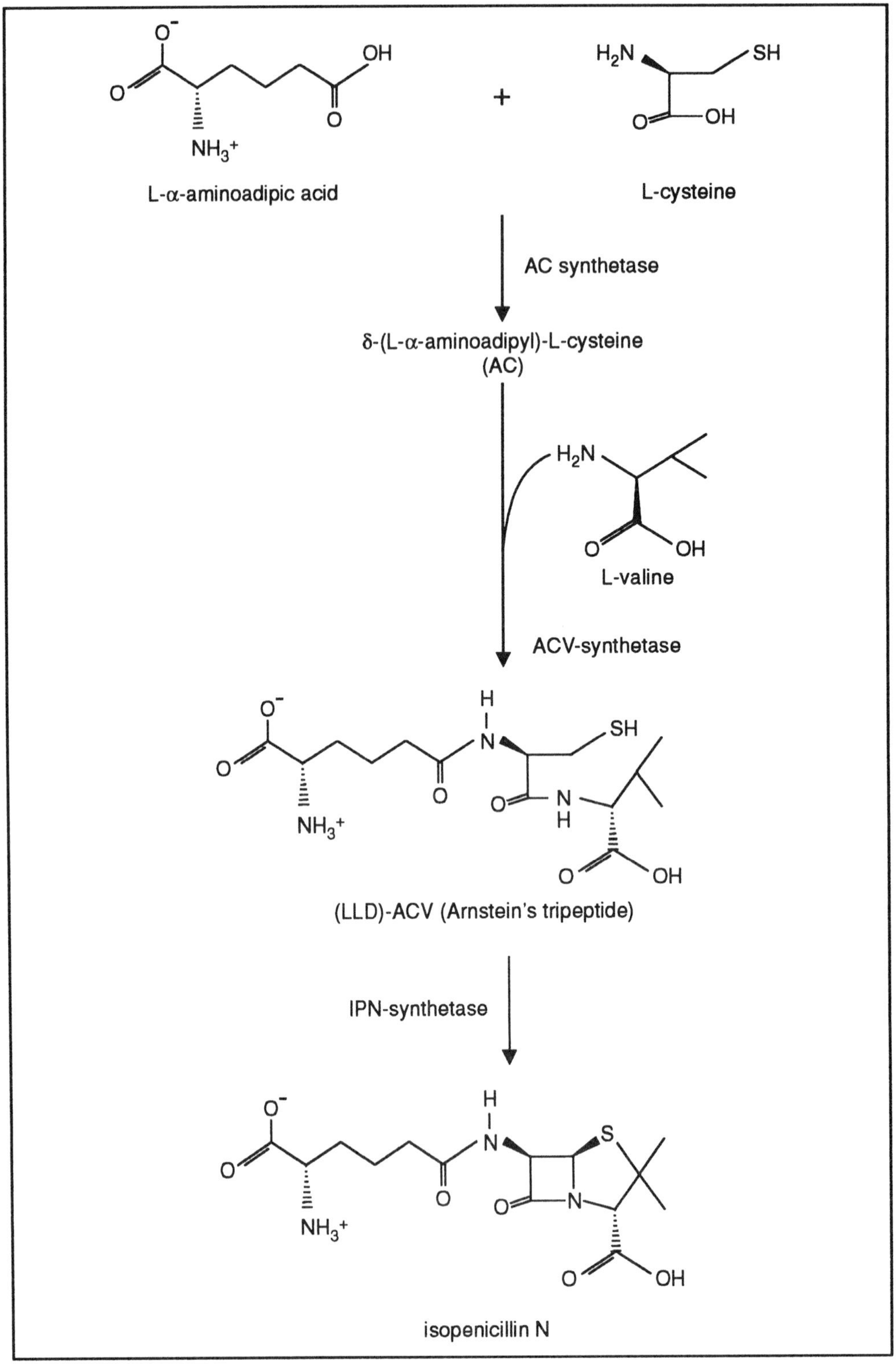

Figure 6.9 Formation of isopenicillin N from its constituent amino acids. After condensation of L-α-aminoadipic acid with L-cysteine, L-valine is coupled. During this transformation, the configuration of the latter amino acid inverts to give D-valine.

Now use Figure 6.10 to follow the rest of the discussion.

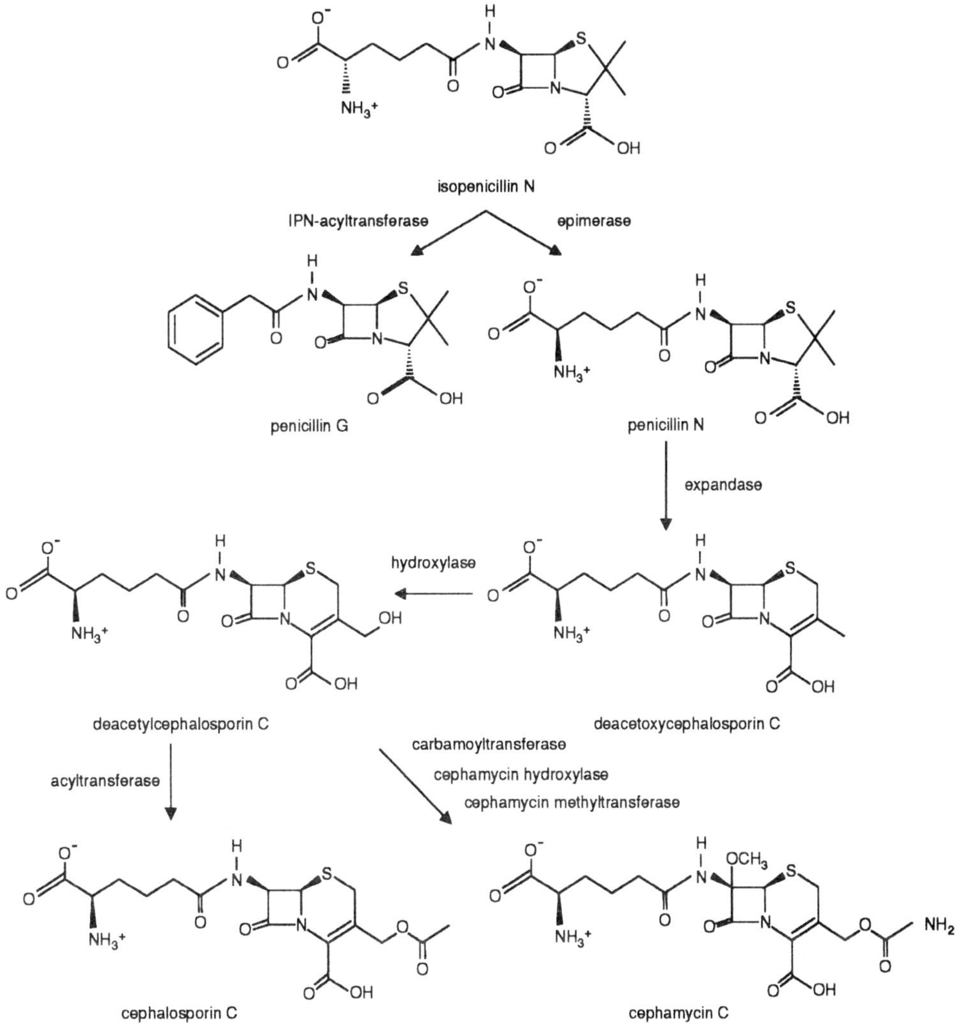

Figure 6.10 Biosynthetic pathways from isopenicillin N to penicillin G and cephalosporin C. Some strains have the ability to convert deacetylcephalosporin C into cephamycin C.

Different strains of micro-organisms are responsible for the production of either penicillins or cephalosporins. In penicillin-producing strains, an acyltransferase enzyme system is present which can remove the side chain from isopenicillin N to give 6-aminopenicillanic acid (6-APA), and which can subsequently acylate 6-APA to generate various penicillins, the most important ones being penicillin G and V(see section 6.3, Table 6.2).

epimerase
expandase
hydroxylase
acyltransferase

Cephalosporin-producing strains are characterised by the presence of the enzyme epimerase, responsible for the conversion of the L-α-aminoadipyl side chain in isopenicillin N, into the D-α-aminoadipyl side chain in penicillin N. The next step is an oxidative ring expansion and involves the loss of two protons, it is catalysed by the enzyme expandase. Unaware of this phenomenon, chemists carried out the ring enlargement of the penicillin skeleton by non-enzymatic means, finding only much later that nature had been doing the same for a long time. Deacetoxycephalosporin C thus obtained is hydroxylated to give deacetylcephalosporin C, using the enzyme hydroxylase (mono-oxygenase). Finally, deacetylcephalosporin C is converted into cephalosporin C with the aid of an acyltransferase.

SAQ 6.4

From what you read in Section 6.3, does the enzyme IPN-acyltransferase exhibit a high degree of specificity? Given reasons for your answer.

6.6 Semi-synthetic penicillins

penicillin G
resistance

In spite of its remarkable therapeutic usefulness and low toxicity, penicillin G appeared to have had its limitations when resistant strains of bacteria emerged. Thus it became of interest to consider producing variants of this molecule with different activities.

Although a range of penicillins could be produced by directed biosynthesis using precursor feeding, this approach is limited by the toxicity of the precursors, the ability of the penicillin producing cells to take up the precursor and by the capability of the acyltransferase to transfer the acyl groups onto the 6-aminopenicillanic acid moiety.

penicillin
acylases
β-lactamases

It was noted that many penicillin-resistant organisms produce enzymes that catalyse the hydrolyses of the amide links in penicillin. These enzymes are penicillin acylases and β-lactamases depending upon the amide links they hydrolyse.

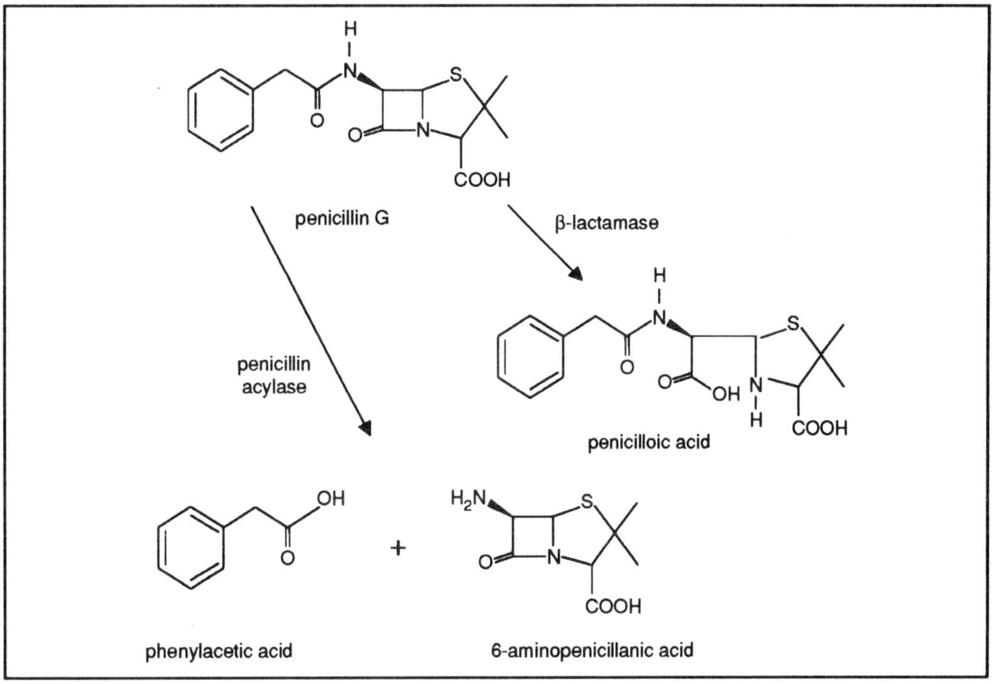

Figure 6.11 Hydrolysis of penicillin G by penicillin acylase and β-lactamase.

Obviously, the production of penicillins that were not sensitive to this hydrolysis would be advantageous.

Π Examine Figure 6.11 and see if you can suggest a strategy that might be adopted to produce modified penicillin.

The strategy we hope you identified is to first produce 6-aminopenicillanic acid, then attempt to add different moieties to the 6-amino group. This can be achieved either chemically or enzymatically. In the following section we will consider the conversion of penicillin G into 6-aminopenicillanic acid and follow this by examining how 6-aminopenicillanic acid may be converted into ampicillin and amoxicillin.

6.6.1 Conversion of penicillin G into 6-aminopenicillanic acid

penicillin acylases

In Figure 6.11 we indicated that penicillin acylases selectively hydrolysed the secondary amide link, releasing 6-aminopenicillanic acid (6-APA). Although these enzymes could be used to produce 6-APA from penicillin G, initially, the vulnerability and high costs of enzymatic deacylation were important reasons to search for alternative, chemical processes.

As can be seen in Figure 6.12, penicillin G contains two amide functionalities, of which the β-lactam linkage is extremely susceptible to basic and nucleophilic attack. Therefore, cleavage of the phenylacetyl side chain could not be performed using classical base hydrolysis. The problem of selectivity was resolved by taking advantage of the fact that the amide bond to be hydrolysed is secondary rather than tertiary.

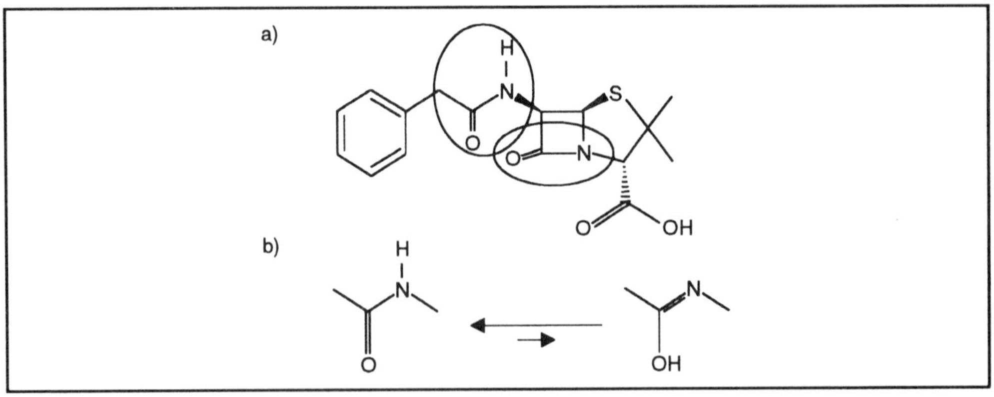

Figure 6.12 Penicillin G contains two amide linkages (circled in a). The amide linkage to the side chain is secondary and exists in two forms (shown in b).

silyl chemistry

Key factor in addressing this problem was the application of silyl chemistry in order to protect the penicillin C-3 carboxyl *in situ*, giving high yields at low cost. The process is illustrated in Figure 6.13. Examine this figure carefully so that you remember, at least in outline, how the chemical conversion takes place. Furthermore, the occurrence of various undesirable side-reactions, so easily met within penicillin chemistry, was successfully avoided by performing reactions with phosphorus pentachloride and alcohol at low temperatures. Thus an inexpensive one-stage process to 6-APA, now one of the world's largest selling β-lactam intermediates, was developed.

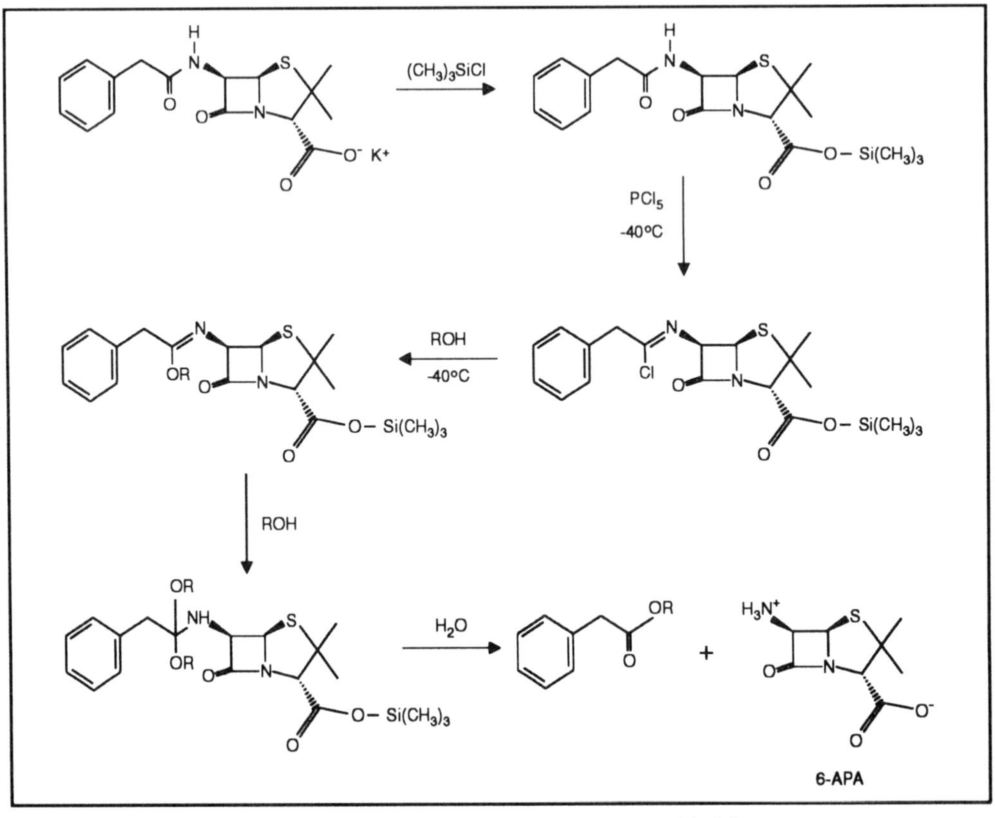

Figure 6.13 Deacylation of penicillin G using phosphorous pentachloride.

Enzymatic approach

limitations of enzymes

Eventually, a growing concern for environmental safety prompted researchers to re-investigate the possible application of the previously described enzymatic deacylation. Application of enzymes in industrial synthesis had long been viewed as being highly impractical since enzymes are unstable at extreme reaction conditions (organic solvents, pH and temperature), sometimes require very expensive co-factors, and are often difficult to recover from complex reaction mixtures. However, as research progresses, it becomes more and more clear that these drawbacks are not always commonplace. As we have illustrated elsewhere in this text, nowadays enzymatic catalysis is known to be very practical in several non-aqueous solvents. Enzymes that function at extreme pH and temperatures have been isolated and certain enzymes, particularly hydrolytic ones, do not need co-factors. Enzymes have become potentially useful catalysts for a wide range of chemical transformations. However, practical industrial application is dictated by economic factors. Cost-effectiveness is still the most important consideration and one always needs to question whether the enzyme of choice can indeed fulfil its catalytic properties. In many examples, enzyme regeneration is not practically applicable and therefore the enzyme needs to be available at low cost in order to compete with traditional chemicals. In other instances, regeneration is a relatively simple concept and the enzyme can indeed be referred to as a catalyst. If this is the case, the cost of the enzyme is of almost no consequence.

enzyme regeneration

water-immiscible solvents

Simple enzyme regeneration techniques can be applied when the product of a given transformation is soluble in a water-immiscible solvent. Since the enzyme is soluble in water, the reaction can be carried out in a two-phase system consisting of water and a water-immiscible solvent. Transformation can only occur at the contacting surfaces of the two phases, and it is for this reason that severe mechanical stirring is often required in order to achieve a sufficient degree of mixing. When the desired transformation has been effected, separation of the phases gives an uncontaminated solution of the enzyme in water which is easily reusable.

immobilised enzyme systems

When the reaction product is soluble in water, enzyme regeneration is difficult to achieve, since the enzyme is often lost during isolation of the product. One way to overcome this problem is application of immobilised enzyme systems. The enzyme is either covalently or ionically attached to an insoluble carrier material or is entrapped in a gel. Depending on the size of the particles used, a simple filtration and washing procedure can be used to separate the immobilised enzyme from the dissolved product. A well-known example of this technique is the industrial production of 6-APA.

Here we will focus on the biochemical aspects. The techniques of isolating enzymes, the process of enzyme immobilisation and the behaviour of immobilised enzyme reactors are discussed in detail in the BIOTOL text "Technological Applications of Biocatalysts", so will not deal with these aspects in detail here. In outline, however, once the desired enzyme is isolated, it is attached to a carrier material. In order to ascertain sufficient accessibility of the enzyme, a bifunctional spacer molecule is attached to the carrier:

functionalisation

Thus, the active functions of the carrier material, which is usually a naturally occurring or synthetically prepared polymer, can be of almost any nature. Introduction of the spacer molecules (functionalisation) is also the phase determining the activity of the immobilised enzyme preparation; the more spacer molecules per unit area, the more enzyme molecules can be attached. Of course, steric hindrance exerts a limit on this number. After the carrier has been functionalised, excess reagent is removed by filtration and washing and the enzyme can be attached to the support. The immobilised enzyme thus obtained is usually stored in an aqueous medium in order to avoid dehydration, which may lead to irreversible deactivation of the enzyme. Just before use, the beads containing the enzyme are collected by filtration, washed, and added to the aqueous solution of substrate. Once the desired conversion has been effected, the beads are removed by filtration, washed, and either stored or reused directly afterwards.

affinity of acylases for antibiotics

6-APA is produced from penicillin G or penicillin V. We remind you that during fermentation leading to these products, a precursor is added which determines the formation of product. When phenylacetic acid is used as precursor, penicillin G is formed. Likewise, when phenoxyacetic acid is used as precursor, penicillin V is formed. Different enzymes show different specificities. Those of bacterial origin have, in general, higher affinity for benzyl penicillin and hence hydrolyse penicillin G more readily than they do penicillin V. In contrast, the penicillin acylases produced by actinomycetes and fungi are generally more active with phenoxymethyl penicillin and therefore more readily hydrolyse penicillin V. We might conveniently regard these two types of enzymes as penicillin G acylase and penicillin V acylase respectively. The reactions they catalyse are illustrated in Figure 6.14.

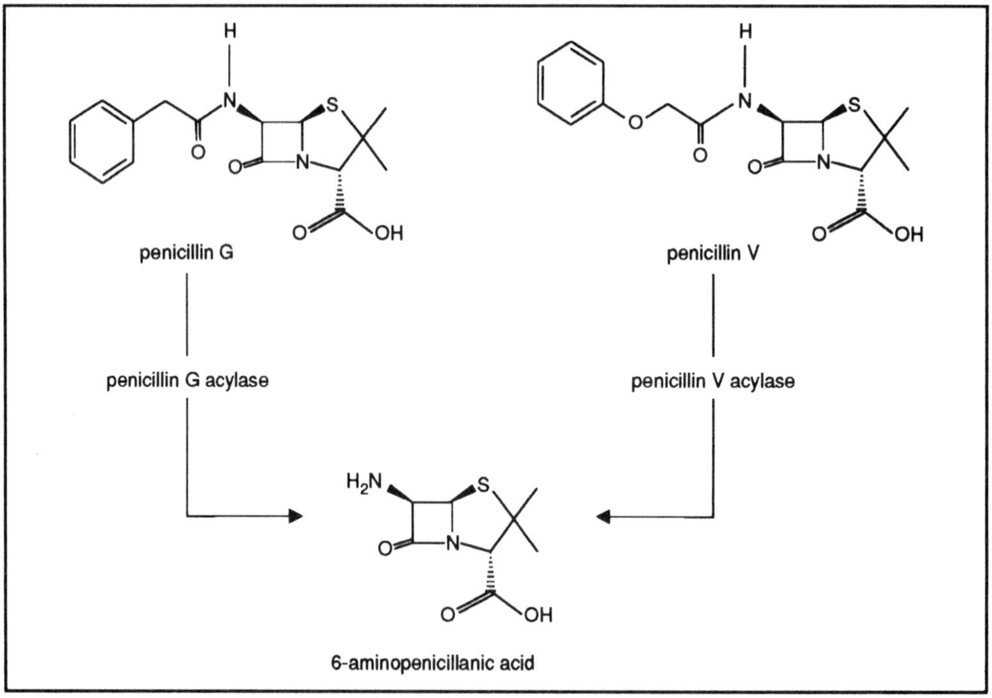

Figure 6.14 Enzymatic side chain cleavage of penicillins. 6-Aminopenicillanic acid, a valuable intermediate for the production of various semi-synthetic penicillins, can be obtained through enzyme-mediated hydrolysis of the phenylacetyl group of penicillin G or the phenoxyacetyl group of penicillin V. The active site of the enzyme recognises the aromatic side chain and the amide linkage, rather than the penicillin nucleus. Chemical entities other than penicillins are therefore often good substrates, as long as they contain the aromatic acetamide moiety.

A variety of enzyme sources have been used. Some examples are given in Table 6.3.

Organism
Bacillus megaterium
Bovista plumbia
Escherichia coli
Kluyvera citrophila
Pseudomonas melanogenum

Table 6.3 Some sources of penicillin acylases used for the large scale production of 6-APA.

Π From the description given previously, see if you can identify a likely sources of i) penicillin G acylase and ii) penicillin V acylase, from the examples given in Table 6.3.

To answer this you would need to have a little knowledge of the taxonony of the organisms listed in Table 6.3. You probably could identify *Bacillus megaterium, Escherichia coli, Kluyvera citrophila* and *Pseudomonas melanogenum* as bacteria and should have predicted, therefore, that these are likely to produce penicillin G acylases. You would have been correct except for the enzyme from *P. melanogenum,* it has a rather different specificity. We will deal with this in section 6.6.2. The enzyme from *E. coli* has found particular industrial application. In some cases whole cells of *E. coli* are entrapped in polyacrylamide gels or the isolated enzyme is entrapped in cellulose triacetate fibres or immobilised onto Sephadex beads or polymethacrylate resin. Immobilised enzyme preparations may, with careful handling, be used for over a hundred batches. Yields of over 90% 6-APA with a purity of over 95% are routinely achieved.

The most commonly used source of penicillin V acylase is the fungus *Bovista plumbia.* Incubation of phenoxymethyl penicillin (penicillin V) with this enzyme produces a yield of about 90-92% 6-APA.

Π What product other than 6-APA is produced from the enzymatic hydrolysis of penicillin G? Has this product any value? If so explain how it may be used.

You should have identified that the acyl moiety, phenylacetic acid, is the second product arising from the hydrolysis of penicillin G. This could be recovered from the reaction mixture and reused as a precursor for penicillin G production. In practice, an integrated process is used. We can represent this in the following way:

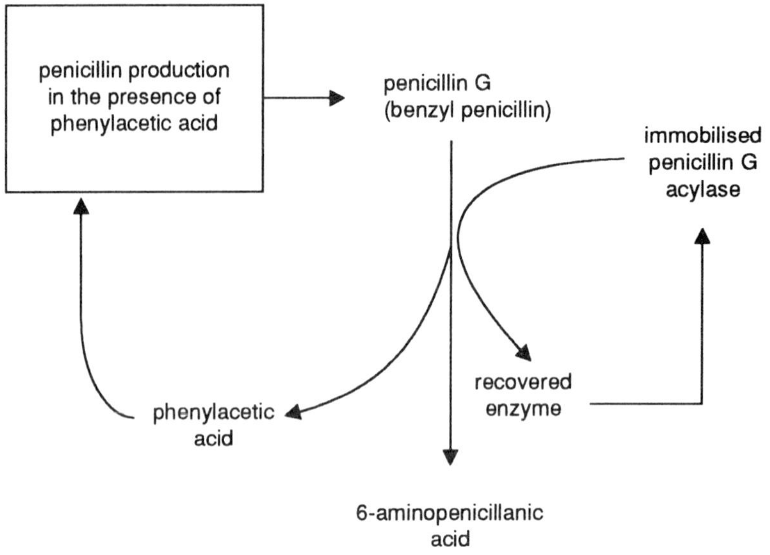

Π In our description we have mentioned that either immobilised whole organisms which produce penicillin acylases or purified enzymes may be used to hydrolyse penicillins. See if you can list two advantages and two disadvantages of using whole cells compared with using purified enzymes (you should recall that the advantages and disadvantages of using whole cells and enzymes in chemical synthesis were considered in Chapter 2).

There are several advantages and disadvantages you might have listed. The principle advantages are:

advantages of whole cells

- using whole cells requires less preparation of the catalytic entity. All that needs to be done is for the cells to be grown, harvested and immobilised. With purified enzymes, substantial post-harvesting processing needs to be carried out;
- with whole cells, the enzyme is held intracellularly and is, to some extent, protected from denaturation arising from the physical and chemical conditions of the reaction mixture;
- chemical attachment of the cells to matrix used for immobilisation may be made between components of the cells (for example the cell walls), rather than directly to the enzyme. There is, therefore, likely to be less damage to the enzyme;
- if entrapment is used for immobilisation, cells are larger and a more open matrix may be used. This, in turn, will offer less resistance to the diffusion into the matrix;
- although not relevant to acylases, you may also have mentioned that for enzymes requiring co-factors, these co-factors will be present within the cells and need not be added.

The principle disadvantages are:

disadvantage of whole cells

- using whole cells means that the substrate has to penetrate both the immobilisation matrix and the cells in order to come into contact with the enzyme. Cell walls and membranes may be a considerable barrier and, therefore, the rates of reactions may be slowed substantially. Of course, if the cells have an active transport mechanism which pumps the substrate into the cells, then the rate will be governed by the rate of diffusion into the immobilisation matrix. We should also consider the diffusion of products away from the enzyme. If these are retained inside the cells, then their accumulation may result in a slowing of the overall reaction (remember, the hydrolytic reaction is reversible);

- inevitably, using whole cells means that the amount of catalyst (the penicillin acylase) per unit volume of mixture will be much lower than that achievable using purified enzyme. The penicillin acylase will represent only a small fraction of the cells used;

- using whole cells may result in the production of a greater variety of end products. Cells will contain enzymes other than the acylases that may catalyse the transformation of the substrate or the products of the acylases.

The latter two points usually tip the balance in favour of using purified enzymes. Ideally the enzyme should be easy to isolate. The penicillin acylase from *Bacillus megaterium* is, for example, an extracelluar enzyme and can be readily absorbed into bentonite.

Before we leave our discussion of preparing 6-aminopenicillanic acid for use as a starting material in the manufacture of semi-synthetic penicillins, we should point out that similar processes are used in the manufacture of semi-synthetic cephalosporins. Here the key intermediate is 7-aminodeacetoxycephalosporanic acid (7-ADCA). We have drawn outline schemes comparing the production of semi-synthetic penicillins and cephalosporins in Figure 6.15. You will see that the two schemes are very similar.

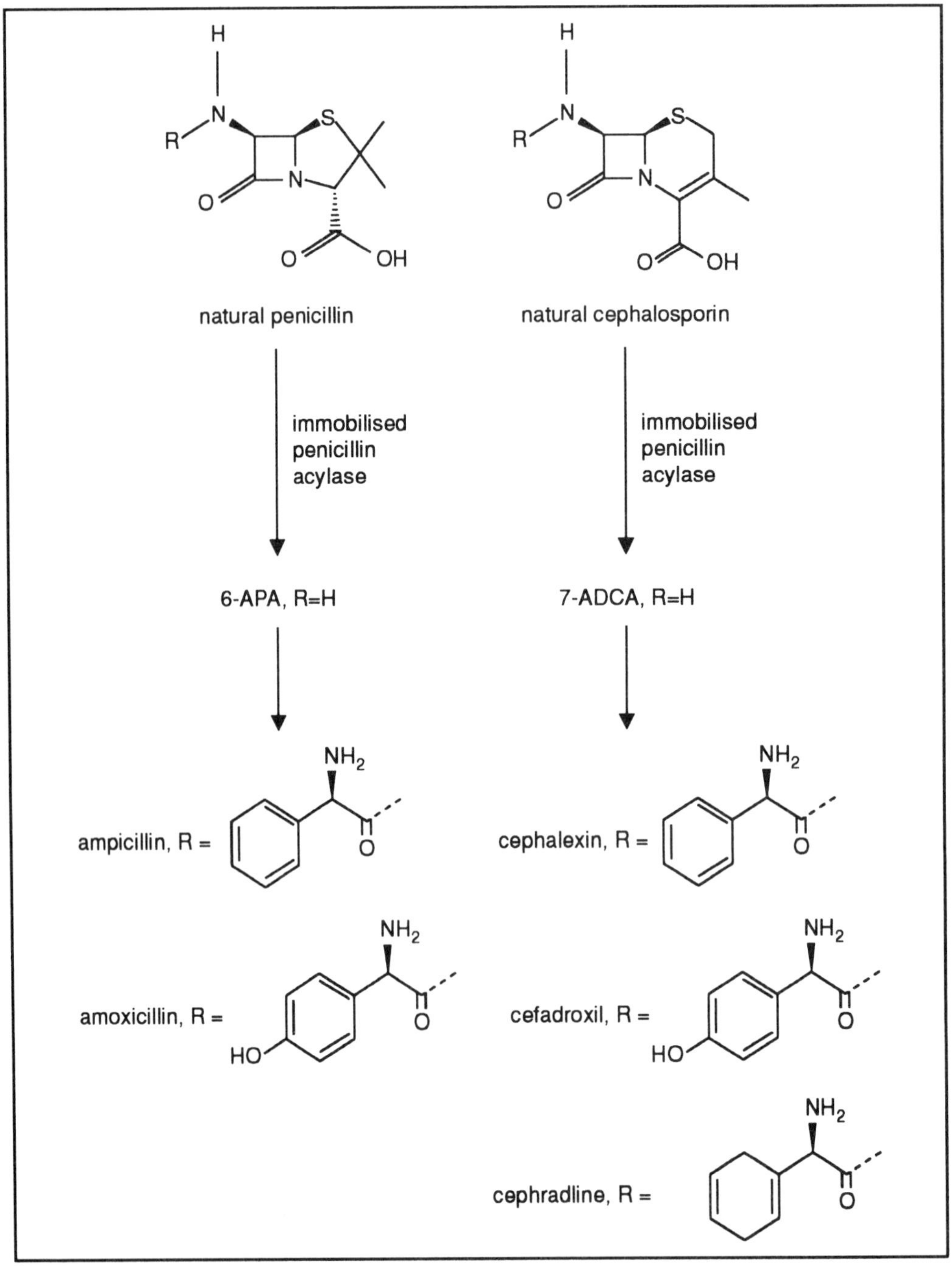

Figure 6.15 Production of semi-synthetic penicillins (left) and cephalosporins (right), from enzymatically obtained intermediates 6-APA and 7-ADCA respectively.

Π We have learnt that the penicillin acylases show some specificity towards the acyl groups attached to the APA moiety. For example, we have distinguished between penicillin G acylases and penicillin V acylases. Some of these enzymes will also work with cephalosporins. What does this tell you about their specificity?

Clearly these enzymes are not highly specific, since they do not distinguish between the ring structures of the penicillins and those of the cephalosporins.

immobilised acetyl esterase cefuroxime

Another important enzymatic process in the production of 7-ADCA, for use in the production of semi-synthetic cephalosporins, is the hydrolysis of 7-aminocephalosporanic acid (7-ACA) by the enzyme acetyl esterase. This process, again using immobilisation techniques, is illustrated in Figure 6.16. The deacylated product can be used, for example, as an intermediate in the production of the important oral cephalosporin cefuroxime. We will return to cephalosporin antibiotics later in this chapter.

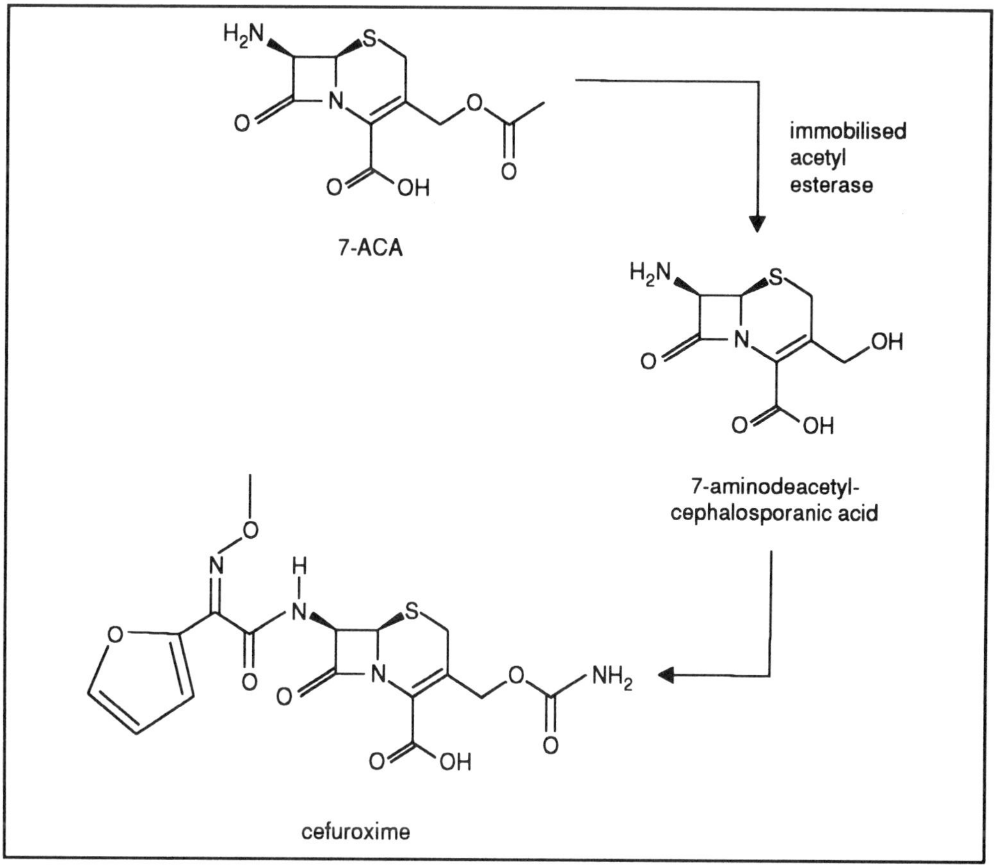

Figure 6.16 Production of 7-aminodeacetylcephalosporanic acid from 7-ACA using an immobilised acetyl esterase. Following enzymatic removal of the acetyl group from 7-ACA, a 3-hydroxymethyl cephalosporin is obtained that can serve as intermediate in the production of cefuroxime.

SAQ 6.5

The introduction of immobilised enzymes has several advantages over the chemical de-acylation of β-lactam. List as many as you can. (You may need to refer back to Figures 6.13-6.16 and the associated text.

6.6.2 Conversion of 6-aminopenicillanic acid into ampicillin and amoxicillin

In section 6.6.1, we described how enzymatic methods have come to dominate the production of the important intermediates used in the manufacture of semi-synthetic β-lactams. In principle, the hydrolytic penicillin acylases may be used in the reverse direction to add acyl groups to 6-APA. For example, a two-step enzymatic process has been described for the preparation of ampicillin (D-(-)-α-aminobenzylpenicillin; structure shown in Figure 6.17).

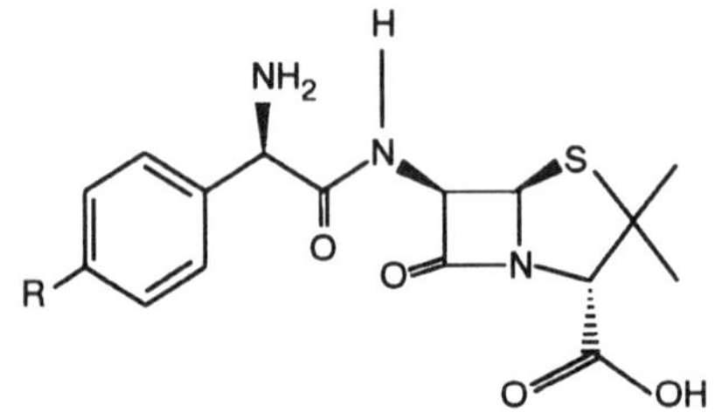

Figure 6.17 Structures of amoxicillin (R=OH) and ampicillin (R=H).

ampicillin

In this process, penicillin G is first hydrolysed to 6-APA with the acylase derived from *Kluyvera citrophila* at a slightly alkaline pH (pH 7.5). Subsequently the 6-APA is incubated with an acylase derived from *Pseudomonas melanogenum* and with DL-phenylglycine methyl ester at pH 5.5. This produces ampicillin in reasonable yields only because of the specificity of the *P. melanogenum* enzyme. This enzyme does not react with penicillin G nor phenylacetic acid.

Efficient synthetic methodologies for N-acylation of 6-APA have been developed. Ampicillin, for instance, can be prepared conveniently by acylation with phenylglycyl chloride hydrochloride under Schotten-Baumann conditions in an aqueous medium. As circumstances require, either ampicillin anhydrate, ampicillin trihydrate or ampicillin sodium can easily be produced on an industrial scale.

amoxicillin

Also illustrated in Figure 6.17 there is another important antibiotic, amoxicillin. Both amoxicillin and ampicillin can be made enzymatically or chemically. Although enzymes are available that can be applied very well for the conversion of 6-APA into a variety of semi-synthetic penicillins, economic reasons are still impeding large scale applications.

A different approach has been used for the synthesis of amoxicillin (Figure 6.18). Based on the application of the inexpensive Dane salt of 4-hydroxyphenylglycine, a process has been developed giving the required compound in almost quantitative yield.

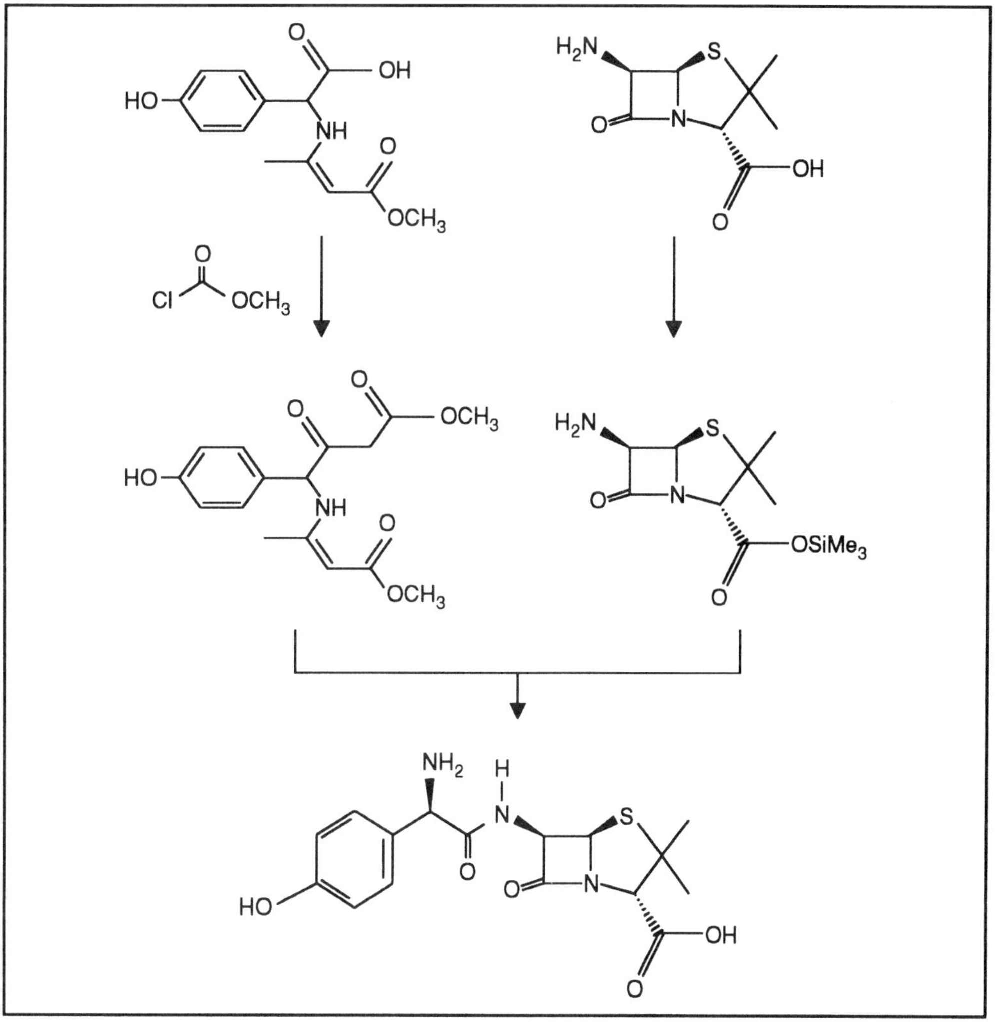

Figure 6.18 Chemical synthesis of amoxicillin from 6-aminopenicillanic acid via the Dane Salt of 4-hydroxyphenylglycine.

chemical approach

We do not expect you to remember the details of this chemistry. What we do, however, hope you realise that just because a process may be carried out enzymatically, this may not be the route chosen by industry. Commercial consideration may still favour the chemical approach. In other words, biotechnological processes have to compete commercially with alternative approaches if they are to be implemented.

6.7 Cephalosporin diversification

6.7.1 Cephalosporin C: discovery, fermentation and production

A highlight in the search for new β-lactam antibiotics was the finding at Oxford (UK) in 1953 by Abraham and Newton that a fungus of the genus *Cephalosporium*, which had been discovered by Brotzu in Sardinia (Italy), produced a number of antibiotics, among them cephalosporin C (Figure 6.19). Interest was aroused when it appeared that this antibiotic was of potential medical importance, after it had shown activity against

Gram-negative bacteria, had displayed resistance to hydrolysis by certain penicillinases and had the low toxicity of the penicillins.

After a strain improvement and development programme similar to, but more complicated than that of penicillin, the D-α-aminoadipyl side chain containing cephalosporin C was obtained by large scale fermentation. However, cephalosporin C could not be isolated as easily as penicillin G or V. Due to its amphoteric nature it is soluble at any pH in the fermentation broth. Several costly isolation procedures involving ion-exchange chromatography have been developed, as a result of which cephalosporin C is much more expensive than penicillin G.

ion-exchange chromatography

6.7.2 Conversion of cephalosporin C into 7-aminocephalosporanic acid

narrow substrate specificty of acylases

It was almost immediately recognised that the deacylated product, 7-aminocephalosporanic acid (7-ACA, Figure 6.16), would be of similar importance as was 6-APA in the development of new penicillins. However, 7-ACA, the cephalosporin equivalent of 6-APA, could not be found in fermentations of *Cephalosporin acremonium*. In Figure 6.15 we have shown that penicillin acylase hydrolyses the acyl residue from natural cephalosporins. Up to a point this is true. These acylases will, however, only work with a limited range of acyl residues. It now seems that nature does not provide for acylases or transacylases that have the capacity to remove or change the D-α-aminoadipyl side chain of cephalosporin C efficiently in a single step. Widespread search for such an enzyme still remains unsuccessful.

chemical side chain cleavage

Fortunately, 7-ACA became readily available after the discovery that cephalosporin C could be converted into 7-ACA by chemical side chain cleavage. Initially, reaction of cephalosporin C with nitrosyl chloride led to a rearranged product that could be easily hydrolysed to give 7-ACA, as outlined in Figure 6.19. However, this method suffered from low yields and eventually a better approach was developed using silyl protection, reaction with phosphorous pentachloride and subsequent alcoholysis. It is well worth mentioning that this chemistry, originally uncovered for cephalosporin C, was successfully applied for many years for the production of 6-APA as outlined before (see Figure 6.13). After the successful isolation of 7-ACA, numerous approaches to obtain new antibiotically active entities were developed. Apart from acylation of the 7-amino function, which is a logical approach in view of similar modifications of 6-APA, a remarkable innovation was accomplished when the acetoxy group at the C-3 position was replaced with various substituents. The smallest of these substituents, hydrogen, gave entry into the very important class of deacetoxycephalosporin intermediates.

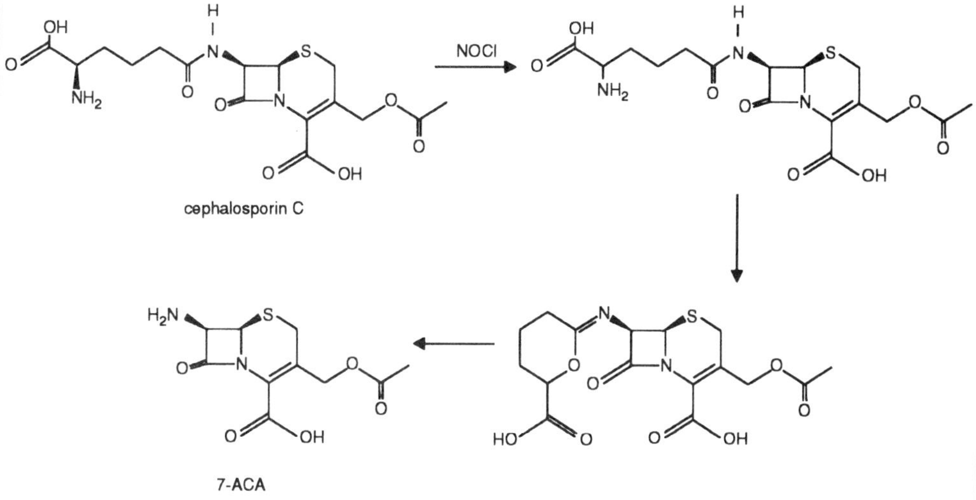

Figure 6.19 Chemical conversion of cephalosporin C into 7-ACA.

We will leave the story of cephalosporin here, since much of the subsequent modifications depend more upon synthetic chemistry than upon biotechnology. It is for example possible to convert deacetoxycephalosporin, exomethylenecephain and demethylcephalosporin derivatives using synthetic chemical procedures. If you wish to follow up this aspect of antibiotic production in more detail, we would recommend Sebek K. O "Antibiotics" in Biotechnology - Volume 6a, edited by Kieslich, K. 1984. Verlag Chemie, Weinheim.

6.8 Alternative strategies for product diversification

In this chapter, by using the examples of β-lactams we have briefly examined how microbial cultures may be used to produce sufficient antibiotics to meet market demands. We have also explained how enzymes (or cells) may be used to biotransform, and thereby diversify, antibiotics. By outlining the history of penicillin production, we explained how analysis and manipulation of culture regimes may be used to enhance the yields of antibiotics (and other secondary products). These studies led to the concept of directed biosynthesis by precursor feeding.

Π This is not the only route by which new antibiotics may be produced in fermentation broth. See if you can identify ways in which new antibiotics or antibiotic-related compounds may be produced.

metabolic inhibitors

One alternative strategy is to use metabolic inhibitors. Although this approach has not been used with the β-lactams, it is worthwhile bearing in mind as a possibility. It has been used in the production of tetracyclines.

chlortetracycline

Streptomyces aureofaciens naturally produces chlortetracycline. This has the structure

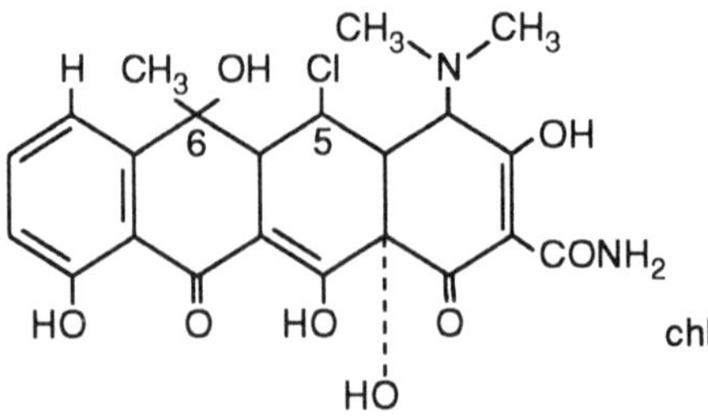

chlortetracycline

By including an inhibitor of the chlorination step in the fermentation broth, the main product formed is tetracycline

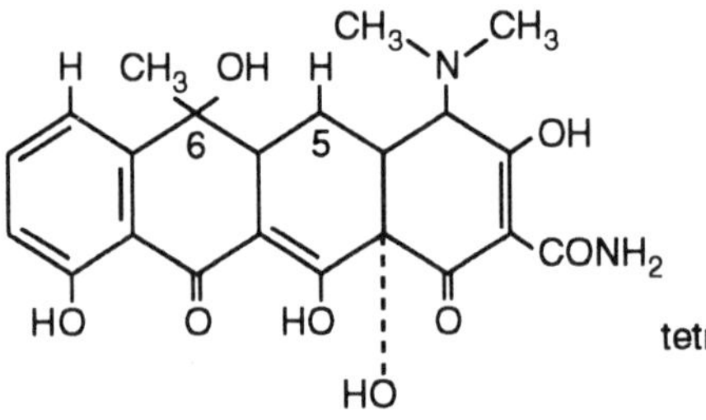

tetracycline

hydrogen at position 5

Notice that, in this case hydrogen, not chlorine, is present at position 5.

use of analogs

Similarly, by adding analogs of L-methionine, the methylation of C6 is inhibited, resulting in the formation of 6-demethylchlortetracycline. The analogs that may be used include D-methionine and ethionine.

use of mutants

An alternative strategy for producing new derivatives by directed biosynthesis is to produce mutants in which particular pathways may be blocked or a new pathway created. Again, we will use a specific example to illustrate this approach.

adnamycin

Many strains of *Streptomyces peucetius* produce daunomycin. These strains often carry a permanently repressed (silent) gene that codes for the enzyme duanomycin 14-hydroxylase. If this is reactivated by mutation, the daunomycin is further metabolised to produce a new antibiotics, 14-hydroxydaunomycin (adnamycin).

neomycin

More frequently, however, mutation is used to block a particular pathway. *Streptomyces fradiae* produces neomycin. 2-Deoxystreptamine is an intermediate in the biosynthetic pathway leading to the production of neomycin (see Figure 6.20).

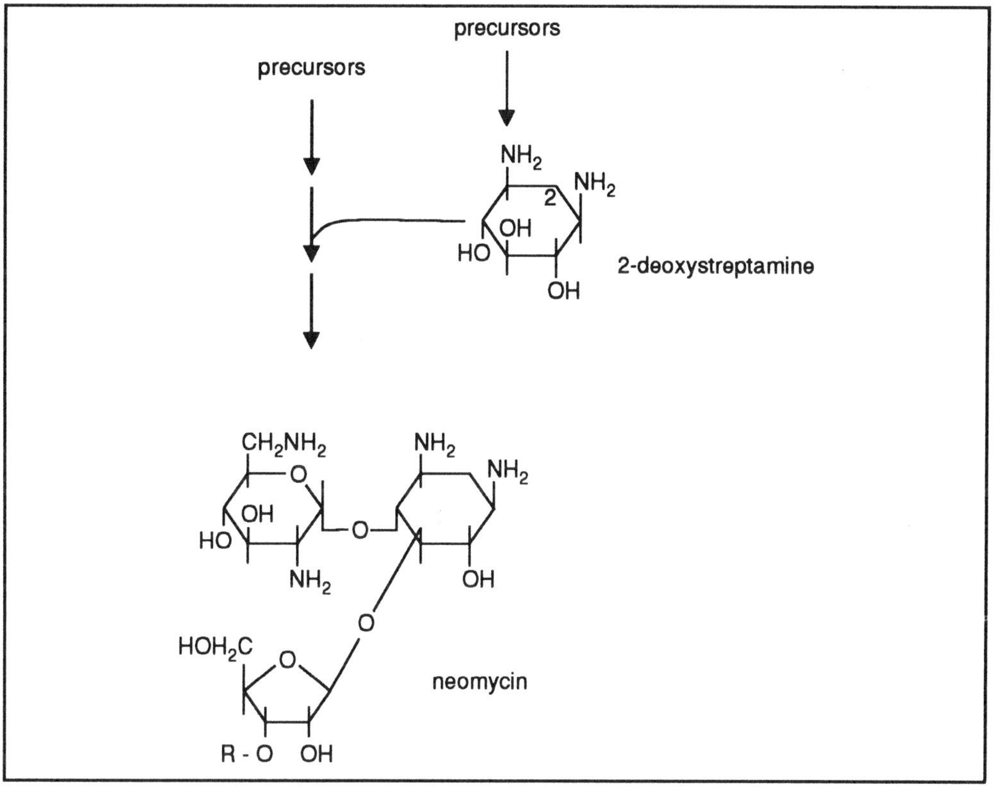

Figure 6.20 Stylised representation of neomycin biosynthesis.

Mutants can be isolated which are unable to produce 2-deoxystreptamine.

Π How may such mutants be used to produce new neomycin analogs?

The obvious solution is to feed these cells with compounds that have structures similar to 2-deoxystreptamine, which would become incorporated into the neomycin. If for example, streptamine is included in the fermentation broth, this is incorporated into the neomycin produced. Thus:

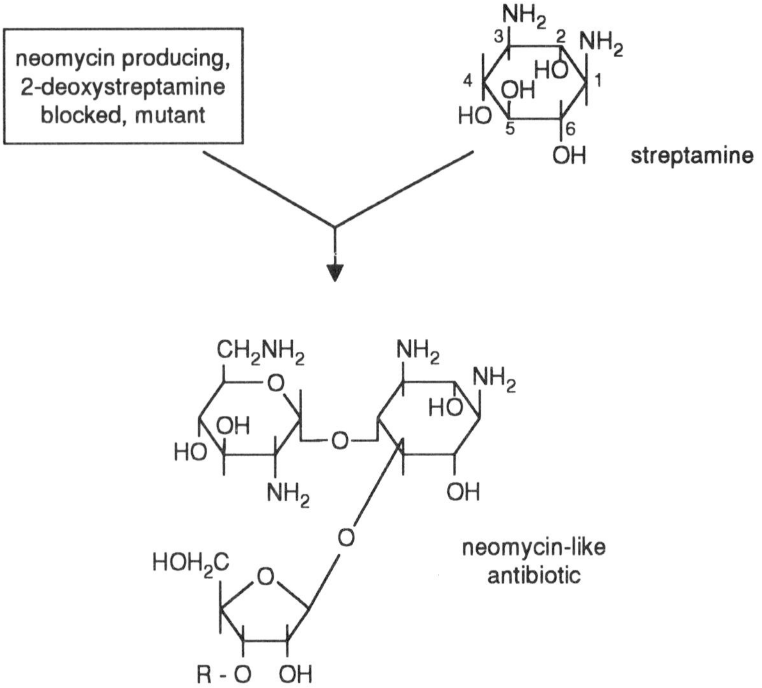

streptamine as precursor

By this approach, a wide variety of structural analogues of 2-deoxystreptamine have been used to produce a range of neomycin-like antibiotics. These structural analogues include methylated and halogenated derivatives of 2-deoxystreptamine.

SAQ 6.6

1) Assume that you have a mutant of a neomycin producing organism that is blocked in the synthesis of 2-deoxystreptamine. You cultivate this organism in the presence of:

   a) 6-*O*-methyldeoxystreptamine

   b) 3-*N*-methyldeoxystreptamine

   c) 5-*O*-methyldeoxystreptamine

   In each case, state whether or not a neomycin-like analogue is likely to be produced and give its probable structure.

2) Explain what assumption you have made in coming to your conclusion to part 1.

We complete this section by reminding you that we have now identified three strategies of using directed biosynthesis to produce diversified products. These are:

- precursor feeding (eg for producing penicillin G, V etc);
- metabolic inhibition (eg for producing various tetracyclines);
- mutation (eg for producing various neomycins).

In the next section, we will expand on the possibilities of using *in vitro* biotransformation to diversify products.

6.8.1 Further biotransformations of antibiotics

In our discussion of the diversification of the β-lactams, we explained how acylases and acylating enzymes may be used in the production of modified (semi-synthetic) β-lactams. However, the potential of using enzymes to modify organic molecules is much wider than this.

Π You should be able to recall from your earlier studies a much wider range of reaction types that might be used. Try to write a list of these before reading on.

The sorts of reactions we hope you would list include:

- oxidations;
- reductions;
- aminations/deaminations;
- esterifications/de-esterifications;
- methylations/demethylations;
- nucleotidylations;
- phosphorylations/dephosphorylations;
- isomerisations;
- hydrations/dehydrations;
- glycosylations.

You may well have included many others.

potential for using enzymes

Many of these have been demonstrated with a range of antibiotics and antibiotic precursors, although relatively few have been applied commercially. We have included a list of published examples in the form of an Appendix at the end of this chapter. We do not expect you to remember the details of this Appendix. It has been included as an illustration of the potential to use enzymes to modify organic molecules like antibiotics. It should be anticipated that, as enzyme technology develops and the search for new antibiotics continues, an increasing number of enzyme-based transformation will find commercial application.

Summary and objectives

In this chapter, we have examined some aspects of the application of biotechnology to the production of antibiotics. We began by briefly describing the range of antibiotics and their modes of action. We described the general strategies used for the production of antibiotics, with particular emphasis on the approaches used to improve yields and to diversify the products synthesised in antibiotic fermentations. We used the β-lactams in particular to illustrate the principles involved, with special emphasis on directed biosynthesis by precursor feeding and *in vitro* modification of fermentation products. We explained how growth and antibiotic synthesis occurs in the phases called the 'trophophase' and 'idiophase' and the implications of this in process design. We also explained the competition between chemical and biological approaches to the production of so called semi-synthetic antibiotics. The use of metabolic inhibitors and mutants to achieve biosynthesis tailored to produce alternative products was also examined. At the end of the chapter we explored briefly the further potential of using biocatalytic systems to modify antibiotics.

Now that you have completed this chapter you should be able to:

- list a wide variety of classes of antibiotics and, in outline, described their modes of action;
- explain how antibiotic-producing micro-organisms may be identified;
- explain, using penicillin production as an example, how culture conditions may be manipulated to improve the yields of antibiotics produced by fermentation;
- explain, by using suitable examples, how directed biosynthesis may be achieved using precursor feeding;
- describe how enzymes may be used to produce 6-aminopenicillanic acid from penicillin G and V and explain how this product may be used in the production of semi-synthetic penicillins;
- apply the principles of using precursor feeding mutants and metabolic inhibitors to producing specific end products;
- explain, using suitable examples, metabolic inhibitors and mutation may be used to re-direct biosynthesis;
- demonstrate an awareness of the potential to use enzymes to modify antibiotics.

Appendix 6.1 Examples of biotransformation of antibiotics (data abstracted from Sebek, O. K. "Antibiotics" in Biotechnology - Volume 6a, edited by Kieslich, K. 1984 Verlag Chemie, Weinheim).

Substrate	Reaction	Product	Micro-organism
β-Lactams			
Benzyl- and Phenoxymethylpenicillins, Ampicillin, Carbenicillin Cephalosporin C Cephaloglycine, Cephaloridine, Cephalothin	Hydrolysis	Corresponding β-lactam ring cleavage products	*Escherichia coli* *Streptomyces albus* *Pseudomonas aeruginosa* *Enterobacter cloacae* *Streptomyces* sp.
Benzyl-, phenoxymethyl- and other penicillins	N-Deacylation	6-APA and the corresponding acyl side chains	*Escherichia coli* *Fusarium semitectum* *Penicillium chrysogenum* *Aspergillus oryzae* Nine different bacteria
N-Acetyldehydroxy-thienamycin (PS-5)		Deacetylated PS-5 (NS-5)	*Streptomyces olivaceus*
Cephalosporin C	*O*-Deacylation	3-Deacetylcephalo-sporin	Various bacteria and actinomycetes
Cephalothin		3-Deacetylcephalo-thin	*Eschericia coli*
Nocardicin C		3-Aminonocardicinic and α-aminoadipic acids	*Pseudomonas schuylkilliensis*
6-APA + phenylacetic acid	Acylation	Benzylpenicillin	*Escherichia coli* *Alcaligenes faecalis*
+ phenoxyacetic acid		Phenoxymethyl-penicillin	*Alcaligenes faecalis*
+ carboxylic acid and esters		Penicillins with the corresponding acyl sides chains	*Kluyvera citrophila* *Pseudomonas melanogenum*
+ phenyglycine esters		Ampicillin	*Kluyvera citrophila*
7-ACA, 7-ADCA, and their organic acid esters		Corresponding cephalosporins (cephalexin, cephaloglycine)	*Kluyvera citrophila* *Xanthomonas citro* and other pseudomonads

Substrate	Reaction	Product	Micro-organism
Aminoglycosides			
Mannosidostreptomycin	Hydrolysis	Streptomycin	*Streptomyces griseus*
Validamycins		Validamycin A and Validoxylamine	Various bacteria and yeasts
Validamycins A and D		Validoxylamine A	*Pseudomonas denitrificans* and other micro-organisms
Validamycin B		Hydroxyvalidamine	*Pseudomonas denitrificans*
Validamycins C, E and F		Validmycin A	*Endomycopsis* spp and *Candida intermedia*
Gentamicins C, C_1 and C_{1a}	Acylation	3-*N*-Acetyl derivatives of the respective substrates	*Escherichia coli* *Klebsiella penumoniae* *Pseudomonas aeruginosa*
Kanamycin A Kanamycin B			*Escherichia coli* *Klebsiella pneumoniae* *Pseudomonas aeruginosa*
Tobramycin			*Escherichia coli* *Klebsiella pneumoniae*
Gentamicin C_1 and C_{1a}		2'-*N*-Acetyl derivatives of the respective substrates	*Providencia* spp.
Sisomicin Tobramycin			*Providencia* spp.
Lividomycins A and B Paromomycin Ribostamycin			*Providencia* spp.
Amikacin		6'-*N*-Acetyl derivatives of the respective substrates	*Pseudomonas aeruginosa*
Butirosins			*Pseudomonas aeruginosa*
Amikacin	Phosphorylation	3'-*O*-Phosphorylated derivatives of the respective substrates	*Staphylococcus aureus*
Butirosin			*Escherichia coli* *Pseudomonas aeruginosa*
Butirosin A			*Escherichia coli* *Pseudomonas aeruginosa*

Substrate	Reaction	Product	Micro-organism
Kanamycin A Neamine			*Pseudomonas aeruginosa* *Escherichia coli*
Neamine Neomycin Ribostamycin Gentamicins A, B, C_{1a}, C_2 Kanamycin B Sisomicin Tobramycin			*Pseudomonas aeruginosa* *Escherichia coli* *Pseudomonas aeruginosa* *Escherichia coli* *Staphylococcus aureus*
Dihydrostreptomycin		3''-*O*-Phosphorylated derivatives of the respective substrates	*Pseudomonas aeruginosa*
Streptomycin			*Escherichia coli* *Pseudomonas aeruginosa*
Gentamicins C_{1a} and C_2	Methylation	Gentamicins C_{2b} and C_1	*Micromonospora purpurea*
N-Demethylclindamycin		*N*-Demethyl-*N*-hydoroxymethyl clindamycin	*Streptomyces lincolnensis*
Anthracyclines			
Adriamycin	Reduction	7-Deoxyadria-mycinone and 7-Deoxyadria-mycinol aglycone	*Streptomyces steffisburgensis*
Daunomycin		Daunomycinol	*Corynebacterium equi* *Mucor spinosus*
		7-Deoxydaunomy-cinone	*Streptomyces steffisburgensis*
Daunomycinol and its aglycone		7-Deoxydauno-mycinol aglycone	*Streptomyces steffisburgensis* *Streptomyces coeruleorubidus* *Streptomyces galilaeus*
Daunomycinone		13-Dihydrodauno-mycinone	*Streptomyces aureofaciens*
Nogalamycin		7-Deoxynogalarol	*Streptomyces nogalater*
Rubeomycin A		Rubeomycin B	*Rhodotorula glutinis*
Auramycinone	Hydroxylation	11-OH-Auramy-cinone and 9-methyl-10-OH-daunomycin	*Streptomyces coeruleorubidus*
Daunomycin		14-OH-Daunomycin (adriamycin)	*Streptomyces peucetius* var. *caesius*

Substrate	Reaction	Product	Micro-organism
ε-Isorhodomycinone ε-Pyrromycinone		1-OH-13-Dihydrodauno mycin (and its *N*-formyl derivative)	*Streptomyces coeruleorubidus*
Daunomycinone derivatives	Demethylation	Corresponding demethylated Carminomycinone derivatives	*Sepedonium chrysospermum* *Beauveria sulfurescens*
Macrolides			
14-*O*-Acetyl-8-*O*-acyl-lankacidin C	Deacylation	Corresponding 8-*O*-Acyllankacidins	*Streptomyces rochei*
Lankacidin A Lankacidinol A		Lankacidin C	Streptomyces rochei, Aspergillus niger Aspergillys sojae *Trametes sanguinea*
Various 8-*O*-14-*O*-di acylated Lankacidin C derivatives		14-*O*-Acylated lankacidin C	*Streptomyces rochei*
Leucomycines A_1 and A_3 Magnamycins A and B Middamycin		Corresponding 4''-*O*-deisovaleryl derivatives	*Cunninghamella elegans* and other fungi *Streptomyces* sp.
Leucomycin A_5		Leucomycin V (4''-debutyrylleu-comycin)	*Actinoplanes missouriensis*
Maridomycin III		4''-*O*-Depropionyl-maridomycin II	*Bacillus megaterium* *Streptomyces pristinaespiralis*
		18-Dihydromarido-mycin III, 4''-Depropionyl-maridomycin III and 18-Dihydro-4''-depropio nylmaridomycin III	*Streptomyces* sp
A23187-Methyl ester and polyvinylpyrrolidone	Hydroxylation	16-Hydroxylated (and N-demethylated) products	*Streptomyces chartreusis*
Josamycin Maridomycin I		3'''-OH-Josamycin 3'''-OH-Maridomycin I	*Streptomyces olivaceus*
Narbomycin		Picromycin	*Streptomyces narbonensis*
Narbonolide		Picronolide	*Streptomyces zaomyceticus*
Ketone formation			
Lankacidinol		Lankacidin A	*Streptomyces rochei* var. *volubilis*
Midecamycin A_1		Midecamycin A_3	*Streptomyces mycarofaciens*

Substrate	Reaction	Product	Micro-organism
Expoxidation			
Carbomycin B Leucomycin A3		Carbomycin A Maridomycin II	*Streptomyces hygroscopicus*
cis-Propenylphosphonic acid		Fosfomycin	*Penicillium spinulosum*
Albocycline	Reductions	2,3-Dihydroalbocycline	*Streptomyces venezuelae*
Carbomycin A		Carbomycin A P1	*Streptomyces halstedii* *Streptomyces lutea* and others
Carbomycin A		Maridomycin II	*Streptomyces hygroscopicus*
Carbomycin B		Leucomycin A_3	
Maridomycin III		18-Dihydromaridomycin III	*Streptomyces* sp.
Midecamycin A_3		Midecamycin A_1	*Streptomyces mycarofaciens*
Tylosin		Relomycin	*Nocardia corallina*

Production and applications of microbial exopolysaccharides

Production and applications of microbial exopolysaccharides

7.1 Introduction

market potetential

Commercial applications for polysaccharides include their use as food additives, medicines and industrial products. Although plant polysaccharides (such as starch, agar and alginate) have been exploited commercially for many years, microbial exopolysaccharides have only become widely used over the past few decades. The diversity of polysaccharide structure is far greater in micro-organisms compared to plants and around 20 microbial polysaccharides with market potential have been described. However, micro-organisms are still considered to be a rich and as yet underexploited source of exopolysaccharides.

chapter overview

Successful commercial application of microbial exopolysaccharides depends on exploiting their unique physical properties. These properties concern the rheology of exopolysaccharides in solution and their ability to form gels at relatively low concentrations. The physical properties of exopolysaccharides arise largely from: their high molecular weight; their molecular confirmations determined by their primary structure; associations between molecules in solution. In this Chapter we will commence by considering the origin, or natural sources of exopolysaccharides, and their molecular composition. We will then consider how the molecular composition determines exopolysaccharide structure and, in turn, how structure determines the unique physical properties of exopolysaccharides. The commercial applications of the unique physical properties of exopolysaccharides will then be discussed. We shall see that microbial exopolysaccharides are widely used as viscosifiers and as gelling agents. They are, for example, used in the food industry as stabilisers, adhesive, thickening agents and foam stabilisers. Other industrial applications include their use as thickening agents in the printing industry and as gels for improved petroleum recovery in the oil industry. The prospect of those approaching commercial use in food, medical and industrial areas will also be considered. Later in the Chapter, the biosynthesis of exopolysaccharides and, in particular, the genetics and regulation of synthesis will be discussed. Much of our knowledge in this area comes from studies on the bacterium *Xanthomonas campestris* the industrial producer of the commercially very important exopolysaccharide xanthan. Our understanding of the genetics of exopolysaccharide synthesis is advancing rapidly and offers opportunities, not only to improve yields of exopolysaccharides, but also to modify their composition and thus their structure and properties, giving rise to new applications. Finally, we will consider industrial fermentation of microbial exopolysaccharides, including medium formulation and product recovery, particular in relation to xanthan.

7.2 Origin and composition

Many different types of carbohydrate-containing molecules are located on the surface of microbial cells. Some of these are components of the microbial cell wall and are limited to certain types of micro-organisms; such as bacterial peptidoglycan, lipopolysaccharides, techoic acids and yeast mannans. Other polysaccharides are not wall components but are either associated with surface macromolecules or are totally dissociated from the microbial cell. These are extracellular polysaccharides, also known as exopolysaccharides, and they show considerable diversity in their composition and structure.

Exopolysaccharides occur widely, especially among bacteria, and include free-living saprophytes and animal and plant pathogens. They are produced by most microalgae but relatively few yeasts and filamentous fungi produce exopolysaccharides. Although plants produce a wide range of polysaccharides, their diversity is considerably less than those produced by micro-organisms. The number of different sugars found in polysaccharides is an indicator of diversity of structure and is eight fold higher (around 200) in those of microbial origin compared to those of plant origin.

diversity

Although exopolysaccharides do not normally have a structural role, they do form structures that can be detected by either light or electron microscopy. Exopolysaccharides may form part of a capsule closely attached to the microbial cell surface, or appear as loose slime secreted by the cell but not directly attached to it. Exopolysaccharide producing cells usually form mucoid colonies on solid media and liquid cultures of these cells may become very viscous. However, growth conditions can influence the composition, physical properties and organisation of exopolysaccharide.

mucoid colonies

7.2.1 Composition

Exopolysaccharides are mainly composed of carbohydrates (see Figure 7.1).

common sugars

The sugars commonly found in microbial polysaccharides are extremely diverse and include most of those found widely in animal and plant polysaccharides:

- D-glucose;
- D-galactose;
- D-mannose;

} pyranose forms, ie

- L-rhamnose;
- L-fucose;

} furanose forms, ie

However, whereas eukaryotic polysaccharides may contain pentoses such as D-xylose and D-ribose, they are only very rarely found in microbial polysaccharides.

Π Draw the ring structures of L-fucose (6-deoxy-L-galactose) and L-rhamnose (6-deoxy-l-mannose). Use the information in Figure 7.1 to help you do this.

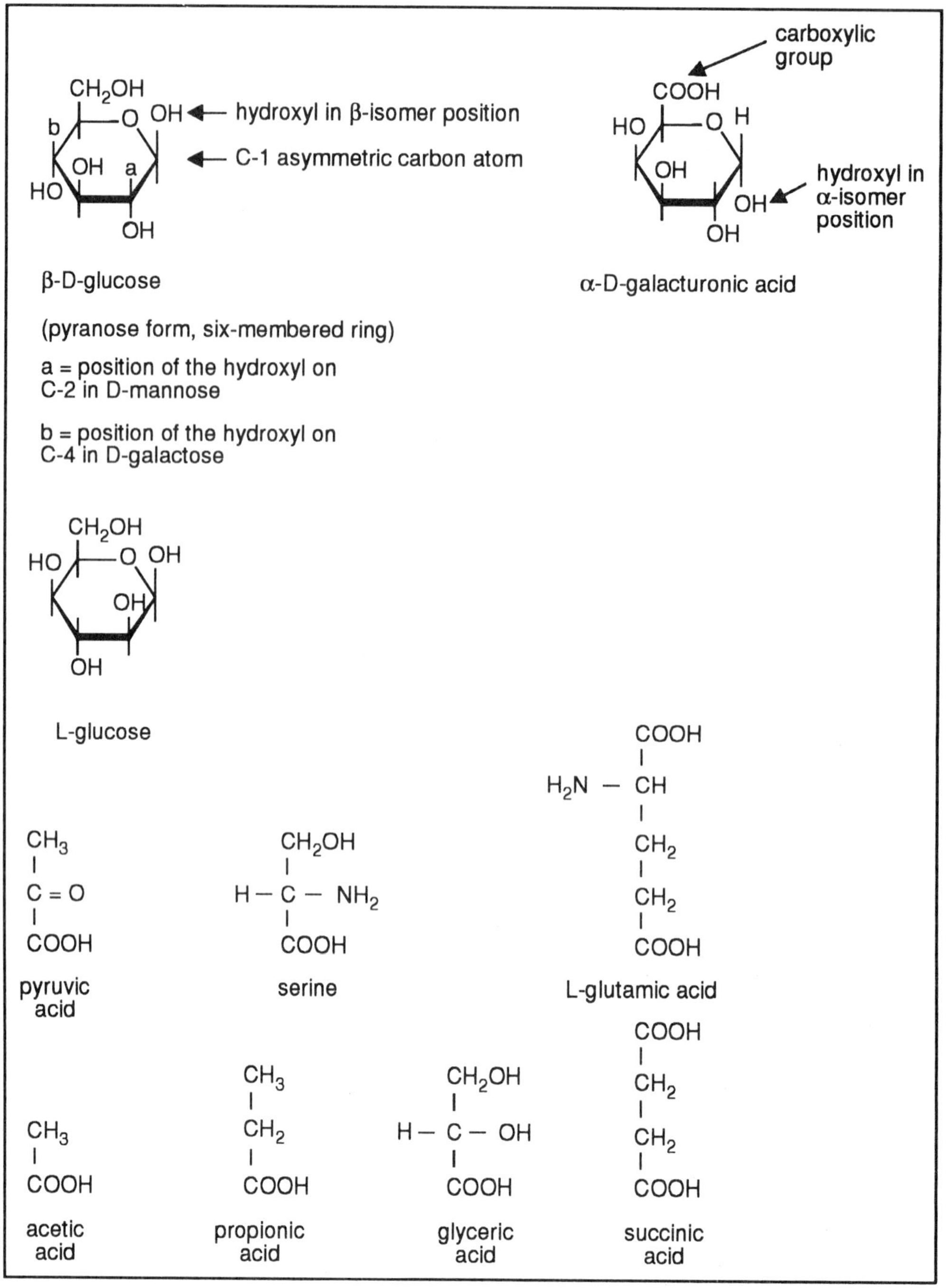

Figure 7.1 Structures of some of the components of microbial exopolysaccharides.

rare sugars

Examples of rare sugars which have been found in some microbial polysaccharides are:

- L-glucose;
- L-galactose;

- N-acetyl-D-glucosamine (an amino sugar);
- D-glucuronic acid;
- D-galacturonic acid.

polyanionic nature

The presence of uronic acids in microbial exopolysaccharides results in their polyanionic nature.

In addition to one or more sugars, exopolysaccharides from prokaryotes commonly contain pyruvate ketals and various ester-linked organic substituents. These are only rarely found in eukaryotic exopolysaccharides.

pyruvate ketals

Pyruvate ketals add to the anionic nature of the exopolysaccharide and are usually present in stoichiometric ratios with the carbohydrate component. Pyruvate is normally attached to the neutral hexoses but may also be attached to uronic acids. In the absence of uronic acids, pyruvate alone contributes to the anionic nature of the exopolysaccharide.

acetate

Acetate is the commonest ester-linked component of exopolysaccharides and does not contribute to their anionic nature. Less common ester-linked components, which may be found along with acetate in some exopolysaccharides, include:

- propionate;
- glycerate;
- succinate;
- 3-hydroxybutanoate.

organic acids

amino acids

The presence of organic acid substituents in exopolysaccharides increases the lipophilicity of the molecule. In addition, for some exopolysaccharides with relatively high organic acid contents, their interaction with cations and with other polysaccharides may be influenced. Several amino acids have also been found in bacterial exopolysaccharides, including serine and L-glutamic acid (Figure 7.1).

phosphate and sulphate

Some microbial exopolysaccharides contain the inorganic substituents phosphate and sulphate. Phosphate has been found in exopolysaccharide from bacteria of medical importance, including *Escherichia coli*. Sulphate is far less common than phosphate and has only been found in species of cyanobacteria. In addition to these inorganic components, which form part of the structure of some exopolysaccharides, all polyanionic polymers will bind a mixture of cations. Exopolysaccharides are, therefore, purified in the salt form. The strength of binding of the various cations depend on the exopolysaccharide; some bind the divalent cations calcium, barium and strontium very strongly, whereas others prefer certain monovalent cations, eg Na^+.

SAQ 7.1

Identify each of the following statements as true or false. If false give a reason for your response.

1) D-xylose is a sugar commonly found in microbial polysaccharides.

2) Pyruvate ketals contribute to the cationic nature of exopolysaccharides.

3) The presence of acetate in exopolysaccharides increases their lipophilicity.

4) Exopolysaccharides are not produced by yeasts and filamentous fungi.

5) An exopolysaccharide containing a high content of D-glucuronic acid will tend to bind cations.

7.3 Exopolysaccharide structure

As with all polysaccharides, microbial exopolysaccharides can be divided into homopolysaccharides and heteropolysaccharides.

Homopolysaccharides contain only one type of monosaccharide, whereas heteropolysaccharides contain more than one.

Many are neutral glucans, being composed of the monosaccharide component D-glucose.

7.3.1 Homopolysaccharides

Three main types of homopolysaccharides are known (Figure 7.2).

1) Single linkage type. Several of these are neutral glucans, eg curdlan. Others are polyanionic homopolymers and, unlike the glucans, also contain acyl groups.

2) Side chain type. Scleroglycan is typical and possesses tetrasaccharide repeating units due to a 1,6-β-D-glucosyl side-chain on every third main chain residue.

3) Branched types. These are dextrans which are composed entirely of α-linked glucosyl residues. In some dextrans the linkage is almost entirely $1 \rightarrow 6$, but up to 50% of the glucose residues may be linked $1 \rightarrow 2$, $1 \rightarrow 3$ or $1 \rightarrow 4$.

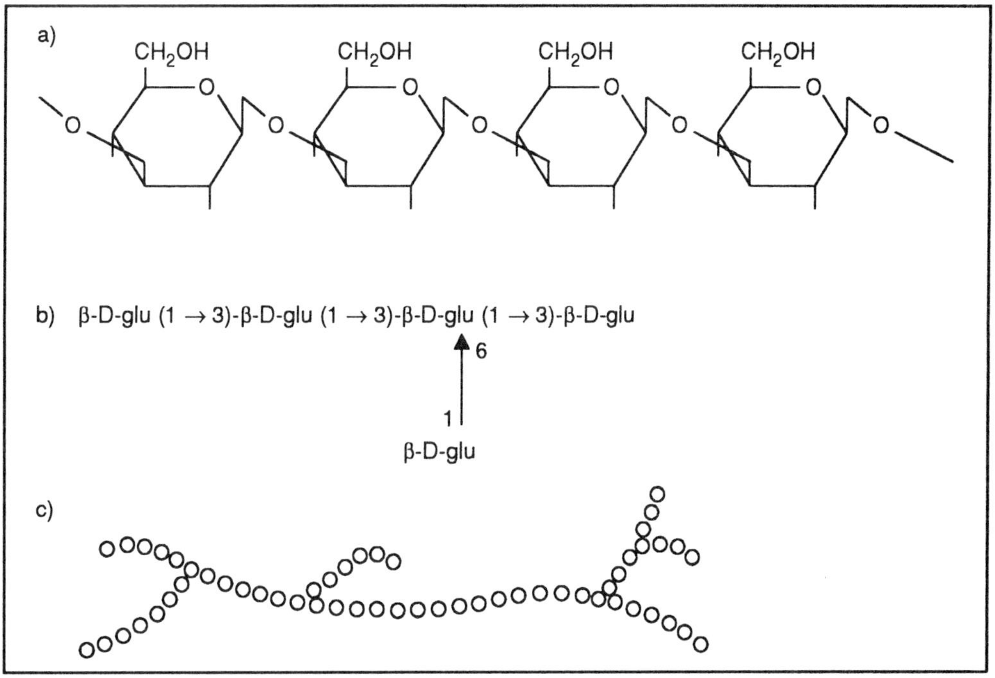

Figure 7.2 Three main types of homopolysaccharides. a) Single linkage type, eg curdlan. b) Side chain type, eg scleroglycan. c) Branched chain type, eg dextran. The figure also show various ways of illustrating exopolysaccharide structure.

7.3.2 Heteropolysaccharides

repeat units

Microbial heteropolysaccharides are almost entirely all composed of repeating units of between two and eight monosaccharides. The units often contain D-glucuronic acid and short side-chains of one to four residues are common. Several different side-chains are found in some heteropolysaccharides. The structure of xanthan illustrates these points and is shown in Figure 7.3. A few bacterial alginates are exceptions.

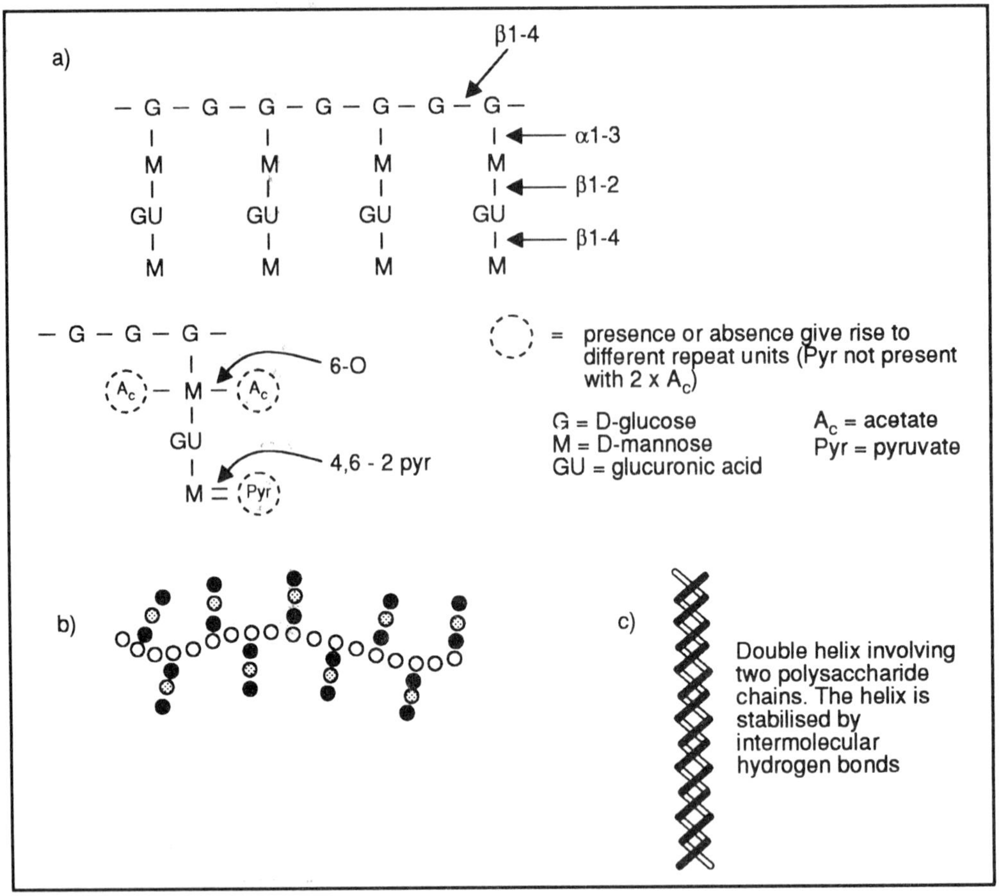

Figure 7.3 The structure of xanthan illustrated in three ways (a, b, c).

SAQ 7.2

Complete the illustration of the repeat unit in xanthan by adding bonds and groups to the molecule (similar to the illustration in Figure 7.2a). Refer to Figures 7.1 and 7.3 for chemical structures.

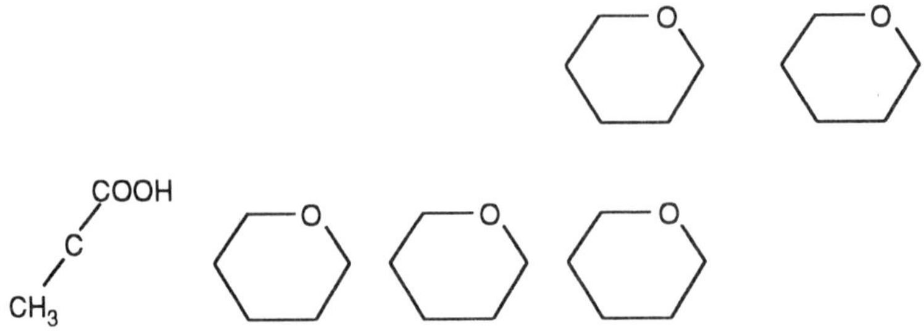

Π Examine Figure 7.3 and give two reasons why xanthan has an anionic nature.

The anionic nature of xanthan arises from the presence of glucuronic acid and pyruvate.

molecular conformation

The unique physical properties of microbial exopolysaccharides (considered in Section 7.7), which determine their commercial importance, arises from their molecular conformation. This, in turn, is determined by the primary structure and from associations between molecules in solution.

bond angles

For most exopolysaccharides their shape is determined by the angle of bonds which governs the relative orientations of adjacent sugar residues in the chain. However, the range of relative orientations of adjacent sugar molecules is limited by steric interactions between molecules along the chain.

Π Which group substituents influence inter-atomic electrostatic repulsion in a glycosyl chain?

The carbonyl (COO^-) group, which carries a full charge, will have the most pronounced effect. Oxygen atoms of hydroxyl groups carry a partial negative charge and, therefore, repel each other.

helical conformation

Exopolysaccharides in solution have an ordered helical conformation, which may be single, double or triple; for example, xanthan forms a double or triple helix (Figure 7.3c). These are stabilised by intermolecular hydrogen bonds. The helical conformation makes the exopolysaccharide semirigid and the molecules can move large volumes of solution. These volumes overlap even at low concentrations of exopolysaccharide, giving rise to relatively high viscosities.

intermolecular interactions

The intermolecular interactions stabilise the helices and greatly influence the properties of exopolysaccharides in solution, ie solubility, viscosity and gel-formation. A strong interaction or 'good-fit' between molecules will lead to insolubility, whereas poor interaction will lead to solubility of exopolysaccharides. The interactions between molecules is influenced by the presence of side-chains. For example, cellulose is insoluble but introduction of a three monosaccharide side-chain into the cellulose chain gives the soluble xanthan. Small changes in the structure of the side-chains can alter the molecular interactions and thus properties of the exopolysaccharide.

These changes may be brought about by:

- choice of production organism;
- conditions of fermentation;
- chemical modification (post fermentation);
- enzymatic modification (post fermentation).

7.4 Medium composition for exopolysaccharide production

Many different types of carbon substrate can be converted by micro-organisms to exopolysaccharides, these include:

- carbohydrates;
- amino acids;

- hydrocarbons;
- fatty acids;
- certain central metabolites (such as TCA cycle intermediates).

carbohydrates

Carbohydrates are the most widely utilised carbon substrate by exopolysaccharide producing micro-organisms and are used as substrate for commercial production. The structure of the exopolysaccharide is generally independent of the carbon substrate. However, choice of carbon substrate can influence both the quantity produced and the extent of acylation of exopolysaccharides. The bacteria that produce dextran are unusually specific in their carbon substrate requirement for exopolysaccharide production: they synthesise dextran only when grown on sucrose and are apparently unable to synthesise the polymer when grown on other substrates, such as glucose.

Π Why do you think carbohydrates are the most widely utilised carbon substrate for commercial production?

Carbohydrates are relatively cheap, available in large quantities and are readily utilisable source of carbon and energy for most micro-organisms. These considerations are particularly important for those exopolysaccharides produced on a large (bulk chemical) scale.

nitrogen sources

Utilisable nitrogen sources for exopolysaccharide producing organisms include:

- ammonium salts;
- amino acids;
- nitrate;
- dinitrogen (nitrogen gas).

Ammonium salts or amino acids are by far the most commonly used nitrogen sources in production media. Nitrate is rarely used. Although most nitrogen fixing micro-organisms do produce exopolysaccharide, their growth and the quantity of the polymer produced is often improved if a fixed source of nitrogen, such as NH_4^+, is supplied.

cations

Most micro-organisms require various cations for optimal growth, in particular K^+, Mg^{2+}, Fe^{2+} and Ca^{2+}. Other cations (trace elements) are required in smaller quantities and, in some culture media, may be present as components of other ingredients. Phosphate is the major anionic requirement of micro-organisms.

K^+ has a role in substrate uptake and during efficient exopolysaccharide synthesis, adequate supplies of this ion is essential for ensuring sufficient intracellular carbon substrate is maintained. Other ions, such as phosphate and magnesium, have roles in the acylation of exopolysaccharides and influence their physical properties.

organic components

Exopolysaccharide production may be improved by the provision of various organic components, other than the main carbon and energy source. These can improve growth of the production organism (growth factors) and/or directly enhance the synthesis of exopolysaccharide. Additions that improve polymer yield include tricarboxylic acid (TCA) cycle intermediates, which are thought to improve metabolic balance between carbon flow from carbohydrate substrate through the catabolic pathways and oxidation through the TCA cycle.

Π The components of a defined mineral salts medium for exopolysaccharide production are given below:

KH_2PO_4	Glutamate	$FeSO_4.7H_2O$
$MnSO_4.7H_2O$	KI	Glucose
H_3BO_4	$MgSO_4.7H_2O$	Na_2MoO_4
$CuSO_4.\ 5H_20$	$CoCl_2.6H_2O$	$CaCl_2.6H_2O$

What is the nitrogen source in the medium? List the cations provided in the medium. What is the main anionic component? How might the addition of fumaric acid and peptone (vegetable infusion) improve the medium?

The nitrogen source in the medium is the amino acid glutamate. There are several cations: K^+ Mn^{2+}, Cn^{2+}, Zn^{2+}, Mg^{2+}, Co^{2+}, Fe^{2+}, Ca^{2+} Mo^{6+}. Phosphate (PO_4^{3-}) is the major anionic component. Fumaric acid is a TCA cycle intermediate and may improve metabolic balance through the catabolic pathways and oxidation through the TCA cycle. Peptone may improve growth through the provision of growth factors (amino acids, vitamins, nucleotides).

limiting nutrient

The balance between carbon substrate and a growth limiting nutrient also influences polysaccharide production. Carbon substrate must be provided in adequate amounts to ensure good yields of exopolysaccharide. However, very high cell density cultures are not necessary during fermentation. It is, therefore, usual to control cell density by limitation of a nutrient other than carbon. Nitrogen has traditionally been used as the limiting nutrient although others, such as sulphur, potassium, magnesium and phosphorous have also been studied. The type of limiting nutrient has been shown to influence both the yield and composition of exopolysaccharide. In the case of potassium limitation, yields are often low because of the involvement of potassium in nutrient uptake. In the case of xanthan, limitation of sulphur, magnesium or phosphate is thought to influence acylation and thus the physical properties of the exopolysaccharide.

industrial production media

Media used for laboratory studies of exopolysaccharide production may vary considerably from industrial production media. In laboratory studies pure substrates such as glucose, sucrose and glycerol, can be used to determine exopolysaccharide yields. In industrial production the main factors that influence the decision as to which substrate to use are:

- cost of pure substrate;
- product yield for the substrate;
- the quality of the product required.

For a high-quality product, the substrate itself must be relatively pure to minimise carryover of impurities to the final product. Most industrial process micro-organisms produce optimal yields of exopolysaccharides from carbohydrate and the most commonly used substrates for industrial production are:

- glucose from cane or beet;
- sucrose;
- starch or starch hydrolysates;
- corn syrup.

Cruder and thus cheaper substrates may be used if less pure products are acceptable. Some of these are waste products from other industries, for example:

- dry milled corn starch;
- cereal grain hydrolysates;
- whey.

For industrial production, the nitrogen source may be a relatively cheap proteinaceous product, such as:

- yeast hydrolysate;
- distillers solubles;
- casein hydrolysate;
- soybean meal.

Local availability as well as cost may well determine the choice of the nitrogen source.

Industrial production media must also contain sources of potassium, phosphorous and magnesium. Trace elements may also have to be added. The water used for medium preparation will be from the public water supply or other readily available source. The quality of the water is carefully monitored because the presence of certain metal salts, for example, calcium, copper and iron, can have adverse effects on both the growth of the organism and the rheological properties of the exopolysaccharides.

SAQ 7.3

A medium for the production of an exopolysaccharide in batch culture has the following components:

Glucose
Succinic acid
Ammonium chloride (limiting substrate)
Yeast hydrolysate
Potassium
Phosphorous
Magnesium sulphate
Iron
Calcium
Trace elements (Cu, Zn, Ni, Mo, Mn, Co).

1) What are the roles of each component?

2) If the growth yield coefficient for ammonia was 10 $g\ g^{-1}$, what concentration of biomass would you expect if ammonia was added to a concentration of 2.5 $g\ l^{-1}$?

3) If the growth yield coefficient for magnesium for the organism grown in the medium described in 2) was 200 $g\ g^{-1}$, what concentration of biomass would you expect if magnesium was added to this medium at a concentration of 0.2 $g\ l^{-1}$?

7.5 Fermentation

steps

The various steps involved in commercial production of exopolysaccharide are:

1) Strain maintenance.

2) Inoculum train.

3) Exopolysaccharide production (50-200 m^3).

4) Enzyme treatment to modify properties, eg filterability.

5) Concentration via precipitation or ultrafiltration.

6) Storage and packaging.

Biocide may be added after steps four or five.

7.5.1 Strain maintenance

standard inoculum

Microbial strains must be maintained in such a way that they do not lose their desirable characteristics. Some strains are maintained by regular subculturing, whereas others are lyophilised (freeze-dried), or frozen under nitrogen, or held at -80°C in a freezer. To ensure that a standard inoculum can be obtained on demand, great care is taken to ensure that the stored cultures are pure and the viability is known.

7.5.2 Inoculum train

The medium used to produce the inoculum should be designed for rapid growth of the production organism without exopolysaccharide production. Production of the latter in the inoculum train can give rise to highly viscous cultures that are difficult to transfer from one vessel to another.

Optimisation of biomass production would require a large inoculum, comprising 10% of each inoculum stage. However, this involves many transfers which increases the risk of contamination.

Π How many transfers are required for inoculation of a 200 m^3 fermentor, using a 10% inoculum train? The initial transfer is the inoculation of a 200 ml shake flask with a lyophilised culture.

The correct answer is seven. In practice, the number of inoculum steps does not usually exceed four to reduce the risk of contamination. This also reduces capital investment and production costs since fewer transfers require fewer vessels for development of the inoculum.

Contamination of the production vessel leads to serious financial penalties and each step in the inoculum train is monitored for contamination. To reduce the risk of contamination during sampling it is usual to take a sample from the residue left in each vessel after its contents have been transferred to the next reactor. Since these contamination checks are retrospective, a heavy reliance is placed on the growth characteristics of the production organism. Kinetic variables such as growth rate and oxygen consumption rate are also used to assess the quality of the inoculum.

contamination checks

7.5.3 Fermentation conditions

pH

The optimum pH for synthesis of exopolysaccharides is normally between 6.0 and 7.5. For xanthan the optimum is pH 7.0 and below pH 5.0 synthesis is severely depressed.

Π Why do you think it is particularly important to control the pH of a xanthan fermentation?

Xanthan itself is an acidic product which, in the absence of pH control, would reduce the pH of the fermentation.

Temperature can influence the characteristics of the product. For example, the succionoglycan from *Agrobacterium radiobacter* grown at 30°C has a lower viscosity than that grown at a temperature of 35°C, although the final concentration is similar.

The amount of oxygen in an aerobic fermentation is used in a few seconds by actively growing cultures and most aerobic microbial industrial processes are oxygen limited. This problem is particularly acute for exopolysaccharide production on a large scale. Figure 7.4 demonstrates the dramatic effect of increasing viscosity on oxygen transfer. Clearly in a system that produces a highly viscous compound (xanthan at 3% (w/v) has a viscosity in excess of 10,000 cp) oxygen transfer severely reduces the product concentration that can be achieved in the medium.

viscosity and oxygen transfer

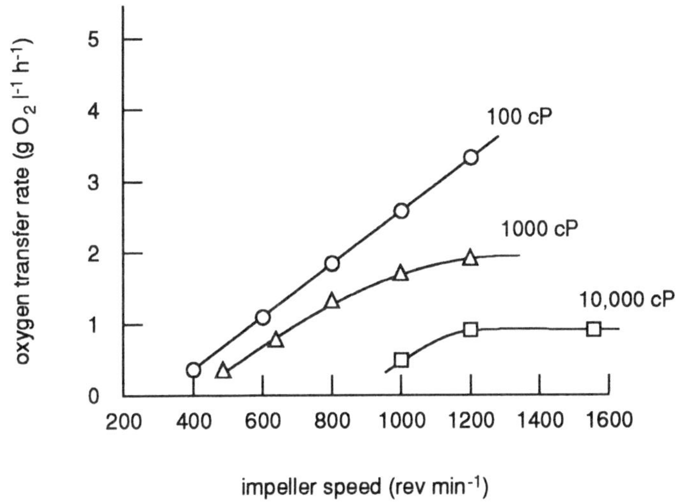

Figure 7.4 The effect of viscosity on oxygen transfer rates. Adapted from Biochemical and Biotechnology Handbook. B Atkinson and F Mavituna, (Eds) 1991, Stockton press.

7.5.4 Production of xanthan gum

From a commercial point of view, xanthan gum is the most important microbial exopolysaccharide currently being manufactured. Therefore, we shall consider the fermentation of this product by *Xanthomonas campestris* in some detail.

complex nutrients

The wild-type strain of *X. campestris* requires complex nutrients, such as yeast extract and several vitamins, to achieve adequate growth and yield of exopolysaccharide.

Π Give reasons why a requirement for 1) yeast extract and 2) vitamins is undesirable for commercial production of exopolysaccharide.

1) Complex nutrients, such as yeast extract, are variable in composition and consequently it is difficult to maintain process reproducibility within the narrow window required to produce a product of consistent quality.

2) Vitamins are expensive.

These limitations have been overcome by the development of strains able to grow well in a minimal salts medium in the absence of complex nutrients and vitamins. The carbon source is usually glucose and a complex nitrogen source, such as glutamate, can replace ammonia.

Π The components of a defined mineral salts medium for xanthan production are given below:

KH_2PO_4	Glutamate	$FeSO_4.7H_2O$
$MnSO_4.7H_2O$	KI	Glucose
H_3BO_4	$MgSO_4.7H_2O$	Na_2MoO_4
$CuSO_4.5H_2O$	$CoCl_2.6H_2O$	$CaCl_2.6H_2O$
$ZnSO_4.7H_2O$		

1) List those ingredients which are added at concentrations below 1 mg l^{-1} (micro-elements).

2) Which ingredients are added at the following concentrations?

$g\ l^{-1}$

25 - 40
3.5 - 7.0
0.68
0.4
0.012
0.011

Our response:

1) The micro-elements are:

$MnSO_4.7H_2O$
$ZnSO_4.7H_2O$
H_3BO_4
Na_2MoO_4
$CuSO_4.5H_2O$
$CoCl_2.6H_2O$
KI

2)

	$g\ l^{-1}$
Glucose	25 - 40
Glutamate	3.5 - 7.0
KH_2PO_4	0.68
$MgSO_4.7H_2O$	0.4
$CaCl_2.2H_2O$	0.012
$FeSO_4.7H_2O$	0.011

submerged fermentation - xanthan

The characteristics of a typical submerged fermentation for xanthan are shown in Figure 7.5. Nitrogen exhaustion limits the cell density to around 4 g dry weight l^{-1}. The growth phase is complete in about 25 hours and exopolysaccharide production ceases soon after. This reflects the high specific exopolysaccharide production rate during growth of the organisms. The sugar concentration and the viscosity of the fermentation is monitored at regular intervals and the production is considered to be complete when all the glucose is exhausted and the desired viscosity has been reached.

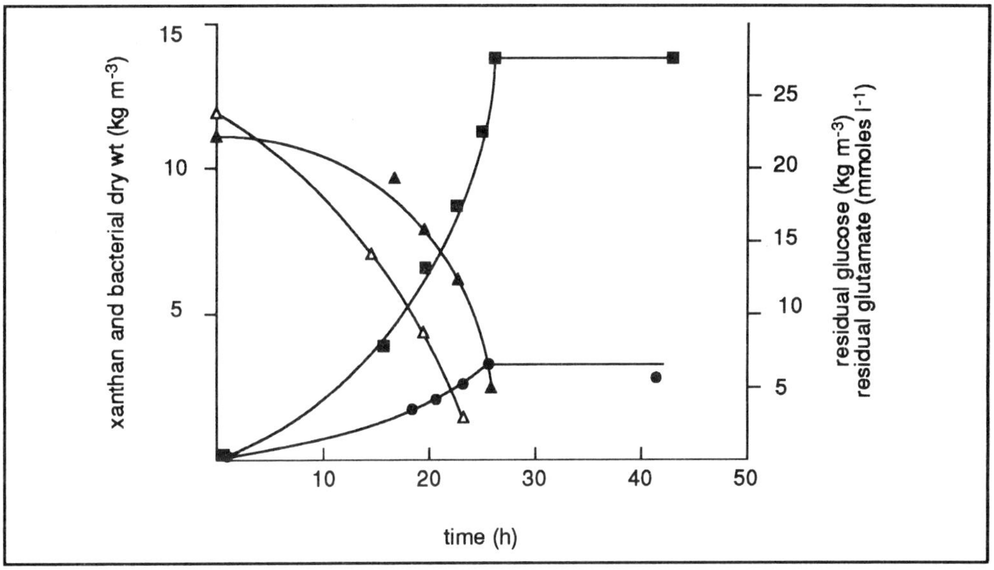

Figure 7.5 Production of xanthan gum in batch culture using *X. campestris*. Bacterial dry weight (●); xanthan gum (■); residual glucose (▲); residual glutamate (Δ). Adapted from Microbial exopolysaccharide, Yenton *et al* pp 217-261. In biomaterials; Novel Materials from Biological Sources, D Byrom (Ed), MacMillan Academic Professional Ltd, 1991.

7.5.4 Production of succinoglycan

Typical batch production of succinoglycan by a *Pseudomonas* sp is shown in Figure 7.6.

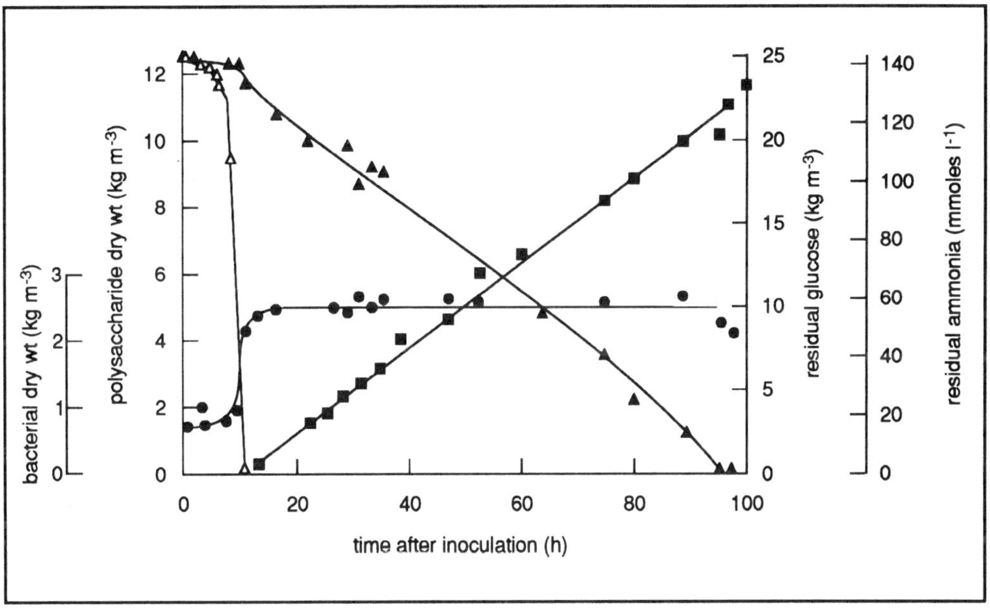

Figure 7.6 Production of succinoglycan in batch culture using a *Pseudomonas* sp. Bacterial dry weight (●); succinoglycan concentration (■); residual glucose (▲); residual ammonia (Δ). Adapted from as Figure 7.5.

In the succinoglycan fermentation, exopolysaccharide production commences after cessation of growth. During the 80 hour production phase of the fermentation, there is a linear increase in product concentration and a linear decrease in residual glucose.

Π Use the data presented in Figure 7.6 to calculate the following kinetic parameters and variables for the fermentation:

1) Y_P, the yield coefficient for product (kg polysaccharide (kg glucose)$^{-1}$);

2) q_P, the specific rate of polysaccharide production (kg polysaccharide h^{-1} (kg dry weight cells)$^{-1}$);

3) productivity (kg polysaccharide m^{-3} h^{-1}); assume that the fermentor was harvested at t =100h and the set-up time for preparation of the fermentor was an additional 24h.

1)
$$Y_P = \frac{12.5 \text{ kg xanthan m}^{-3} \text{ produced}}{25 \text{ kg glucose m}^{-3} \text{ consumed}} = 0.5 \text{ kg polysaccharide (kg glucose)}^{-1}$$

2)
$$q_P = \frac{\text{slope of polysaccharide accumulation line}}{\text{biomass concentration}} = \frac{\left(\frac{12.0 \text{ g m}^{-3}}{83 \text{ h}}\right)}{2.5}$$

$$= 0.058 \text{ kg polysaccharide h}^{-1} \text{ (kg dry wt cells)}^{-1}$$

3)

$$\text{Productivity} = \frac{\text{product concentration at end of fermentation}}{\text{total fermentation time}}$$

$$= \frac{12 \text{ kg m}^{-3}}{124 \text{ h}} = 0.10 \text{ kg polysaccharide m}^{-3} \text{ h}^{-1}.$$

SAQ 7.4

Refer to Figures 7.5 and 7.7.

1) Explain briefly why the rate of xanthan production increases during the production phase while that of succinoglycan remains constant.

2) Explain why it would be inadvisable to use the carbon source as growth limiting nutrient in the succinoglycan fermentation.

3) What limits growth in the xanthan fermentation?

4) What limits succinoglycan production and why is it necessary to do so?

5) In the succinoglycan fermentation, what is the growth yield coefficient for ammonia? (in units of g g^{-1}; molecular weight of ammonia = 17).

7.6 Product recovery

There are two main methods for the recovery of exopolysaccharides from fermentation liquors:

- solvent precipitation followed by spray drying;
- ultrafiltration.

The method used is governed by the market application of the exopolysaccharide. In general, the food industry has a requirement for a dry powder, whereas for several other applications, such as enhanced oil recovery, a liquid product is required and the ultrafiltration concentrate is preferred.

7.6.1 Solvent precipitation

polar organic solvents

Polar organic solvents readily precipitate exopolysaccharides from solution. The solvents commonly used are acetone, methanol, ethanol and propan-2-ol. Cation concentration of the fermentation liquor influences the amount of solvent required for efficient product recovery. In the case of propan-2-ol, increasing the cation concentration can lead to a four-fold reduction in the volume of solvent required to precipitate xanthan gum. Salts such as calcium nitrate and potassium chloride are added to fermentation broths for this purpose.

Further reduction in the volume of solvent required can be achieved by heat treatment of the exopolysaccharide at 100-130°C for 1-15 minutes before solvent precipitation.

distillation

The solvent can be recovered by distillation and the precipitated product is freed from excess liquid by centrifugation or pressing. Continuous drying by vacuum or with an inert gas, to reduce the dangers associated with the flammable organic solvents, gives the dry polymer. This is finally milled to the required mesh size, which forms an off-white powder. The powdered exopolysaccharide must be stored in a cool, dry atmosphere to prevent adsorbtion of moisture which would lead to biological degradation of the product. The dry powder remains stable if stored at temperatures below 25°C.

For food and pharmaceutical applications, the microbial count must be reduced to less than 10,000 viable cells per g exopolysaccharide. Treatment with propylene oxide gas has been used for reducing the number of viable cells in xanthan powders. The patented process involves propylene oxide treatment for 3 h in a tumbling reactor. There is an initial evacuation step before propylene oxide exposure. After treatment, evacuation and tumbling are alternated and if necessary the reactor is flushed with sterile nitrogen gas to reduce the residual propylene oxide level below the Food and Drug Administration permitted maximum (300 mg kg^{-1}). The treated polysaccharide is then packaged aseptically.

propylene oxide

The main disadvantage of precipitation with a polar (water-soluble) solvent is the need for a costly distillation stage to recover the relatively large volumes of solvent used. Another disadvantage is the precipitation of proteins, salts and, in some cases, pigments which reduces the purity and leads to discoloration of the product. To overcome these problems, precipitation using less polar solvents, such as methylethylketone, has been proposed. Only 23% (w/v) methylethylketone is sufficient to saturate the aqueous phase and precipitate exopolysaccharides quantitatively.

methylethylketone

Π What are the advantages of having two distinct phases (aqueous/organic) during precipitation?

1) It is relatively easy to recover organic solvents.

2) Pigments are extracted into the organic phase resulting in the production of a pure white precipitate.

7.6.2 Ultrafiltration

pseudoplastic ploymers

The alternative large scale recovery method to precipitation is **ultrafiltration**. For concentration of viscous exopolysaccharides, ultrafiltration is only effective for pseudoplastic polymers (shearing reduces effective viscosity; see section 7.7). Thus, pseudoplastic xanthan gum can be concentrated to a viscosity of around 30,000 centipoise by ultrafiltration, whereas other polysaccharides which are less pseudoplastic, are concentrated only to a fraction of this viscosity and have proportionally lower flux rates. Xanthan gum is routinely concentrated 5 to 10-fold by ultrafiltration.

membranes

Commercially available plate- and frame- type ultrafiltration equipment are used for exopolysaccharide concentration. The membranes are polysulphone or polyvinylidine fluoride with molecular weight cut-off between 20-60,000. There is a relatively low energy requirement (1-2 kWh m^{-3}) for pumping the fluid through the filtration unit at the desired pressure. Pressure difference across the membrane is of the order 2-14 atmospheres.

advantages of ultrafiltration

Ultrafiltration has the advantage that there is removal of low molecular weight fermentation products and medium components during concentration of the exopolysaccharide. In addition, biological degradation is minimised because fluid is held only for a short time during the filtration process. Other advantages lie in the fact that there is no requirement for solvent recovery and the process is carried out at ambient (not elevated) temperature.

antimicrobial agents

The concentrate derived from ultrafiltration is usually a thick colourless gel containing about 4-8% solids. This must contain an antimicrobial agent to inhibit microbial growth and biological degradation. The type of antimicrobial agent used depends on the particular application for the exopolysaccharide. For example, the nature of the antimicrobial agent is less critical for industrial applications, such as enhanced oil recovery, than for use in cosmetics.

Π Make your own list of advantages and disadvantages of the two main methods of recovery of exopolysaccharides from fermentation liquors, based on the information provided in this Section (7.6).

SAQ 7.5

List the following stages involved in recovery of exopolysaccharides by solvent precipitation and subsequent packaging in an appropriate order.

1) Addition of solvent.
2) Centrifugation.
3) Vacuum drying.
4) Aseptic packaging.
5) Propylene oxide treatment.
6) Cool dry storage.
7) Heat treatment.
8) Addition of cations.
9) Distillation.
10) Sterile nitrogen gas treatment.
11) Milling.

7.7 Physical properties

Microbial exopolysaccharides are widely used in industry as viscosifiers and as gelling agents. In this section we will consider, in general, the rheology of exopolysaccharides in solution and their ability to form gels. Specific properties of individual microbial exopolysaccharides and applications which exploit these characteristics are considered later in this chapter.

7.7.1 Viscosity

pseudoplastic flow

Many microbial polysaccharides show pseudoplastic flow, also known as shear thinning. When solutions of these polysaccharides are sheared, the molecules align in the shear field and the effective viscosity is reduced. This reduction of viscosity is not a consequence of degradation (unless the shear rate exceeds 105 s^{-1}) since the viscosity recovers immediately when the shear rate is decreased. This combination of viscous and elastic behaviour, known as viscoelasticity, distinguishes microbial viscosifiers from solutions of other thickeners. Examples of microbial viscosifiers are:

viscoelasticity

- xanthan gum;
- succinoglycan;
- welan;
- scleroglucan.

shear rate

The viscosity of a solution of microbial exopolysaccharide must, therefore, be defined as a function of the shear rate (see Figure 7.7).

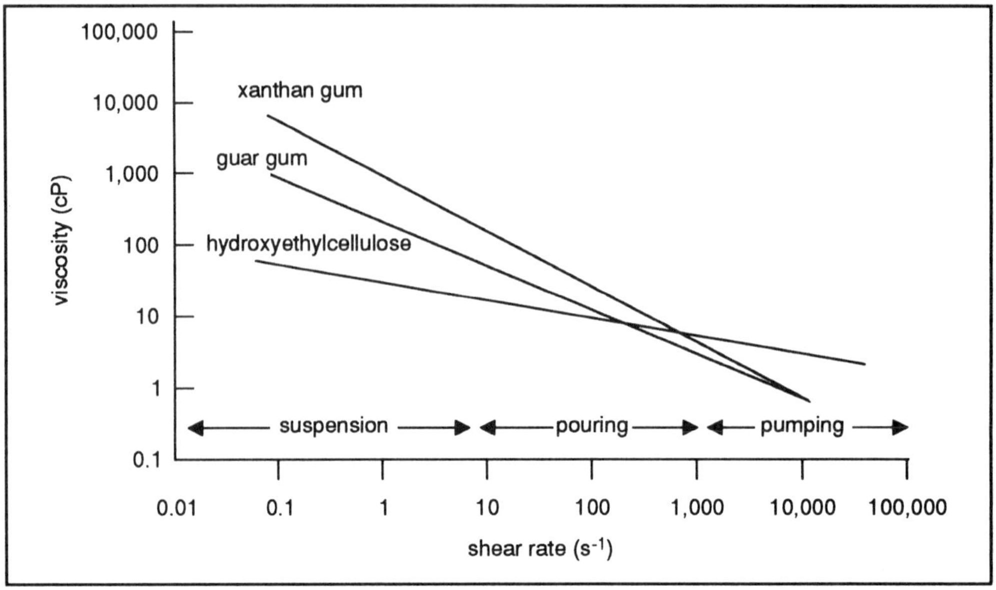

Figure 7.7 Viscosity verses shear rate profiles (polymer concentration, 2.25 kg m^{-3}). Adapted from as Figure 7.5.

yield stress

We can see that although the rate of flow may be very low in very viscous solutions, there is no yield stress, ie a stress that must be exceeded for flow to commence. However, some exopolysaccharides do display yield stress characteristics.

Π In practice, the pseudoplastic flow behaviour and elasticity are important characteristics. Try to think of useful properties that these characteristics would confer on exopolysaccharides.

High viscosity at low shear rate is essential to inhibit particle sedimentation or to give good cling of a film of exopolysaccharide, whereas reduced viscosity under conditions of high shear rate maintains good pumping and spraying characteristics.

7.7.2 Gel formation

mechanisms

One of the most striking and useful properties of exopolysaccharides is that they can form gels at relatively low concentration (typically around 1%). Gels are distinct from viscous solutions that flow readily and are used widely in the food industry and in some personal care products. The mechanism of gel formation depends on the type of microbial exopolysaccharide:

- polymers that require the presence of ions for gel formation, eg alginate and gellan;
- polymers that form gels without the involvement of ions, eg curdlan;
- polymers that require the presence of another polysaccharide, eg xanthan.

curdlan

Curdlan gel formation is heat dependent. This polysaccharide is not soluble in water, but when an aqueous suspension is heated it becomes clear at about 54°C. Further heating leads to gel formation. The gels are stable over a wide range of pH (3 to 9.5) and do not melt at temperatures below 100°C. Curdlan gels are formed by cross-linking, involving conformational ordering of the exopolysaccharide molecules to give a triple helical structure.

triple helix

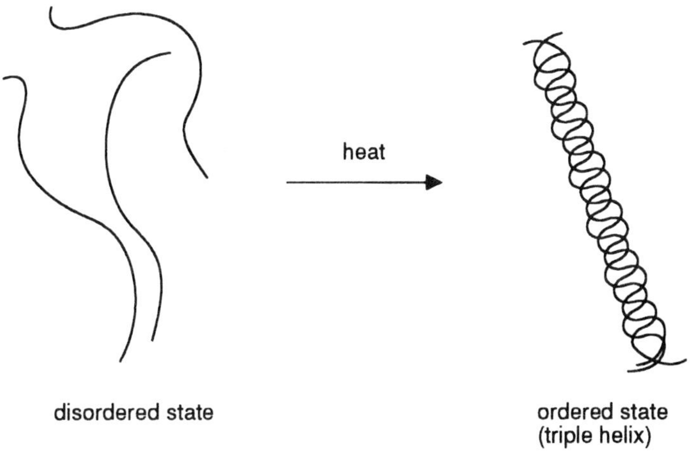

alginate
binding of Ca^{2+}

Alginate forms gels by the selective co-operative binding of divalent cations. Cross-linking of alginate gels involves the binding of Ca^{2+} within alginate ribbons of polyguluronate sequences, giving rise to the so-called 'egg-box' model. The egg-box arrangement ends when the polyguluronate sequence gives way to polymannuronic acid or mixed sequences. The physical properties of the alginate gels thus depend on the ratio of D-mannuronic acid and L-guluronic acid.

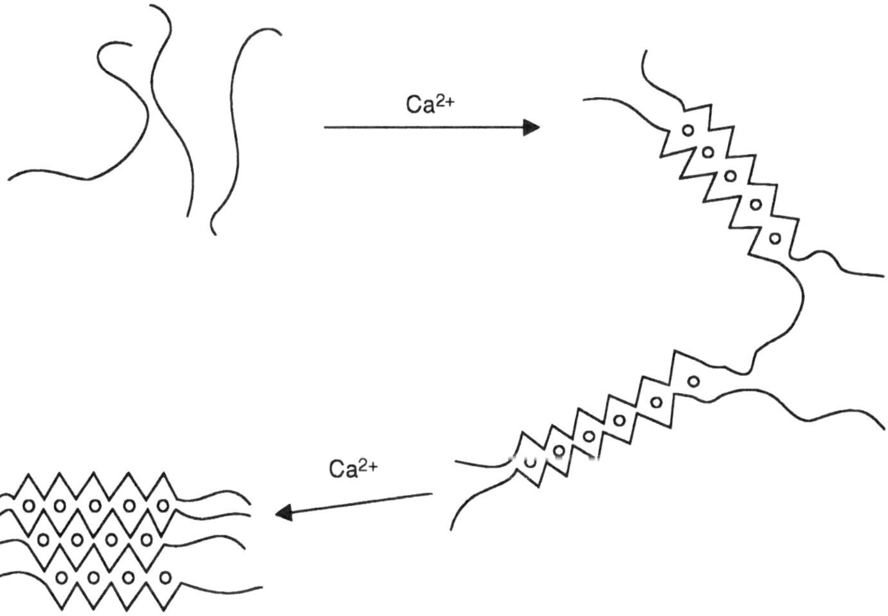

xanthan

Xanthan does not in itself form gels, despite the strong intermolecular interactions which occur in solution. However, some of the rheological properties of xanthan have been explained by non-covalent association of xanthan with relatively small amounts of other polysaccharides to form three-dimensional networks (weak gels).

7.7.3 Stability

melting temperature

Microbial polysaccharides in solution lose their ordered conformation on heating. The temperature at which the polymer 'melts' to a disordered state is known as the melting temperature (T_m) and is determined by a variety of factors:

- polymer structure;
- nature and concentration of ions in solution;
- nature and concentration of miscible solvent (if present).

ionic strength

At low ionic strengths, T_m increases exponentially with ion activity. The effect of high concentrations of salts or miscible solvents depends on the influence they have on hydrogen-bonding and may increase or decrease T_m. In the case of xanthan gum, the value of T_m can be adjusted from ambient to over 200°C by the addition of appropriate salts. Table 7.2 presents T_m values for some industrial viscosifiers.

Polymer	ApproximateT$_m$ (°C)
Xanthan	120
Succinoglycan	70
Scleroglucan	150
Welan	150

Table 7.2 Appropriate T_m values for microbially derived viscoifiers in hard tap water (or sea waters).

At temperatures below T_m, there is relatively little change in viscosity of microbial exopolysaccharides (Figure 7.8).

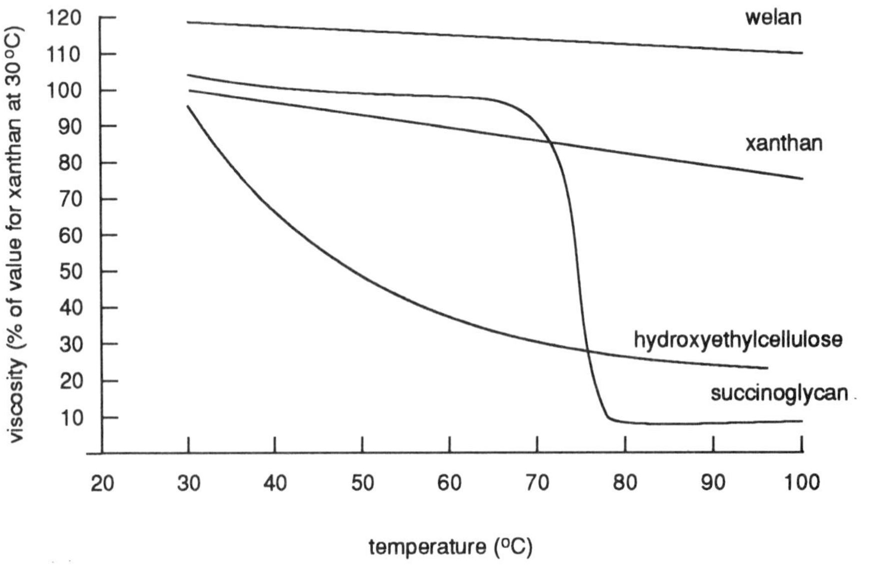

Figure 7.8 The effect of temperature on viscosity. Adapted from as Figure 7.5.

viscosity above T_m

At temperatures above T_m the change in viscosity with temperature depends on the type and concentration of polymer in solution. At low polymer concentrations, the more flexible disordered conformation reduces both the viscosity and the pseudoplastic behaviour. However, at concentrations exceeding a few percent the viscosity of some polymers will show a small increase compared to their viscosity below T_m. Conversely, some exopolysaccharides such as succinoglycan show a virtual loss of all viscosity at temperatures above T_m (Figure 7.8). On cooling to below T_m some or all of the viscosity observed before heating will be regained, or gelling will occur.

degradation

At temperatures above T_m, chemical and enzymatic degradation of microbial exopolysaccharides is enhanced. The apparent enhanced stability of microbial exopolysaccharides in their ordered confirmation is thought to be due to the glycosidic bonds in the backbone of the polymer which raises the activation energy. This restricted movement would also restrict access of enzymes and chemicals to the backbone.

SAQ 7.6

1) Complete the following statements:

a) ______ flow is also known as shear thinning.

b) _______ behaviour describes recovery of viscosity with reduction in shear rate.

c) The _________ conformation of a microbial exopolysaccharide is lost at temperatures at or above its ______ temperature.

d) Chemical and enzymatic degradation of microbial exopolysaccharides is enhanced at temperatures ____________ T_m.

2) Explain why brittle gels are obtained from alginate with a high guluronic acid content and more flexible gels from alginate if the D-mannuronic acid content is high.

7.8 Biosynthesis

cell membrane

Exopolysaccharides, in almost every case, are synthesised at the cell membrane and then exported from the cell. The only exceptions that have been recorded to date are the homopolysaccharides levan and dextran which are synthesised extracellularly.

sugar nucleotide phosphates

The building blocks for the exopolysaccharides are usually sugar nucleotide diphosphates with some monophosphates. The sugar nucleotide diphosphates provide energy for the synthesis of the oligosaccharides and are readily interconverted (Figure 7.9). The energy released from the sugar nucleotide diphosphates (32 kJ per mole) is greater than that released by sugar monophosphates (20 kJ per mole) and provides up to 70% of the energy requirement for the synthesis.

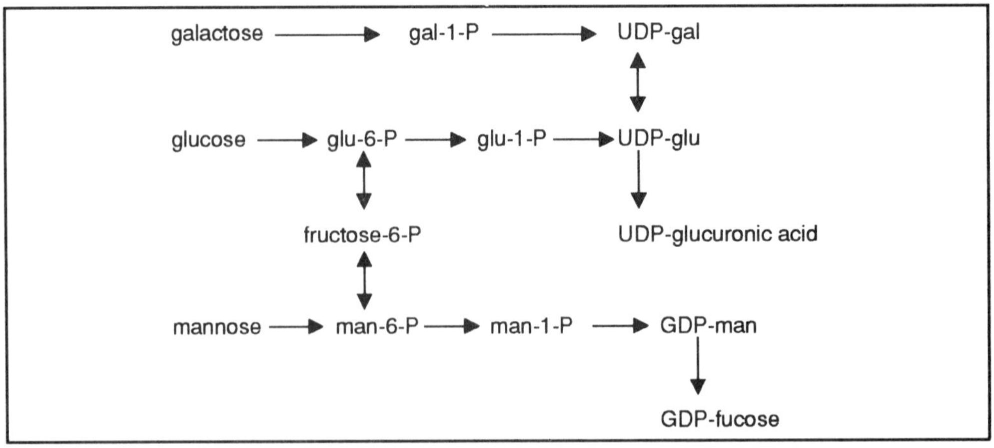

Figure 7.9 Interconversion of sugar phosphates and sugar nucleotide phosphates. Adapted from "Biotechnology of microbial exopolysaccharides". I W Sutherland, Cambridge University Press, 1990.

acetyl CoA

transferases

Other groups are added to modify the basic structure of the polysaccharide. These commonly include acetate, pyruvate, succinate, 3-hydroxy butanoate, phosphates and, in cyanobacteria, sulphate. Acetyl CoA is the source of acetyl groups in xanthan synthesis and is likely to be the source for most other exopolysaccharides. Methyl groups are derived from methionine or S-adenosylmethionine, as has been demonstrated for cell wall polysaccharides. There has as yet been no definitive study of the mechanism for introducing amino acids or inorganic substituents. It is probable that by anology with peptidoglycan (bacterial cell wall structural polysaccharide) synthesis, amino acids are added to the sugar nucleotides by a specific transferase requiring ATP and Mn^{+} and the phosphate groups are derived from the sugar nucleotides.

heteropolysaccharides

Heteropolysaccharide biosynthesis involves four stages:

- synthesis of the sugar nucleotide diphosphates;
- assembly of the repeat unit on a C_{55} lipid carrier (undecaprenyl phosphate);
- pyruvylation and addition of other substituents;
- transfer of the polysaccharide to the new subunit.

The completed exopolysaccharide is then transported to the cell surface through membrane adhesion zones.

The synthesis of xanthan has been studied in some detail and is depicted in Figure 7.10.

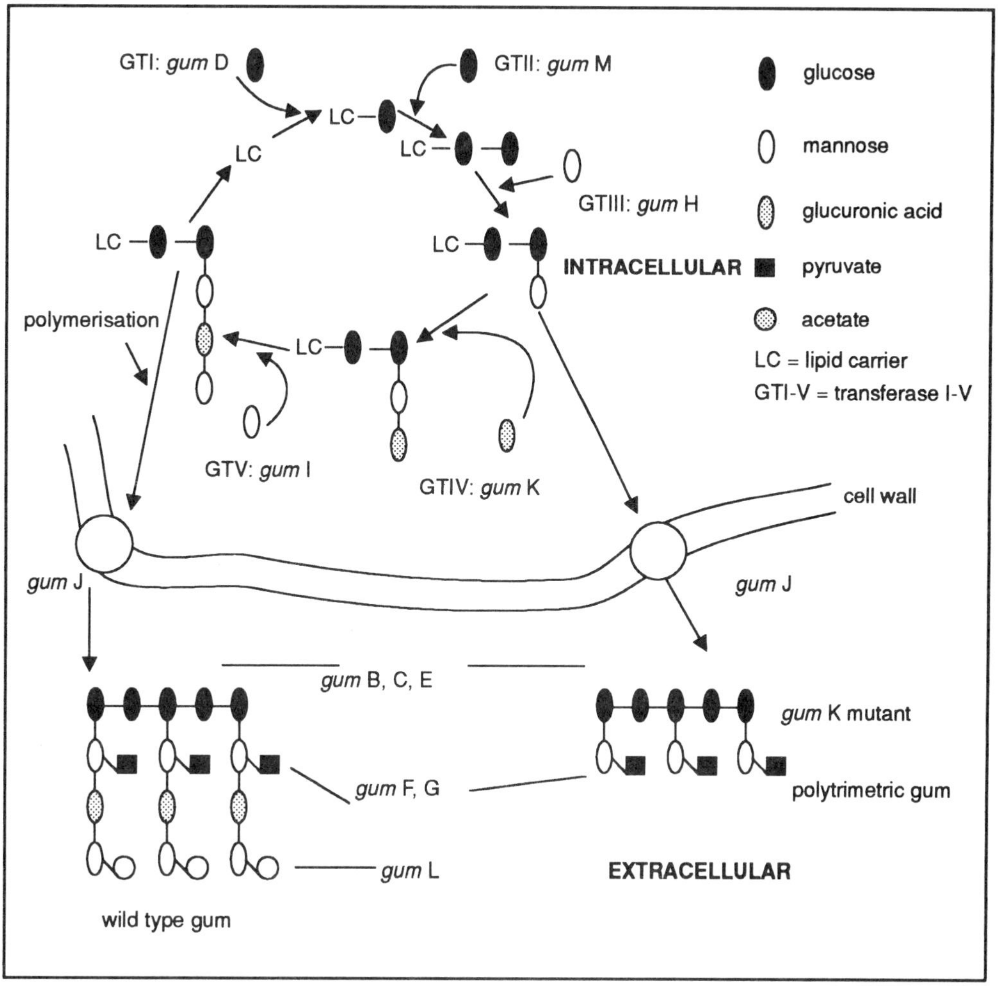

Figure 7.10 Simplified scheme for the biosynthesis of xanthan. Adapted from Exopolysaccharides in Plant Bacterial Interactions. J A Leight and D L Coplin. Annu Rev Microbial 1992 46 pp 307-46 (see text).

lipid carrier

polymerase

Glycosytransferases I to V add glucose, glucose, mannose, glucuromic acid, mannose respectively. Pyruvate and acetate are added and the growing xanthan molecule is transferred to a single repeat unit on a second lipid carrier. The acetylation of xanthan is variable and this is probably explained by other processes requiring acetyl CoA having priority when the intracellular concentrations are low. This process continues until the macromolecule is formed. The polymerase is relatively unspecific and will polymerise 'imperfectly formed' units synthesised by the *gum*K and I mutants resulting, in polytrimeric and polytetrameric gums respectively; the genes involved in xanthan synthesis are considered further in sections 7.8.1 and 7.8.2.

homopolysaccha rides

Homopolysaccharides are synthesised by relatively few specific enzymes and are not constructed from subunits. The commercially important homopolymer dextran is synthesised extracellularly by the enzyme dextransucrase. In *Leuconostoc mesenteroides* the enzyme is induced by the substrate sucrose. This is cleaved to release free fructose and link the glucose to the reducing end of the acceptor dextran chain, which is bound to the enzyme. The product from this bacterium is composed almost exclusively of

1,6-α-linked glucose with 1,2, 1,3 and 1,4 linkages as minor components. The mechanism for branching is unclear but there is no specific branching enzyme or requirement for sugar nucleotides. Dextran synthesised by other bacterial species differs in the extent of branching.

Alginates are unusual in that although they are heteropolysaccharides they are synthesised as homopolymers in the cell. Final epimerisation from mannuronic acid to glucuronic acid then occurs extracellularly.

SAQ 7.7

Using a simple illustration show how the completed pentasaccharide subunits of xanthan are assembled into the finished polymer.

SAQ 7.8

Using the word list below fill in the blanks in the following paragraph concerning exopolysaccharide biosynthesis.

Exopolysaccharides are synthesised ____________ at the cell membrane and then exported. They are formed predominantly from sugar nucleotide ________________, the cleavage of which provides much of the ___________ required for polymer synthesis. They are constructed on _________ carrier molecules. Heteropolysaccharides are made as _____________ which are then added to the growing exopolysaccharide molecule. This process requires several specific ____________ systems to transfer each sugar nucleotide diphosphate to the subunit etc. _____________. are synthesised by relatively few specific enzymes and are not constructed from subunits.

Word list:

bisphosphates, lipid, subunits, intracellularly, enzyme, homopolysaccharides, energy.

7.8.1 Genetics of exopolysaccharide synthesis

gene clusters

A common feature is the arrangement of genes responsible for the control and regulation of synthesis of specific exopolysaccharides in tight gene clusters or 'cassettes'. This is exemplified by the arrangement in *Xanthomonas campestris* which is used to produce the commercially important xanthan gum. Mutation studies have revealed a number of strains that produce xanthan with alterations in the pattern of acetylation and pyruvylation, as well as mutants which show increases in yield, rates of production and composition of the repeat sub-unit. The genes are clustered in a 16kb DNA sequence containing 12 genes, *gum*B to M (Figure 7.11).

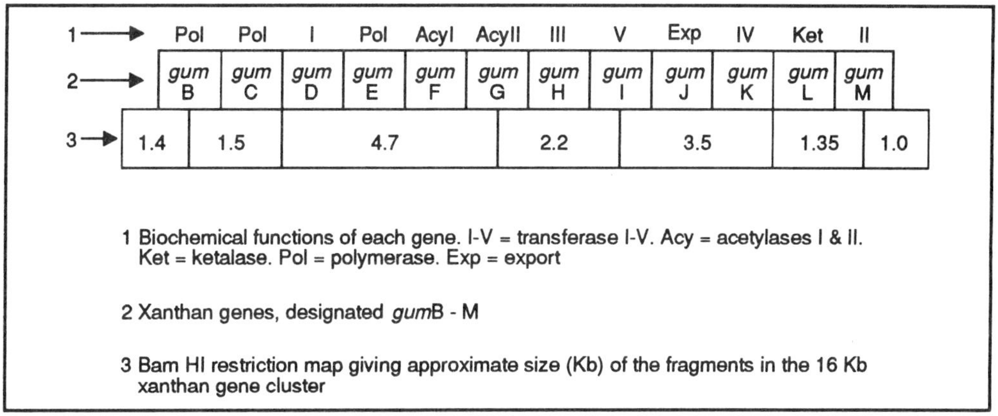

Figure 7.11 Restriction map of *Xanthomonas campestris* xanthan gene cluster. Adapted from R W Vanderslice *et al.* Genetic engineering of polysaccharide structure in *Xanthomonas campestris*. In Biomedical and Biotechnological Advances in Industrial Polysaccharides, 1989, Gordon and Breach N Y.

gum genes

These genes code for all the transferase activities, three polymerases and a gene (*gum*J) which appears to control the export of the polysaccharide. Mutations in *gum*J are lethal but blocking xanthan synthesis suppresses the lethality. This indicates that mutants deficient in *gum*J cannot export the xanthan, which accumulates and kills the cells. Polytetramer and polytrimer products are synthesised by mutants unable to add the side chain terminal β-mannosyl (*gum*I) and the terminal disaccharide β-mannosyl β-glucuronic acids (*gum*K) respectively.

large operon

Genetic analysis has shown that there is only one promoter and no internal termination sequences have been found, indicating that this gene cluster is one very large operon. There is a second large gene cluster, about 35kb, also involved in the synthesis of xanthan. The precise function is unknown but it is thought to encode for proteins involved in the synthesis of sugar nucleotide diphosphates.

SAQ 7.9

Cell lysates from mutant strains of *X. campestris* were incubated with radiolabelled UDP[^{14}C] glucose or GDP[^{14}C] mannose, the other sugar nucleotide substrates being unlabelled. The reaction mixture was then divided into lipid and soluble fractions. Where would you expect the radiolabel to be found and what product, if any, would you expect from strains with deficiencies in the following genes?

*Gum*D, transferase I
*Gum*M, transferase II
*Gum*K, transferase IV

SAQ 7.10

Briefly outline the major organisational features of genes encoding the enzymes responsible for exopolysaccharide synthesis, using *X. campestris* as the example.

7.8.2 Regulation of exopolysaccharide synthesis

Several of the important exopolysaccharides are produced by bacteria that exploit plants either as pathogens (*X. campestris*) or symbionts (*Rhizobium* spp). Exopolysaccharide production is often essential for plant/bacterial interaction and in *X. campestris* is co-ordinately regulated with other pathogenicity factors. Non-pathogenic mutants defective in extracellular protease and polygalacturonic acid lyase also have very low amounts of exopolysaccharide. Seven positive regulatory genes have been identified (*rpf*A-F regulation of pathogenicity factors) that act co-ordinately on the degradative enzymes, exopolysaccharides and pathogenicity. This system consists of an environmental sensor protein located in the membrane which detects changes in the external conditions and an effector protein in the cytoplasm which transmits this signal to the regulatory genes.

pathogenicity factors

sensor protein

The *rpf* system is balanced by a negative regulatory system; mutants defective in this chromosomal locus are pathogenic with elevated levels of exopolysaccharide and extracellular enzymes, whilst multiple functional copies have a repressive effect. An additional more specific regulation of exopolysaccharides has also been identified, involving a two component effector/sensor system, and there is also a general regulatory mechanism of which at least one component is the catabolite activating protein. Mutants deficient in catabolite activating protein have reduced amounts of exopolysaccharide.

negative regulatory system

catabolite activating protein

The synthesis of the exopolysaccharide colanic acid in *E. coli* is limited by the availability of an activator protein (*rcs*A) which acts on the exopolysaccharide gene cassette (*cps* loci). This is normally rapidly degraded by a protease (lon protease). A two component effector/sensor system also regulates the *cps* loci. The sensor component (*rsc*C), is likely to be a phosphorylase/kinase that activates the effector protein (*rsc*B). This in turn activates the *cps* genes either as a *rcs*B dimer or as *rcs*B-*rcs*A. *Rcs*C is itself negatively regulated by *rcs*D apparently through a modification of signal perception.

colanic acid

cps genes

SAQ 7.11

Complete the following diagram by identifying the components to give a generalised scheme for the regulation of exopolysaccharide synthesis based on the examples from *X. campestris* and *E. coli*.

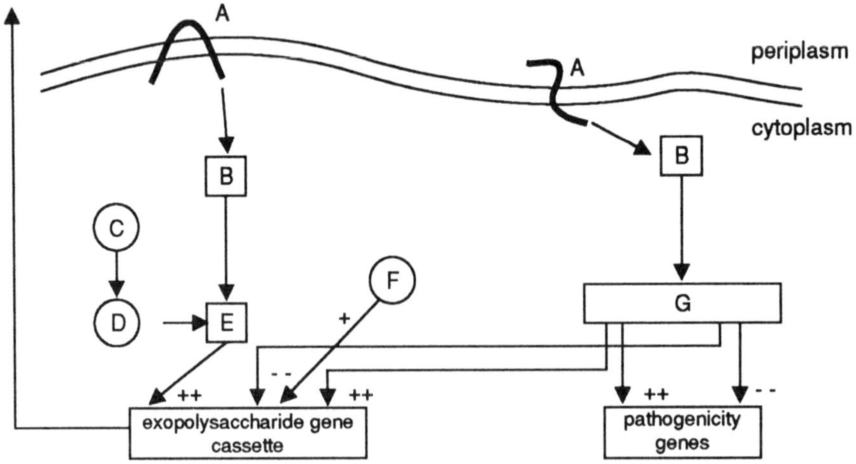

Components

activated complex
effector protein
activator protein
sensor
catabolite activator protein
pathogenicity regulating genes
degrading enzyme

7.9 Applications of microbial exopolysaccharides

powders and concentrates

The unique properties of exopolysaccharides are exploited in a wide range of industrial applications. They are sold either as powders or as concentrates (approx 8% w/v). Although more expensive to transport, concentrates offer several advantages: they can be pumped; any lumps formed during make up of the solid are avoided; the viscosity at a given concentration is higher than from solids.

xanthan

Xanthan has some unique properties and high activity at low concentrations and is commercially the most important exopolysaccharide. Food products account for approximately 60% of xanthan use, 15% is accounted for by toothpaste, textiles and crop protection products, 10% in the oil industry and the remainder in miscellaneous industrial/consumer applications.

7.9.1 Food industry applications

The properties of exopolysaccharides utilised in the food industry are presented in Table 7.3. Alteration of the food texture by thickening or gelling is one of the more important uses. This in turn affects less easily defined parameters that are nevertheless crucial in food stuffs, such as 'mouth feel'. The different properties of exopolysaccharides mean that a number of different gel types are available for use in the food industry.

Property	Food	Exopolysaccharide
adhesive	icings, glazes	gellan
binding agent	pet foods	gellan, alginates
coating	confectionery	xanthan
emulsifying agent	salad dressing	xanthan, alginates
film formation	sausage case	pullulan, elsinan
gelling agent	pastry, filling, jelly	gellan, curdlan
inhibit crystal formation	frozen food, pastilles	alginate, xanthan
stabiliser	ice cream	alginate, gellan, xanthan
syneresis inhibitor	cheeses	xanthan
synergistic gels	synthetic meat gels	xanthan
thickening agent	jams, sauces	xanthan, gellan, alginate
foam stabiliser	beer, dough	xanthan, alginate*

Table 7.3 Examples of the use of exopolysaccharides in the food industry.
* Alginate used in the propylene glycol form; this makes it less susceptible to precipitation by acid and can be used in food and beverages at pH < 3.0.

alginates

Alginates form non-thermoreversible gels and are useful when shape retention on heating is desired. Alginates form gels with calcium ions that are bound between sequences of polyguluronyl residues. The ion binding, gel strength and sensitivity for calcium, barium and strontium ions (which are more strongly bound than other cations) increases with increasing guluronosyl content but is reduced if the mannuronic acid residues are heavily acetylated.

gellan

Gellan is stable to heat and gives a very clear, thermoreversible gel which sets at lower concentrations and more rapidly than most other polysaccharides. Gellan is a fermentation product of *Pseudomonas elodea* comprised of glucose, rhamnose and glucuronic acid residues with 3-4.5% O-acetyl groups. The native exopolysaccharide does not gel but removal of the acetyl groups by heating at pH 10 results in the low acetyl form which gives clear gels in the presence of divalent cations (Mg^{2+}). The properties of this exopolysaccharide make it suitable for use in glazes, jellies and icings. Gellan also has superior flavour release characteristics than most other polysaccharides used in gels.

evaluation

Before microbial exopolysaccharides can be used in foodstuffs they must be evaluated against an industrial standard to ensure that product quality is maintained or improved. The tight deadlines imposed on industry mean that product evaluation must be effective and logical to ensure a rapid result. Therefore, clear objectives and criteria are set against which new products are tested.

The criteria that must be satisfied for quick setting jelly desserts are as follows:

1) One stage make up from powder - the product (jelly dessert) is already multi-component, the introduction of any further stage is undesirable.

2) The powder must be easily dispersible. If vigorous stirring is required, aeration of the product may occur reducing its clarity. Also as hot water is used, difficult systems to mix can be dangerous for the user.

3) Setting time. Must be within 60 minutes.

4) Texture. Must not be dramatically different to gelatine since the nature of the product will change unacceptably.

5) Flavour release properties. Must be similar or superior to the currently used compounds (gelatine).

Π From Table 7.4 determine which of the gel-systems is the most suitable for inclusion in quick-set jellies.

	Stages	Powder dispersal	Setting time	Texture	Flavour release	Relative cost
carrageenan	1	moderate	>60	+++	+++	5
carrageenan + LBG	1	moderate	>60	++++	+	4
gellan	1	moderate	<<<60	+	++++	20
alginate	1	vigorous	<60	+	+	6
gelatine (CWS)	1	moderate	>>60	+	+	2
xanthan + LBG	2	moderate	<60	+	+	6
pectin	2	moderate	<60	+	+	5

Table 7.4 Comparative features of gel-systems evaluated for use in quick setting jelly products. CWS = cold water soluble, LBG = locust bean gum.

Pectin and xanthan/locust bean gum fail on the first criterion as they would only work if a two stage make up process was used. Alginate is too difficult to mix, and cold water soluble gelatine takes an excessive time to set. The remaining three all show variations from the standard gelatine. Carrageenan and the carrageenan/locust bean gum mixture only just meet the criteria for setting time and gellan imparts a poorer texture. However, gellan scores heavily in the other tests: very rapid setting, excellent flavour release and can be used at lower concentrations than the others. The overriding consideration is of course the price. Gellan is currently ten times the price of gelatine, nevertheless the advantages gained by using gellan justify the extra cost in certain products.

synergistic gelling

Less polysaccharide can also be used in products by taking advantage of the synergistic gelling of xanthan/galactomannan mixtures which forms thermoreversible gels at lower concentrations than if each is used separately.

curdlan

Another promising exopolysaccharide that may come to replace some of the traditional setting agents is curdlan. This is an α-1,3 linked glucan made by *Alcaligenes faecalis* which retains its shape in cooked food and only needs temperatures of between 55 and 80°C for preparation.

Π This property of curdlan has an important consequence for use in food stuffs. What do you think this might be?

Flavour compounds and other heat labile constituents will not be lost as extensively in the preparation, thus improving the product and reducing the cost of production.

Another feature of this particular exopolysaccharide is that gel strength depends upon the temperature used. It is constant between 60-80°C, increasing in strength from 80-100°C and finally changing structure from a single to a triple stranded helix at temperatures over 120°C. This makes it particularly well suited for use as a molecular sieve, immobilised enzyme support and a binding agent.

flavour release

emulsion stabiliser

Different polysaccharides change the perception of flavour, thus xanthan is superior to gum guar in the perception of sweetness. Mixtures of xanthan and locust bean gum have improved flavour release and texture when used in pies and patés compared to starch. Many foods are emulsions, examples being soups, sauces and spreads. Exopolysaccharides are used to stabilise these emulsions and prevent the phases from separating when dried foods are reconstituted. Xanthan is particularly suited for this purpose as it is stable over a wide pH range and can accommodate the low pH's found in several relishes and dressings. Alginates are less useful at lower pH values but conversion to the propylene glycol form makes it less susceptible to precipitation by acid or metal ions and extends its range of applications.

7.9.2 Other industrial uses

Xanthan gum is the most important exopolysaccharide used commercially, its unique properties making it ideal for an extraordinary range of applications Table 7.5.

Usage	Physical properties required
explosives (package gels)	compatibility with $Ca(NO_3)_2$; water resistance (for dynamite)
fire fighting	foam stabilisation
flowable pesticides	suspension and drift control
hydraulic fracturing	viscosity and cross-linking
jet printing	
laundry chemicals	suspension of starch
liquid fertilisers and herbicides	suspension
liquid feed supplements	suspension
oil-drilling muds	shear-thinning and viscosity control
paper finishing	suspension of clay coatings
thixotropic paints	stabiliser
water clarification (or extraction)	flocculant

Table 7.5 Industrial applications of xanthan.

drilling fluids

Xanthan is used as a drilling fluid, either as a mixture with the traditional bentonite clays or alone, as a clear 'mud'. In this application, xanthan acts as a lubricant and removes cut material. Important factors are the ability to pump (pseudoplastic flow), suspend particles plus its resistance to relatively high temperature, pH and its salt compatibility. Succinoglycan has also been used, having the advantage that it interacts less with $CaBr_2$, which is used to increase the density in drilling fluids.

improved oil recovery

There has been some success in using xanthan to improve oil recovery. The water used can often push through the more viscous oil, therefore failing to drive the oil before it out of the well. Increasing the viscosity of the flushing liquid with xanthan can reduce this problem and thus enhance oil recovery. Cross-linking of xanthan with chromium ions has also been used to seal very permeable rocks - so called thief zones - where oil would otherwise be lost. The use of xanthan also has the advantage that the gel can be removed with oxidising agents such as sodium hypochlorite. The temperature limit for this application is 80°C. There are some problems encountered when using xanthan, however, and biocides are often included to prevent microbial degradation during the initial dilution and injection phase. Free radical scavengers are also added to reduce chemical attack and maintain the exopolysaccharide structure at higher temperatures. Furthermore, despite the prolonged stability of xanthan at temperatures of 95°C the temperatures in some oil wells can reach 150°C.

welan

rhamsan

Several new exopolysaccharides such as welan and rhamsan produced by *Alcalingenes* spp may supercede xanthan for some industrial applications. These are based on the same repeat tetrasaccharide backbone of glucose, glucuronic acid, glucose and rhamnose but differ in the substituents: rhamsan has a disaccharide side chain and welan a monosaccharide. Both are stable at high temperature and have excellent pseudoplastic properties.

elastic modulus

The higher elastic modulus (a measure of structure in solution) of rhamsan suggests that it should be superior to xanthan as a stabiliser. Rhamsan also has improved salt compatibility and is used in fertiliser suspension (high polyphosphates) and explosives (high ammonium nitrate).

Π From Figure 7.8 and Table 7.2, which of the alternative gums do you think is the most suitable for use in oil field applications?

One of the most important criteria for choosing gums for use in oil recovery etc is their stability to extended periods of high temperature. Although succinoglycan has a high viscosity, this deteriorates at high temperatures over extended periods. Welan and xanthan are used in oil well drilling fluids. Welan is particularly useful in this application since it has high viscosity which is maintained at an almost constant level at 121 °C for extended periods and can tolerate temperatures as high as 149°C.

thickening agents

Alginates and xanthan are used as thickening agents in the printing industry to control the spread of the dye through fabrics and hold it in place until the dye is fixed. These exopolysaccharides are compatible with a variety of dyes and are also easily washed off. Xanthan and succinoglycan, amongst other exopolysaccharides, are included in paint formulations to stabilise suspensions of the pigment and also confer superior spraying and pumping characteristics.

7.9.3 Enzyme technology applications

purification of biomolecules

The extraction of biological material often requires a relatively gentle method to avoid conditions that would otherwise cause losses through precipitation and denaturation. Aqueous two-phase systems with very high water content (>80%) and low interfacial tension provide such a method. They are extremely useful for the separation of enzymes, nucleic acids and even cells. Polyethylene glycol (PEG), dextran, water mixtures will form separate phases determined by polymer concentration, molecular weight of the polymers and temperature. Purification of biomolecules can be enhanced by the attachment of affinity ligands to PEG (upper phase) which accumulates the target molecule whilst the contaminating molecules partition into the dextran phase. This system can also be used for the simultaneous production and purification of bioproducts from cells or enzymes, for instance to purify α-amylase from *B. subtilis*. In a PEG/dextran system the cells partition into the lower phase with 80%+ of the enzyme product in the upper phase.

immobilised cells or enzymes

Polysaccharides can also be used to immobilise cells or enzymes, permitting the re-use of the catalyst and continuous flow systems. Alginates have the advantage that gel formation occurs under mild conditions, therefore cells remain viable and enzymes are not denatured but calcium gradually leaches out and the gel dissolves. Gellan or other combinations may prove superior for this application.

emulsan

emulsifying agent

A strain of *Acinetobacter calcoaceticus* produces an unusual polysaccharide called emulsan. It is a complex polymer comprising about 15% fatty acyl esters and 20% protein. This structure enables it to act as an emulsifying agent, stabilising hydrocarbon/water emulsions at very low concentrations (0.1-1.0%). This property, coupled with its low toxicity makes it ideal as a cleaning agent and it is incorporated into hand-cleaners to remove compounds used in paints, plastics and in ink manufacture. Other possible industrial uses include cleaning of storage tanks, reducing crude oil viscosity, dispersing slurries and pigments.

7.9.4 Medical applications

In medical applications some important biological properties - immunogenic, anti-tumour and anti-viral - can be exploited, as well as the established functional properties based on rheology and gel formation.

dextrans

Dextrans are particularly useful and are employed as a plasma substitute. A concentration of about 6% dextran (50,000-100,000 relative molecular weight) has equivalent viscosity and colloid-osmotic properties to blood plasma. Dextran can also be used as non-irritant absorbent wound dressings, an application also suited to alginate gels.

sephadex

Dextran can be produced in a range of molecular weights and crossed-linked or substituted with a variety of functional groups. These products (Sephadex) are routinely used in the purification of proteins and pharmaceutical and other medically important compounds.

encapsulated drugs

Exopolysaccharides are used in lotions and gel formation is exploited in encapsulated drugs. The latter application also takes advantage of the mouth feel and flavour neutrality, qualities also vital for the food industry.

anti-viral activity

Many polysaccharides of eukaryotic origin show non-specific anti-viral activity and this property may be shared by some of the exopolysaccharides. The structural requirements for activity are not immediately evident as the polysaccharides exhibiting this activity are very diverse.

curdlan

Curdlan possesses anti-tumour activity similar to that shown by fungal β-D-glucans, a property which appears to be related to the ability to form triple helices.

vaccines

There are a number of practical problems involved with using polysaccharides as vaccines as there are frequently too many different chemotypes for it to be practicable to prepare a vaccine. In some cases a limited number of serotypes are the dominant cause of infection and it may then be possible to produce vaccines. A major problem is the poor immune response elicited by polysaccharide antigens, which may in some cases be improved by chemical modification. This is the case for vaccines for *Haemophilus influenzae* type b (a causative agent of meningitis), where the antigenicity of the polysaccharide can be increased by coupling to proteins.

SAQ 7.12

Use the information given in section 7.9 to complete the 'Important Physical Properties' column of the following table.

Exopolysaccharide	Application	Important Physical Properties
welan	enchanced oil recovery	
emulsan	cleaning agent	
dextrans	plasma substitute	
curdlan	anti-tumor agent	
polyethylene glycol/ dextran water mixtures	purification of biological materials	
alginates	immobilising enzymes	
gellan	use in glazes, jellies and icings	

Summary and objectives

In this chapter we have seen that microbial exopolysaccharides have a wide range of applications, from fine medical to large scale industrial. They are composed mainly of carbohydrate and their molecular conformation is determined by primary structure and associations between molecules in solution. This in turn determines their physical properties and ultimately their commercial importance. Properties that have been exploited commercially include viscoelasticity, thermostability and gel formation. Exopolysaccharides are generally produced by submerged fermentation with high carbon conversion efficiency, although the yields from oxygen are inherently poor. High viscosities of fermentation liquors can adversely affect mixing, oxygen transfer and subsequent downstream processing. A common feature of genes responsible for the control and regulation of synthesis of exopolysaccharides is their arrangement tight gene clusters; most is known for *X. campestris* and *E. coli*. Genetic manipulation of exopolysaccharide producing organisms has not, as yet, been exploited commercially.

Now that you have completed this chapter you should be able to:

- describe broadly the chemical composition and structure of microbial exopolysaccharides and explain how these influence their physical properties;
- relate applications of named microbial exopolysaccharides to their physical properties;
- describe media and fermentation conditions for microbial exopolysaccharides, with particular reference to xanthan and succinoglycan;
- list benefits and limitations of solvent precipitation and ultrafiltration as methods of recovery of exopolysaccharides;
- describe the biosynthesis and regulation of microbial exopolysaccharides with particular reference to xanthan and colanic acid.

Industrial production of amino acids by fermentation and chemo-enzymatic methods

Industrial production of amino acids by fermentation and chemo-enzymatic methods

8.1 Introduction

essential amino acids

non-proteinogenic amino acids

precursors for fine chemical synthesis

Amino acids have always played an important role in the biology of life, in biochemistry and in (industrial) chemistry. There are several reasons why they are of commercial interest. Firstly, amino acids are the building blocks of proteins and they play an essential role in the regulation of the metabolism of living organisms. Large-scale chemical and microbial production processes have been commercialised for a number of essential amino acids. The use of glutamic acid, lysine and methionine as food and feed additives is well established nowadays. Secondly, current interest in developing peptide-derived chemotherapeutics has heightened the importance of rare and non-proteinogenic pure amino acids. For example, D-phenylglycine and D-*p*-hydroxy-phenylglycine are building blocks for the broad spectrum β-lactam antibiotics ampicillin and amoxycillin, respectively. The natural amino acid L-valine is used as feedstock in the fermentative production of the cyclic peptide cyclosporin A, which has immuno-suppressive activity and is used in human transplant surgery. Thirdly, amino acids are versatile chiral (optically active) building blocks for a whole range of fine chemicals. In the last two decades, there has been a growing public awareness and concern with regard to the exposure of man and his environment to an ever increasing number of chemicals. The benefits, however, arising from the use of therapeutic agents, pesticides, food and feed additives, etc are enormous. Hence there is still an ever increasing demand for more selective drugs and pesticides which are targeted in their mode of action, exhibit less toxic side-effects and are more environmentally acceptable. To this end a central role will be played by chiral compounds, as nature at the molecular level is intrinsically chiral. Consequently, this provides an important stimulus for companies to market chiral products as pure optical isomers. This in turn results in an increasing need for efficient methods for the industrial synthesis of optically active compounds.

Amino acids are, therefore, important as nutrients (food and feed), as seasoning, flavourings and starting material for pharmaceuticals, cosmetics and other chemicals. They can be produced in a variety of ways (see Table 8.1):

- chemical synthesis;
- isolation from natural materials (extraction);
- amino acid fermentations (using micro-organisms);
- chemo-enzymatic methods.

In this chapter we consider amino acid production by fermentation and by chemo-enzymatic methods. We first consider the stereochemistry of amino acids and the importance of chirality in chemical synthesis. General approaches to amino acid fermentation and recovery of amino acids from fermentation broths are then dealt with, followed by a detailed consideration of the production of L-phenylalanine by direct fermentation. Later in this chapter, chemo-enzymatic methods of amino acid fermentation are examined. We first consider general aspects of the approach, followed by more detailed case studies. We have again selected L-phenylalanine for detailed consideration, since this important amino acid can be produced by direct fermentation and by a variety of chemo-enzymatic methods. This allows comparisons between the two different approaches to be made, including a consideration of economic aspects of large scale production of the amino acid.

Two appendices are included at the end of this chapter. The first is intended to serve as a reminder, for those of you who might need it, of the nomenclature and representation of stereoisomers. The second appendix contains descriptions of various chemo-enzymatic methods of amino acid production. This appendix has been constructed largely from the recent primary literature and includes many new advances in the field. It is not necessary for you to consult the appendix to satisfy the learning objectives of the chapter, rather the information is provided to illustrate the extensive range of methodology associated with chemo-enzymatic approaches to amino acid production. It is therefore available for those of you who may wish to extend your knowledge in this area. Where available, data derived from the literature are used to illustrate methods and to discuss economic aspects of large-scale production.

amino acid	**chemical synthesis**	**extraction**	**fermentation**	**enzymatic catalysis**
L-alanine		+		+
L-arginine		+	+	
L-aspartic acid		+		+
L-cystine		+		
L-cysteine	+			
L-glutamic acid (Na)		(+)	+	
L-histidine (.HCl)		+	+	
L-isoleucine		+	+	
L-leucine		+		
L-lysine (.HCl)			+	+
L-methionine				+
L-phenylalanine	(+)	(+)	(+)	+
L-proline		+	(+)	
L-serine		+	+	
L-threonine		+	+	
L-tryptophan			+	+
L-tyrosine		+		
L-valine		+	(+)	+

Table 8.1 Production methods of proteinogenic amino acids.

8.2 Essential and nonessential amino acids

An amino acid is defined as a compound that possesses both amino and carboxyl groups. Some amino acids are iminocarboxylic acids, such as proline while others are sulphur containing amino acids, such as cysteine and methionine (Table 8.2). Over 100 amino acids have been isolated and identified from natural sources to date. The great majority of these naturally occurring amino acids have the amino group attached to the

α-carbon

carbon α to the carboxylic acid. With very few exceptions, the α-carbon also bears a hydrogen atom. The fourth bond of the α-carbon is joined to a group which has over 100 variations. Thus, most of the naturally occurring amino acids differ only in the structure of the organic residue attached to the α-carbon.

L-configuration

An interesting and important fact is that almost all amino acids isolated from proteins have the L-configuration at the α-carbon, although some amino acids isolated from microbiological sources are the mirror image isomers, ie in the D-configuration. We shall consider amino acid stereochemistry in more detail in section 8.3.

Of the amino acids isolated from living material, only about 20 are naturally occurring components of proteins. Some of these are shown in Table 8.2. The remainder, non-proteinogenic amino acids, are found as intermediates or end products of metabolism.

One of the amino acids commonly found in protein hydrolysates is called cystine; it has the following structure:

$$\mathrm{HOOC-\underset{\displaystyle NH_2}{\underset{|}{CH}}-CH_2-S-S-CH_2-\underset{\displaystyle NH_2}{\underset{|}{CH}}-COOH} \qquad \text{(cys-cys)}$$

cystine

dimer

It is clearly a dimer of cysteine, where the thiol groups have been oxidised to form a disulphide linkage. The dimer actually results because of two monomers at widely spaced intervals in the polypeptide are joined together by a disulphide bridge. Thus the basic amino acid is cysteine and consequently, the dimer is not included here.

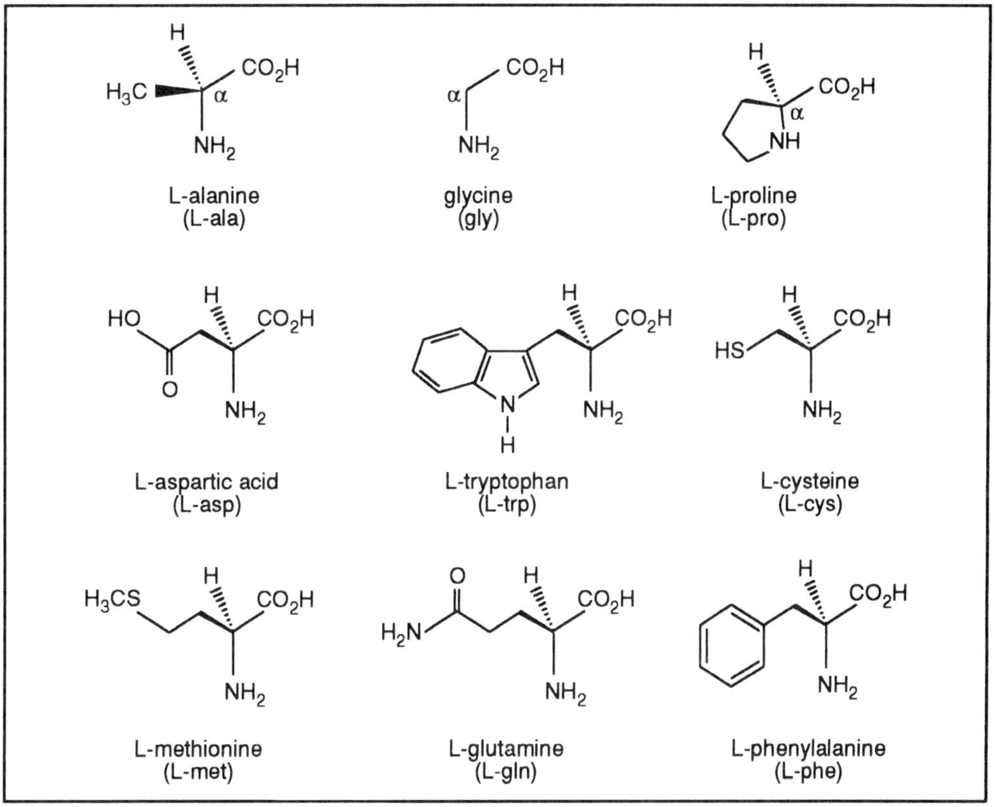

Table 8.2 Structure of some proteinogenic α-amino acids.

essential amino acids

All living species are able to synthesise amino acids. Many species, however, are deficient in their ability to synthesise within their own metabolic system all the amino acids necessary for life. The eight amino acids with this special significance for the human species are called essential amino acids, these are:

- L-valine;
- L-leucine;
- L-isoleucine;
- L-threonine;
- L-methionine;
- L-phenylalanine;
- L-tryptophan;
- L-lysine.

They are essential not because they are the only amino acids required for human functioning, but because they are essential in the diet of the human species.

SAQ 8.1

1) Name *three* sulphur containing amino acids.

2) Name *five* of the eight essential amino acids.

3) Name *two* amino acids that contain a heterocyclic ring.

4) Name the amino acid with the simplest structure.

5) Name the amino acid considered to be a dimer.

6) Name an amino acid that is produced industrially only by enzymatic catalysis.

7) Name an amino acid that is produced industrially only be chemical synthesis.

8.3 Stereochemistry of amino acids

The French physicist Biot discovered during the early nineteenth century, that a number of naturally occurring organic compounds rotate the plane of polarisation of an incident beam of polarised light. In the latter part of the nineteenth century, it was found that many pairs of compounds seemed to have an identical structure and identical physical properties, such as melting point and solubility. Compounds in each pair were differentiated by the fact that even in solution they rotated polarised light in equal amounts but in opposite direction. Such compounds are called optical isomers and are described as being optically active. Optical activity requires, and is explained by an asymmetric arrangement of groups around a tetrahedral carbon atom. The geometric properties of a tetrahedron are such that if there are four different substituents attached to a carbon atom, the molecule does not contain a plane of symmetry, and there are two kinds of geometrical arrangements which the molecule can have. These two arrangements (configurations) are different in that it is not possible to simultaneously superimpose all the atoms of one figure on the like atoms of the other. The two configurations are, in fact non-superimposable mirror images.

optical isomers

tetrahedral carbon atom

An illustrative example is given in Figure 8.1.

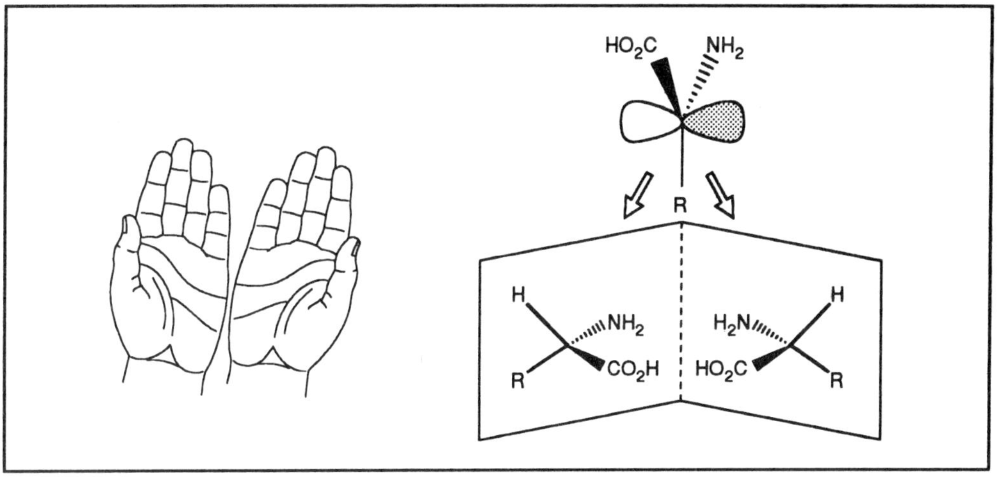

Figure 8.1 Non-superimposable mirror images.

asymmetric carbon

Such molecules result when the four groups attached to the carbon atom are all different and a molecule of this kind is said to be asymmetric, or to contain an asymmetric carbon.

enantiomers

Molecules that are not superimposable on their mirror images are chiral. If two compounds are related as non-superimposable mirror images, they are called enantiomers.

Π Can you explain why the amino acid alanine is optically active, whereas glycine is not (refer to Table 8.2)?

We can see from Table 8.2 that the α-carbon of alanine is asymmetric (four different groups attached), whereas that of glycine is not. Optical activity requires an asymmetric carbon atom.

racemic mixtures

If two enantiomers are mixed together in equal amounts the result is a racemic mixture. We meet a number of enantiomeric items in daily life. The left hand, for example, is the mirror image of the right hand and they are not superimposable (see Figure 8.1). This becomes obvious if we try to put a right glove on a left hand. Similarly, a pair of shoes is an enantiomeric relationship while the stock in a shoe store constitutes a racemic mixture.

Representation of and the nomenclature for stereo-isomers are given in Appendix 1.

8.3.1 Importance of chirality

enantiomeric purity

If we consider natural synthetic processes, enzymes are seen to exert complete control over the enantiomeric purity of biomolecules (see Figure 8.2). They are able to achieve this because they are made of single enantiomers of amino acids. The resulting enantiomer of the enzymes functions as a template for the synthesis of only one enantiomer of the product. Moreover, the interaction of an enzyme with the two enantiomers of a given substrate molecule will be different. Biologically important molecules often show effective activity as one enantiomer, the other is at best ineffective or at worst detrimental.

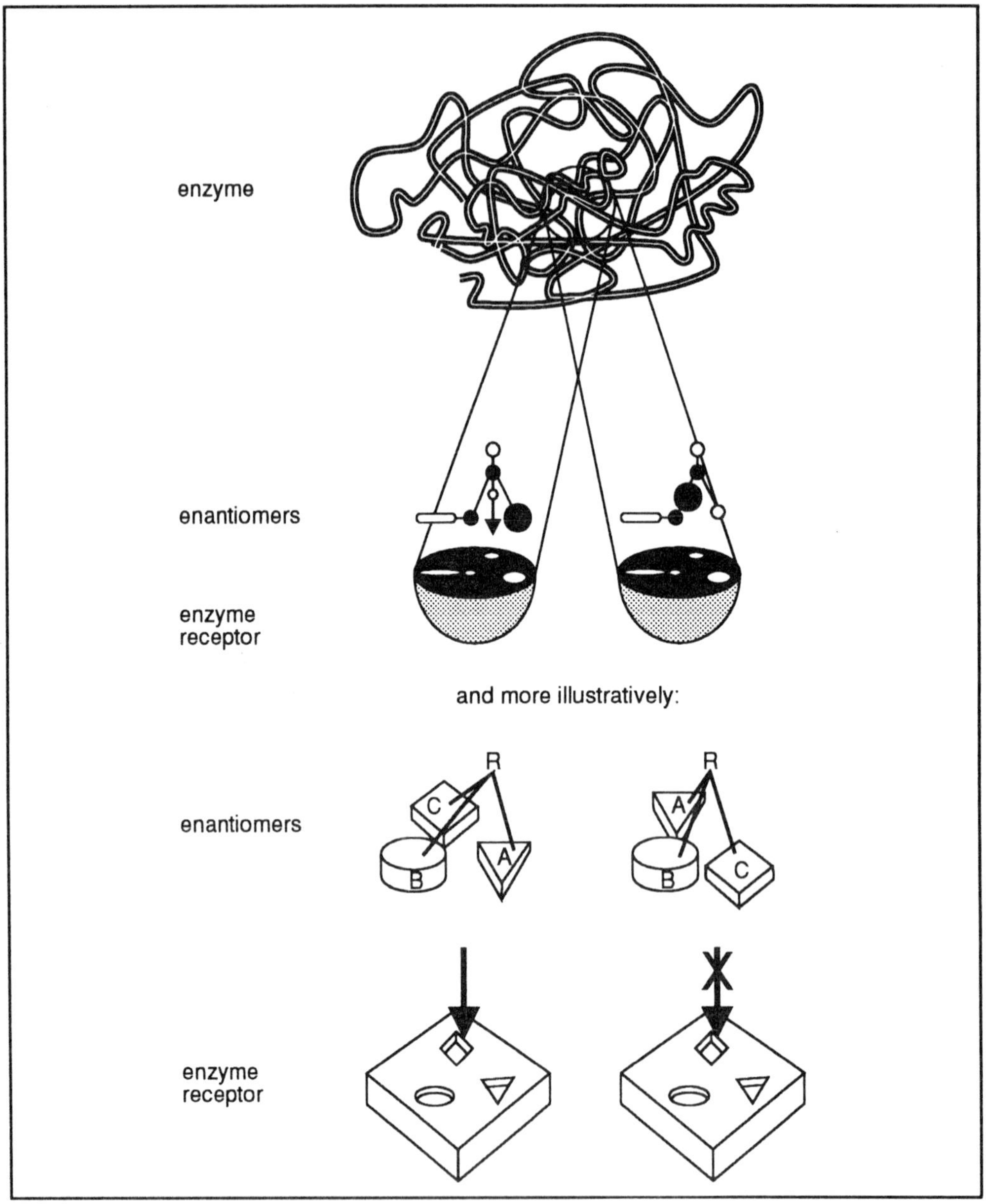

Figure 8.2 Enzyme interaction with two enantiomers of a given substrate molecule.

specific drug action

In some cases the unwanted enantiomer can perturb other biological processes and cause catastrophic side effects. The use of enantiomerically pure compounds thus permits more specific drug action and the reduction in the amount of drug administered. Even in the cases where the other enantiomer is inactive, the work involved in its metabolism before secretion can be avoided.

Numerous examples of the different biological effects of enantiomers are available. One of the enantiomers of limonene smells of lemons, the other of oranges; one of carvone smells of caraway, the other of spearmint. These differences obviously have important consequences for the perfume and flavour industries. Both enantiomers of sucrose are equally sweet, but only the naturally occurring D-enantiomer is metabolised, making the synthetic L-enantiomer a potential dietary sweetener. In the protection of crops from insects, one enantiomer of a compound may be a repellant while the other is an attractant, and the racemic mixture is ineffective.

One enantiomer of penicillamine (D-) exhibits antiarthritic properties but the other is highly toxic (Figure 8.3). The teratogenic effects of thalidomide were induced by one enantiomer, the other exhibited the beneficial effects against morning sickness.

importance of homochiral compounds

Different optical enantiomers of amino acids also have different properties. L-asparagine, for example, tastes bitter while D-asparagine tastes sweet (see Figure 8.3). L-Phenylalanine is a constituent of the artificial sweetener aspartame (Figure 8.3). When one uses D-phenylalanine the same compound tastes bitter. These examples clearly demonstrate the importance of the use of homochiral compounds.

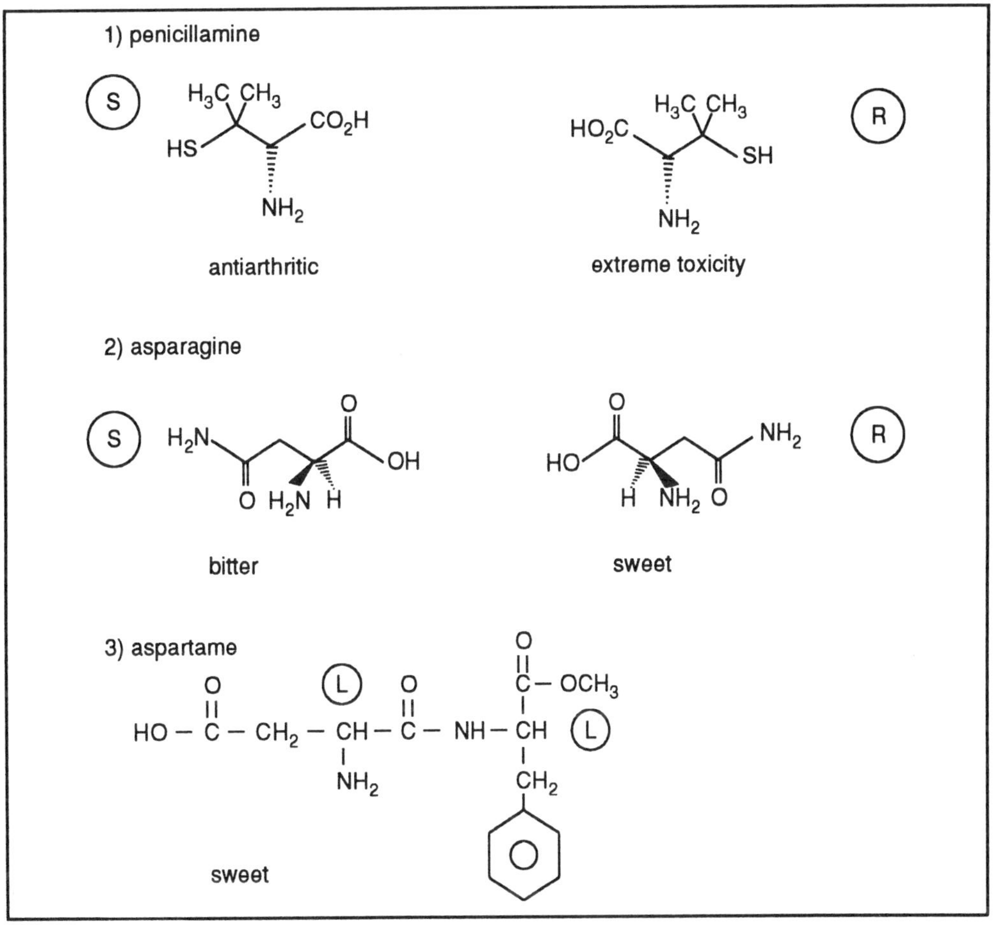

Figure 8.3 Examples of different biological effects of enantiomers. S and R refer to a particular system of nomenclature used to describe chiral carbon. (see Appendix A8.1)

SAQ 8.2

List possible advantages of using enantiomerically pure compounds as drugs, as opposed to using racemic mixtures.

8.4 Amino acid fermentation

wild strains

Many micro-organisms accumulate amino acids in culture media. Indeed, wild strains have proved to be effective producers of amino acids like alanine, glutamic acid and valine.

negative feedback control

Since amino acids are used as essential components of the microbial cells and their biosynthesis is regulated to maintain an optimal level, they are normally synthesised in limited amounts and are subject to negative feedback control. The main problem using wild strains is, therefore, the production of minor amounts of amino acids at an early stage in the fermentation, giving rise to feedback control. To achieve overproduction of amino acids the following procedures can be used:

- improvement of the uptake of the raw material (starting material);
- hindrance of the side reactions;
- stimulation of the enzymes that are involved in the synthesis;
- inhibition of the degradation of the desired amino acid;
- stimulation of excretion of the amino acid that is produced.

strategies for overproduction

The most successful way to achieve overproduction is to make use of mutants. Another way to overcome feedback regulation is to make use of a kind of semi-fermentation process called precursor addition fermentation; this will be considered later in this chapter.

Amino acids produced by fermentation on an industrial scale are listed in Table 8.3.

amino acids	tonnes/year	applications
aspartic acid	ca. 8, 000	aspartame (sweetener) enzymatic synthesis of alanine and phenylanine
glutamic acid	ca. 270, 000	flavours, pharmaceuticals
lysine	ca. 90, 000	dietary
phenylalanine	ca. 8, 000	aspartame (sweetener)
threonine	ca. 500	dietary
tryptophan	ca. 100	pharmaceuticals, dietary

Table 8.3 Amino acids industrially produced by fermentation.

Π Examine the list of procedures to achieve overproduction (shown above) and identify which ones could be achieved by mutation of a wild strain.

Since all the procedures listed involve enzymes they all could be achieved by mutation. This emphasises the potential of using mutation for amino acid production.

8.4.1 Wild strains

Normally amino acid synthesis will just satisfy the metabolic demand. In some cases, when the amino acid occurs in both biosynthetic and energy production pathways, overproduction of the amino acid can take place. This is the case, especially for L-glutamic acid, with *Corynebacterium*, *Brevibacterium*, *Microbacterium* and *Arthrobacter*.

characteristic influencing overproduction

These micro-organisms have some common features: Gram-positive, non-spore forming, non motile, cocci or rod and biotin-requiring for growth. The following general characteristics have been found to influence the overproduction of L-glutamic acid:

- nutritional requirement for biotin;
- intracellular levels of phospholipids;
- lack of, or low activity of, α-oxoglutarate dehydrogenase.

In the first two cases, the permeability of the cell membrane to L-glutamate is altered through changes in the fatty acid composition of the cell membrane. In the third case, the degradation of the amino acid is inhibited, resulting in accumulation.

Other amino acids produced by wild strains include L-valine, DL-alanine and L-proline.

Π Why are amino acid synthesised in only small amounts in wild strains?

The small overproduction of amino acids by wild type strains in culture media is the result of regulatory mechanisms in the biosynthetic pathway. These regulatory mechanisms are feedback inhibition and repression.

These mechanisms are considered fully in other BIOTOL texts. However, details of the mechanisms are not essential for an understanding of the material in this chapter. In this section we will, therefore, only briefly describe the mechanisms.

Feedback inhibition

In the metabolic pathway to an amino acid several steps are involved. Each step is the result of an enzymatic activity. The key enzymatic activity (usually the first enzyme in the synthesis) is regulated by one of its products (usually the end product, eg the amino acid). If the concentration of the amino acid is too high the enzymatic activity is decreased by interaction of the inhibitor with the regulatory site of the enzyme (allosteric enzyme). This phenomenon is called feedback inhibition.

Repression

In repression the enzyme concentration is regulated at the DNA level.

The genetic information for a certain enzyme is located on the genetic material in the cell, the DNA. In general this consists of several integrated regions working together (an operon).

If the information can be translated into messenger RNA, and this information can be transferred to the ribosome, the enzyme can be synthesised. If not, the key enzyme in a metabolic route to the wanted amino acid is not synthesised. This might involve a repressor, consisting of an apo-repressor (ie a protein from a regulatory gene) and the co-repressor (normally the end product of the pathway) which binds to the operator gene and prevents translation of the operon. Normally this is not an absolute effect. If the concentration of the end product is high the end product blocks the operator gene, so the m-RNA cannot be formed and the enzyme cannot be produced. The concentration of the enzyme is kept low in such a case.

apo-repressor and co-repressor

It is obvious that in the case of overproduction of amino acids the above mentioned regulatory mechanisms are not wanted. One way to overcome these regulations is to make use of mutants.

8.4.2 Mutant strains

Two types of mutants have been used for amino acid overproduction: auxotrophic and regulatory mutants. In some cases, mutant strains have been further improved through DNA-recombination.

Auxotrophic mutants

Auxotrophic mutants are mutants that miss one or more of the enzymes used in the biosynthetical pathway for one or more amino acids. In practice this means that the mutant needs one or more key metabolites which it cannot synthesise for growth in its growth medium. For example, consider Figure 8.4.

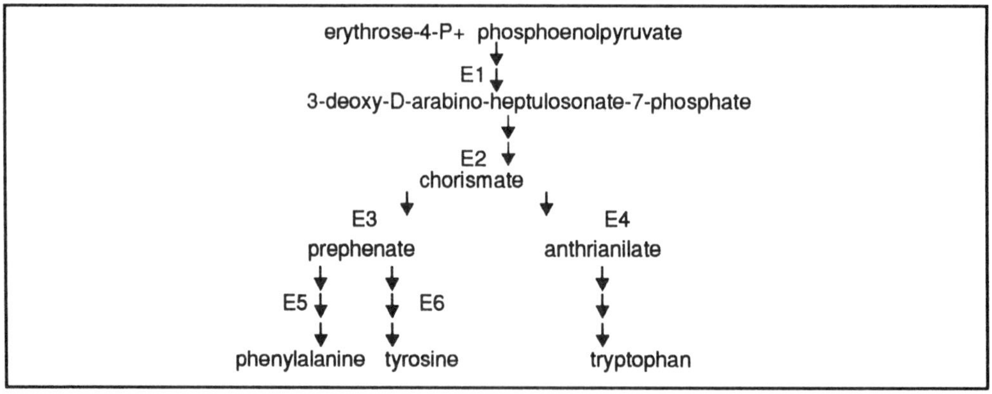

Figure 8.4 Skeleton pathway leading to L-phenylalanine, tyrosine and tryptophan in *Escherichia coli*.

In the case of a tyrosine auxotrophic mutant, the mutant does not produce at least one of the enzymes to synthesise tyrosine (E6 in Figure 8.4).

In the literature this is described as Δtyr. This means that a small amount of tyrosine has to be added to the culture medium, just enough to support growth, because the micro-organism is not capable of producing tyrosine. In the wild strain, E1 is controlled by feedback inhibition involving tyrosine. So, accumulation of tyrosine slows the rate of flow through the pathway. In the mutant strain, however, this does not occur because tyrosine is not produced, so overproduction of phenylalanine will occur. The genetic marker of this particular organism is reported in literature as Tyr^-.

phenylalanine overproduction in Tyr^-

branched pathways

Auxotrophic mutants are used in the production of end products of branched pathways, ie pathways leading to more than one amino acid at the same time. This is the case for L-lysine, L-methionine, L-threonine and L-isoleucine in *Brevibacterium flavum* and *Corynebacterium glutamicum*.

Regulatory mutants

There are two possible sites that are genetically inactivated in regulatory mutants:

- the regulatory site;
- the functions of repressor and co-repressor.

These mutants lack feedback inhibition and are used for the production of many amino acids.

selection using toxic analogues

Selection of these regulatory mutants is often done by using toxic analogues of amino acids; for example *p*-fluoro-DL-phenylalanine is an analogue of phenylalanine. Mutants that have no feedback inhibition or repression to the amino acid are also resistant to the analogue amino acid. They are therefore selected for and can be used to overproduce the amino acid. Some amino acid analogues function as false co-repressors, false feedback inhibitors or inhibit the incorporation of the amino acid into the protein.

The best amino acid producers are organisms that are both auxotrophic and regulatory mutants.

Π What advantage do regulatory mutants have, when compared to auxotrophic mutants, for amino acid fermentation?

Unlike auxotrophic mutants, regulatory mutants can be grown in inexpensive, complex media and they do not require careful control of growth conditions.

DNA-recombinant micro-organisms and combinations

Another way to enhance the production of an amino acid is to make use of DNA-recombinant technology, often in combination with the mutations already described. In this way the negative features of the micro-organisms are avoided. To help explain this, we will consider a well known fermentation of L-phenylalanine using *Escherichia coli*. We have already seen the metabolic pathway leading to the production of L-phenylalanine in Figure 8.4.

regulation by tyrosine, phenylalanine and tryptophan

Conversion of erythrose-4-P and phosphoenol pyruvate to 3-deoxy-D-arabinoseheptulosonic acid (DAHP) is catalysed by DAHP synthetase. In *E.coli* there are three isoenzymes of the enzyme; these are known as *aro*F (regulated by tyrosine), *aro*G (regulated by phenylalanine) and *aro*H (regulated by tryptophan). In each case, regulation is both at the level of enzyme formation (repression) and enzyme activity (feedback inhibition). Another site regulated through phenylalanine controls the expression of the structural genes (*phe*A) for chorismate mutase-prephrenate dehydratase (E3 and E5 in Figure 8.4). The mechanism is based on the phenylalanine mediated repressor from the *phe*A regulator gene that binds to the *phe*A operator site. The effect of this regulation (on the *phe*A level) and feedback inhibition (on the *aro*G level) is low levels of the enzyme chorismate mutase-prephrenate dehydratase and low activities of the enzyme DAHP synthetase. This results in low levels of phenylalanine.

To achieve overproduction of phenylalanine, the micro-organism should be derepressed at the *phe*A level and free of inhibition at the *aro*G level. Both genes are located on the chromosomal DNA of the micro-organism and, by means of amino acid analogues such as *p*-fluoro-DL-phenylalanine, it is possible to make (phenylalanine) feedback resistant mutants of *E.coli* (*phe*A^{FR} and *aro*F^{FR} mutants). The following procedure can be used:

feedback resistant mutants

use of plasmid DNA

- the genes are isolated from the chromosomal DNA and put into a plasmid (circular extra-chromosomal piece of DNA);
- regulation of the *phe*A^{FR} structural gene in the plasmid is altered by using a strong promoter (several very strong promoters are nowadays used, eg the *lac* promoter);
- the total effect (ie overproduction of phenylalanine by deregulation) can be enhanced by using more than one *phe*A^{FR} genes in the plasmid, in a so called multicopy recombinant.

In a good production strain, the plasmid should be present in a stable way and should not be lost from the micro-organism after a few generations.

Π The procedure described above is just one way to get to a good phenylalanine production strain. Briefly outline another way to achieve the goal.

The most obvious alternative approach is to deregulate the *aro*F gene, which is subject to tyrosine regulation. This, of course, could be achieved by DNA-recombinant techniques or by mutation.

SAQ 8.3

Provide brief explanations for the following:

1) Wild strains of *E. coli* are not used for L-phenylalanine production by direct fermentation.

2) Auxotrophic mutants of *E. coli* are particularly useful for the production of L-phenylalanine by direct fermentation.

3) Regulatory mutants improve the rate of L-phenylalanine overproduction by *E. coli*.

4) An *E. coli* mutant, Try^-, is likely to enhance L-phenylalanine overproduction.

8.4.3 Methods of fermentation

fermentation conditions

The growth of micro-organisms used in the production of amino acids is done in a well balanced environment. The conditions required are:

- a controlled pH of the fermentation medium (approximately neutral);
- rich growth media;
- highly aerobic conditions;
- sterile conditions.

Slight changes in the fermentation conditions can greatly affect amino acid production.

These variations are sometimes caused by changes in sterilising conditions, agitation and aeration, temperature, pH, pressure and liquid level. This is why the parameters have to be controlled in a proper way using precise equipment.

methods of amino acid fermentation

We can differentiate between three possible methods of amino acid fermentation:

- batch fermentation;
- fed-batch fermentation;
- continuous fermentation.

1) Batch fermentation

Batch fermentation is the most widely used method of amino acid production. Here the fermentation is a closed culture system which contains an initial, limited amount of nutrient. After the seed inoculum has been introduced the cells start to grow at the expense of the nutrients that are available. A short adaptation time is usually necessary (lag phase) before cells enter the logarithmic growth phase (exponential phase). Nutrients soon become limited and they enter the stationary phase in which growth has (almost) ceased. In amino acid fermentations, production of the amino acid normally starts in the early logarithmic phase and continues through the stationary phase.

short fermentation time preferable

For economical reasons the fermentation time should be as short as possible with a high yield of the amino acid at the end. A second reason not to continue the fermentation in the late stationary phase is the appearance of contaminant-products, which are often difficult to get rid off during the recovery stage. In general, a relatively short lag phase helps to achieve this. The lag phase can be shortened by using a higher concentration of seed inoculum. The seed is produced by growing the production strain in flasks and smaller fermenters. The volume of the seed inoculum is limited, as a rule of tumb normally 10% of the fermentation volume, to prevent dilution problems.

2) Fed-batch fermentation

Fed-batch fermentations are batch fermentations which are fed continuously, or intermitantly, with medium without the removal of fluid. In this way the volume of the culture increases with time.

low residual substrate concentration

One of the advantages of the fed-batch fermentation is the fact that the residual substrate concentration may be maintained at a very low level. This may result in a removal of catabolite repressive effects and avoidance of toxic effects of medium components.

control of oxygen demand

Another advantageous effect is on the oxygen balance. The feed rate of the carbon source (mostly glucose) can be used to regulate cell growth rate and oxygen limitation, especially when oxygen demand is high in the exponential growth phase.

3) Continuous fermentation

advantages of continuous fermentation

In general, there are several advantages of continuous fermentation as compared to batch fermentations. These include higher productivity, operation for a very long period of time, and lower installation and maintenance costs. The maintenance costs are particularly important. In batch cultures, oxygen demand, pH control requirements and amount of cooling required, changes throughout the whole fermentation run, whilst in continuous fermentations these factors are constant.

production strain variability

A disadvantage of using continuous fermentation is the chance of contamination by other micro-organisms during the long fermentation runs (sometimes several weeks). Although this sometimes happens, it should not be overemphasised, since most of the time the fermentation conditions are such that a special niche is created, and only a limited number of other micro-organisms can grow. A more serious problem is the occurrence of variants of the parent production strain by back mutation or loss of genetic elements (plasmids). In addition, phage infections are more likely to occur.

multi-stage continuous system

Despite the advantages of continuous cultures, the technique has found little application in the fermentation industry. A multi-stage system is the most common continuous fermentation and has been used in the fermentation of glutamic acid. The start-up of a multi-stage continuous system proceeds as follows. Initially, batch fermentation is commenced in each vessel. Fresh medium is introduced in the first vessel, and the outflow from this proceeds into the next vessel. The overall flow rate is then adjusted so that the substrate is completely consumed in the last vessel, and the intended product accumulated. The concentration of cells, products and substrate will then reach a steady state. The optimum number of vessels and rate of medium input can be calculated from simple batch experiments.

8.4.4 Problems in the fermentation of amino acids

The following problems in the industrial fermentation of amino acids may arise:

- contamination of the culture by other micro-organisms during fermentation;
- bad fermentation reproducibility due to differences in raw material;
- back mutation or loss of genetic material of the production strain;
- infection of the culture by bacterial viruses (phages).

Although the first two possibilities can lead to severe problems in the fermentation of amino acids, these problems can be prevented by: using proper plant design; maintenance of hygienic conditions throughout the operation; reservation of large batches of raw material with uniform qualities. Much more severe (and much more difficult to control) are the last two possibilities which will now be discussed in more detail.

Back mutation and loss of genetic material

Back mutation (reversion) of the mutant production strain to the wild type (prototrophic form) can occur.

Since auxotrophic mutants and regulatory mutants are widely used in the overproduction of amino acids, this can be a severe problem. In nature, mutation always takes place but this takes some time. However, in fermentation many generations are produced in a relatively short period of time and the chances of back mutation are enhanced.

Π Make a list of possible ways of solving the problem of reversion.

The main ways to solving the problem of reversion are:

solutions to the problem of 'reversion'

- make use of fresh starting material (inoculum) for each run;
- make use of antibiotics that inhibit growth of the wild type (or revertant) but not the mutant;
- start with mutants which are very stable (many companies spend a lot of money in the isolation of stable mutant strains).

Loss of genetic material (for example the constructed plasmid containing the genetic information necessary for overproduction of the amino acid) can also occur.

This will happen in situations where there is an energetically favourable advantage for the micro-organism not to have the plasmid in the cell.

Possible ways to overcome the problem are:

- construction of a strain in such a way that it is energetically advantageous to overproduce the required amino acid, thus keeping the construct in the cell;
- introduction of a genetic marker into the construct (usually antibiotic resistance encoded by plasmid genes) and operation of the fermentation under selective pressure by adding the relevant antibiotic to the medium. Only the micro-organisms carrying the marker would survive, while others die off.

Phage infection

Phage infections sometimes occur in big fermentation plants and cause severe damage.

stages in phage infection

Phages are bacterial viruses composed of a nucleic acid core (DNA or RNA) in a protein envelope. Infection proceeds in the following way:

- adsorption onto the bacterial cell followed by introduction of genetic material into the bacterium;
- multiplication within the cell making use of the genetic apparatus of the bacterial cell;
- liberation of the infective phages by lysis of the cell resulting in the death of the bacterial cell.

There are several different groups of phages, each with there own characteristics and host range specificity.

Phages are widely distributed in nature (soil, air, raw materials etc), they are rather stable and can be introduced to the fermentation easily through the air. Phage infections can be recognised if fermentation characteristics change. The slowing down of the process (for example delay in onset of the fermentation) and decrease in amino acid production yield are the first signs of a moderate form of phage infection. If lysis of the cells occurs in the growth phase, accompanied by a rise in pH and increase in foaming, this suggests that there is a severe phage infection. Confirmation of a phage infection can be done by electron microscopic observation.

latent phage infections

An additional problem is the presence of latent forms of phage infections, in which the production strain carries the phage genome integrated in the chromosome and only in specific situations (after induction) releases the phages. These may then be changed to the virulent form and infect other micro-organisms present. Prevention of phage infections is difficult and there is no absolute preventative measure. However, phages have their own optimum conditions (pH and temperature) for infection and replication. By choosing the proper conditions it is possible to create an environment that is favourable for bacterial growth but not for phage multiplication.

Other ways to prevent phage contamination or multiplication are:

- plant hygiene (sterilisation of raw material; disinfection of equipment by temperature or chemical agents; using proper air filters, etc);
- isolation of phage resistant strains - however, the problem is that after a while other types of phages become capable of infecting the strain;
- inhibition of phage multiplication by prevention of phage adsorption to the cell wall using chemical agents during fermentation (chelating agents capture calcium ions necessary for adsorption; use of surfactants like Tween).

SAQ 8.4

Give possible reasons for selecting each of the following operating conditions for a hypothetical amino acid fermentation.

1) Fed-batch mode.

2) Monoseptic.

3) Antibiotics added to medium.

4) Fresh inoculum used for each run.

5) Chelating agent added to medium.

8.5 Recovery of the amino acid from the fermentation broth

Methods of product recovery are considered in detail in the BIOTOL text entitled 'Product Recovery in Bioprocess Technology'. In this chapter we will briefly review the methods applicable to recovery of amino acids.

general criteria for economic feasibility

General criteria for the economically feasible recovery of amino acids from the fermentation broth are:

- the recovery should be as simple as possible;
- the recovery yield of the processes should be high;
- process steps that give rise to loss of product should be avoided as much as possible;
- the process should be easy to scale-up industrially.

Π Amino acid fermentations do have several advantages over other kinds of fermentations, like for example the production of enzymes, in respect of recovery of desired product. Make a list of the advantages that you can think of.

The main advantages are:

- normally the production strain is constructed in such a way that overproduction of the desired amino acid is obtained and no, or only minor concentrations of, unwanted contaminants appear;
- optical resolution steps are not necessary (as in the case of most chemical-processes) since only the L-form is synthesised;
- the required amino acid can be relatively easily separated from cells and protein impurities.

By making use of the physicochemical properties of the amino acid that is required, it is possible to obtain highly purified and/or concentrated product. This is done by a combination of several processes, the number of which is dictated by the final application for the product and by economical feasibility.

Specific methods to separate the amino acid required from its contaminant products, such as medium components and other amino acids, are preceeded by removal of the cells and proteineous material from the culture broth.

methods for amino acid recovery

Following cell and proteineous material removal, one or more specific methods of amino acid recovery are applied. These are:

- crystallisation;
- ion exchange;
- electrodialysis;
- solvent extraction;
- decolorisation;
- evaporation.

We will now consider each of these methods in turn.

8.5.1 Separation of cells and proteins from the fermentation broth

Two methods are commonly used, these are centrifugation and filtration. Both methods are used on a large scale and are economically feasible.

centrifugation

Centrifugation can be operated on a semi-continuous or continuous basis and there are several different types of centrifuges. Large scale tests have to be performed to choose the proper centrifuge (unloading speed, capacity, separation performance etc).

Sometimes poor centrifugation behaviour of cells can be improved by adding flocculation agents. These agents neutralise the anionic charges (carboxyl and phosphate groups) on the surface of the microbial cells. Examples of flocculation agents are alum, calcium and ferric salts, tannic acid etc.

filtration

Filtration is another possibility to remove the cells from the amino acid containing broth. Factors to be considered are:

- properties of the filtrate (solid/liquid ratio);
- nature of the solid particles (type of micro-organism);
- adequate pressure (or vacuum) to obtain adequate flow rates;
- negative effects of antifoaming agents on filtration.

filteraids

Sometimes filtration can be improved by using filteraids. These filteraids, which are based on diatomaceous earth, improve the porosity of a resulting filter cake leading to a faster flow rate. Before filtration a thin layer is used as a precoat of the filter (normally standard filters). After that a mix is made with the harvest broth and filtration is started.

8.5.2 Crystallisation

pH

Crystallisation is often used as a method to recover the amino acid. Because of the amphoteric character (contains both acidic and basic groups) of amino acids, their solubility is greatly influenced by the pH of the solution and usually show minima at the isoelectric point (zero net charge).

temperature

Since temperature also influences the solubility of amino acids and their salts, lowering the temperature can be used in advance as a means of obtaining the required product.

salts

Precipitation of amino acids with salts, like ammonium and calcium salts, and with metals like zinc are also commonly used. This is followed by acid (or alkali) treatment to obtain the free or acid form of the amino acid.

8.5.3 Ion exchange

Ion exchange resins have been widely used for the extraction and purification of amino acids from the fermentation broth.

cation and anion exchange resins

The adsorption of amino acids by ion exchange resins is strongly affected by the pH of the solution and by the presence of contaminant ions. There are two types of ion exchange resins; cation exchange resins and anion exchange resins. Cation exchange resins bind positively charged amino acids (this is in the situation where the pH of the solution is lower then the isoelectric point (IEP) of the amino acid), whereas anion exchange resins bind negatively charged amino acids (pH of the solution is higher than IEP). Elution of the bound amino acid(s) is done by introducing a solution containing the counterion of the resin.

Anion exchange resins are generally lower in their exchange capacity and durability than cation exchange resins and are seldom used for industrial separation. In general, ion exchange as a tool for separation is only used when other steps fail, because of its tedious operation, small capacity and high costs.

Examples given in literature are:

1) strongly basic anion exchange (OH^-) for the separation of:

- L-glutamic acid from broth;

- basic amino acids from neutral and acidic amino acids;
- separation of aromatic amino acids (phenylalanine and tyrosine) from other amino acids.

2) strongly acidic cation exchange (H^+; NH_4^+) for the separation of:

- L-proline adsorbtion and elution with diluted HCl;
- L-threonine adsorbtion and elution with aquous ammonia.

8.5.4 Electrodialysis

net zero charge

This method is based on the principle that charged particles move towards electrodes in an electric field. A mixture of the required amino acid and contaminant salts can be separated at a pH where the amino acid has a net zero charge (at the IEP).

The salt ions are captured by the ion exchange membranes that are present. The applications are limited to desalting amino acid solutions, eg removal of HCl from L-glutamic acid solution.

8.5.5 Solvent extraction

only small distribution coefficients

This method also has only limited applications. The reason for this is that the distribution coefficients of amino acids between organic solvent and water phases are generally small. There are some possibilities given in literature which are based on the alteration of the amino acid.

Examples are:

- cyclisation of L-glutamic acid and extraction with alkyl and aromatic alcohols;
- conversion of contaminant organic acids (like acetic acid) to the ester form and extraction of the ester;
- extraction of basic amino acids (like L-lysine) from aqueous solution with water immiscible solvents containing higher fatty acids;
- conversion of L-isoleucine/L-leucine mixtures to their cobalt salts and subsequent extraction of L-isoleucine by low molecular weight alcohols;
- conversion of amino acids to their aldehyde derivative and subsequent extraction etc.

8.5.6 Decolorisation

activated charcoal

This step is almost always performed to get rid of the coloured impurities in the fermentation broth. The method is based on the fact that amino acids (especially the non-aromatic amino acids) do not adsorb onto activated charcoal. Although the treatment is very effective, some of the amino acid is lost during this step.

Alternative ways are:

- use of porous ion exchange;

- addition of cationic surfactants, high molecular synthetic coagulants or some phenolic compounds;
- washing of crystals with weakly alkaline water as in the case of glutamic acid.

8.5.7 Evaporation

quick but expensive

Evaporation of the amino acid containing solution is a quick but commercially unattractive way (high energy costs) to obtain amino acids from solution. This can only be used when the total amount of contaminant products is very low, since these compounds are not removed and appear in a concentrated form in the product.

Π Make a list of the main merits and limitations of the various methods of recovery of amino acids from fermentation broth.

Method	Merit	Limitation
Centrifugation		
Filtration		
Crystallisation		
Ion exchange		
Electrodialysis		
Solvent extraction		
Decolorisation		
Evaporation		

SAQ 8.5

In the recovery of amino acids from fermentation broths, several methods can be applied. Match each of the following statements to one or more separation methods.

Statement

1) Separation efficiency is related to a distribution coefficient.
2) Separation relies on the amphoteric nature of amino acids.
3) Separation may be improved by addition of calcium salts.
4) Separation involves adsorption onto activated charcoal.
5) Separation method can only be used if total amounts of contaminating products is very low.

8.6 Case study: fermentative production of L-phenylalanine from glucose

In this section we will examine the fermentative production of L-phenylalanine in detail. This will serve to illustrate some of the general principles of fermentation already considered.

summary reaction

Phenylalanine production from a plasmid harbouring auxotrophic mutant of *E. coli* has been studied in batch cultures. Phenylalanine is produced in the stationary phase due to the release of feedback inhibition after depletion of tyrosine. The fermentation can be summarised as:

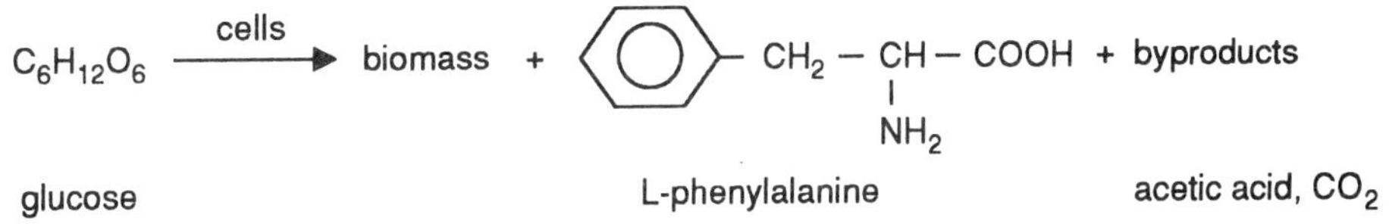

The characteristics of the strain are:

Δtrp, Δtyr /pJN6 carrying the genes *aro*F, *phe*A^{FR}, Tc^R and Ap^R.

Explanation:

Δ = auxotrophic for tryptophan and tyrosine;

FR = feedback resistant to phenylalanine;

R = resistant to the antibiotics tetracycline and ampicillin.

medium composition

The composition of the basic medium is (g l^{-1}):

NH_4Cl:	5.0	Nitrogen source
K_2SO_4:	0.8	S-component
KH_2PO_4:	0.5	P-component/buffer
Na_2HPO_4:	1.0	P-component/buffer
Na-citrate:	2.5	
$FeCl_3.6H_2O$:	0.01	
$CaCl_2.2H_2O$:	0.02	
$MgCl_2.6H_2O$:	0.8	
tyrosine:	0.05	
tryptophan:	0.025	
tetracycline:	0.010	
glucose:	35	Carbon source
trace elements:	1 ml l^{-1}	

These elements are necessary for maintenance and growth of the organism and for product synthesis.

sterilisation

The pH of the medium is set around 7.0 (neutral conditions) and sterilisation is carried out by autoclaving at 121°C. $CaCl_2.2H_2O$, $MgCl_2.6H_2O$ and glucose are sterilised separately. In the first two cases to prevent precipitation and in the last case to prevent caramelisation (decomposition of glucose).

Tyrosine, tryptophan and tetracyline are filter-sterilised to prevent decomposition by heat sterilisation.

inoculation

Three 1 litre baffled flasks, each containing 100 ml medium, are inoculated with cells from one agar plate suspended in 10 ml saline and incubated at 30°C on a rotary shaker (for optimum supply of oxygen). This provides sufficient biomass to inoculate the bioreactor.

A 3 litre bioreactor with a working volume of 2 litre is inoculated with the three shaking flasks. The pH is maintained at 5.5 by automatic titration with 5mol l^{-1} NH_4OH and the temperature is held at 37°C.

The agitation speed is set a 600 rev min^{-1} and the air flow rate at 1.5-1.6 litre min^{-1}.

monitoring the fermentation

The following parameters are measured more or less continuously:

- dissolved oxygen tension, using an *in situ* oxygen electrode;
- phenylalanine, by test kit;
- glucose concentration, enzymatically;
- the byproduct acetic acid, by gas chromatography;
- biomass, by optical density and dry weight;
- temperature, using temperature probe;
- CO_2, by infrared gas analysing system.

Measurement of all these parameters provides sufficient information to evaluate the fermentation and leads to the economical production of the amino acid. We will now discuss several relevant parameters seperately with the help of the data obtained (see Figure 8.5).

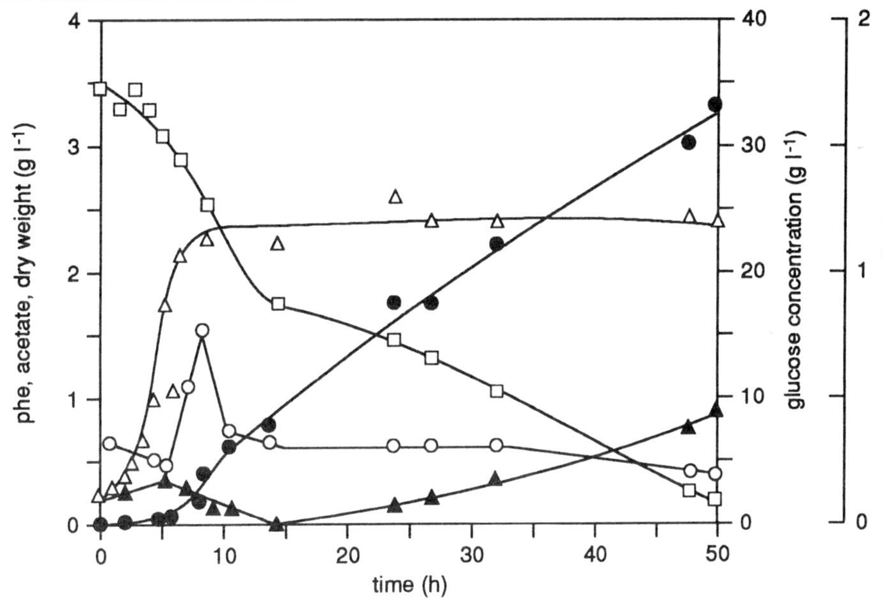

Figure 8.5 The production of phenylalanine from *E. coli*.

Biomass yield

stages on batch growth cycle

Biomass production is determined by measuring the optical density at 620 nm and/or weighing after drying to constant weight. The typical form of the growth curve is given in Figure 8.5 (open triangles). We can recognise at least three stages:

- The lag phase (circa 2 hours) during which the cells are not capable of multiplying;
- The exponential (log) phase (up to 8-10 hours) where cells multiply most rapidly;
- The stationary phase where growth ceases (10-25 hours);
- The death phase where some cells are lysed (after 25 hours; not obvious in Figure 8.5).

We can also see from Figure 8.5 that glucose is consumed rapidly during growth (open squares) and that almost all glucose is consumed by the end of the fermentation.

Phenylalanine production and byproducts

depletion of tryosine

Production of phenylalanine starts after depletion of tyrosine at about 6 hours. This is logical since the micro-organism needs a certain amount of tyrosine, for example to synthesise key enzymes, but synthesis of L-phenylalanine is feedback regulated if tyrosine is present.

L-phenylalanine production is largely independent of the growth stage (see Figure 8.5, closed circles), although maximum specific rate of phenylalanine production (ie the amount of L-phenylalanine produced per unit of time) is at the end log phase of the growth (open circles). For economic reasons it is worthwhile to know the amount of L-phenylalanine that is produced per amount of raw material (glucose, carbon source). This can be calculated from Figure 8.5 after 14 hours during constant product formation and is called yield of L-phenylalanine from glucose. As in most fermentations, byproducts are formed: part of the carbon flow is directed to the tricarboxylic acid cycle and acetate is formed in low amounts.

product yield from glucose

Carbon balance

As we already know, glucose is the major carbon input. Carbon from tyrosine, tryptophane and citric acid is negligible and can be excluded from the calculation.

Carbon from the substrate glucose is converted into the carbon of the cells, phenylalanine, carbon dioxide and byproducts. Carbon balance calculations thus give us more understanding of the amount of carbon in glucose used for cell mass production, for synthesis of the wanted product, maintenance energy and byproduct formation.

In the stationary phase, assuming that the total cell mass in the reactor is constant and that only L-phenylalanine and acetate are produced, the following stoichiometric equations are used:

reaction stoichiometry

$$2C_6H_{12}O_6 + NH_3 + 2O_2 \rightarrow C_9H_{11}O_2N + 8H_2O + 3CO_2$$

(glucose) (phenylalanine) (Equation 8.1)

$$C_6H_{12}O_6 + 2O_2 \rightarrow 2CH_3COOH + 2H_2O + 2CO_2$$

(glucose) (acetate) (Equation 8.2)

From the literature the following stoichiometric equation for *E. coli* growing aerobic on glucose is well known.

$$C_6H_{12}O_6 + 0.1NH_3 + 8.457O_2 \rightarrow 0.5CH_{1.77}O_{0.49}N_{0.24} + 5.5CO_2 + 5.7\ H_2O$$

(Equation 8.3)

glucose yield coefficient

Using these equations it is possible to calculate theoretical values for glucose yield coefficients (Y_G). Y_G values can be calculated for cell mass, phenylalanine and acetic acid and is simply the amount of product produced per unit amount of glucose consumed. So the theoretical yield of phenylalanine on glucose, for example using equation 8.1, is:

0.5 mol phenylalanine (mol glucose)$^{-1}$ or 0.47 g phenylalanine (g glucose)$^{-1}$.

Of course this is an absolute maximum value since part of the glucose that is supplied is used for byproducts such as acetate. In such a case it is more realistic to combine both equations (8.1) and (8.2):

0.33 mol phenylalanine (mol glucose)$^{-1}$ or 0.31 g phenylalanine (g glucose)$^{-1}$.

experimentally derived data

Now consider the experimental results shown in Figure 8.5. From these graphs data can be extracted and presented in tabular form. Table 8.4 presents some data extracted from Figure 8.5, along with other data (for carbon dioxide) that were not shown in the Figure.

Time (h)	2	4	6	8	10	14	24	32	48
glucose (g l^{-1})	34	33	30	28	26	18	15	12	4
dry weight (g l^{-1})	0.3	0.6	1.7	2.0	2.2	2.3	2.3	2.3	2.2
phenyla-lanine (g l^{-1})	0.0	0.02	0.1	0.3	0.6	0.8	1.6	2.2	3.0
acetate (g l^{-1})	0.1	0.2	0.3	0.2	0.2	0.0	0.1	0.3	0.8
CO_2 outlet	-	-	0.25	0.5	0.6	0.7	0.7	0.6	-
culture volume (l)	2	2	2	2	2	2	2	2	2

Table 8.4 Data obtained from Figure 8.5.

Π Use the data given in Table 8.4 to calculate the experimental yield of phenylalanine on glucose [g phenylalanine (g glucose)$^{-1}$] for the interval between 14 and 32 hours:

yield of phenylalanine

For the interval between 14 and 32 hours: glucose consumed = 18-12 = 6 g l^{-1}; phenylalanine produced = 2.2-0.8 = 1.4 g l^{-1}. So, yield of phenylalanine on glucose = 1.4/6 = 0.23 g phenylalanine (g glucose)$^{-1}$.

The maximum yield, as can be seen from Figure 8.5, is in the late exponential phase. Using the data given in Table 8.4 for the interval between 14 and 24 hours, we obtain a value of 0.27 g phenylalanine (g glucose)$^{-1}$ for the maximum yield coefficient.

Π Now compare the maximum experimental Y_G value for phenylalanine (0.27 g g^{-1}) with the theoretical Y_G value (0.31 g g^{-1}). How can we account for this discrepancy?

maintenance energy requirements

This discrepancy is due to the fact that other products such as formate, are formed in very small amounts as byproducts of the metabolic routes leading to L-phenylalanine and polymer synthesis. Of course, part of the glucose is also used for the metabolic activities in the micro-organism necessary to maintain the cells in a viable state, this is termed the maintenance energy requirement.

carbon flow

To find out where the carbon flow goes to we can simply investigate Figure 8.5. It is obvious that:

- in the early exponential phase most of the glucose is used for biomass production;

- in the mid-exponential phase part of the glucose is used for the production of phenylalanine. This portion increases entering the late exponential phase;
- in the stationary phase there is no cell growth, so most of the glucose is used for the production of L-phenylalanine.

SAQ 8.6

1) From the data given in Table 8.4 calculate the observed glucose consumption, in moles, in the interval between 14 and 32 hours (relative molecular mass of glucose = 180).

2) Now calculate the number of moles of glucose necessary to maintain the reactions in this interval. What can you deduce from these calculations?

Hint: Consider the number of moles of phenylalanine and acetic acid formed and the stoichiometric equations 8.1 and 8.2.

(RMM phenylalanine = 165, acetic acid = 60).

8.6.1 Process economics of the fermentative production of L-phenylalanine

The ultimate goal of process development is to achieve feasibility: where it is possible to produce amino acids on a large scale at a production cost per kg of amino acid comparable to, or cheaper than, the processes currently used by other companies. If we presume that the technical performance (fermentation and recovery) are sorted out on a laboratory scale and scaling up looks promising, then it is time to find out whether it is possible to operate economically on a large scale.

cost price evaluation

The thing to do first is a cost price evaluation on the particular process. It is obvious that one should know as much as one can of the process in order to come to a realistic cost price evaluation. Since these data are not always available in the literature (especially not for ongoing commercial fermentation processes) we have to make use of those general data that are available in the literature and patents.

First we start with a general prescription of the process. The aim, for example, is to make 1,000 tonnes of L-phenylalanine a year (1 million kg per annum).

Presume that a yield of 20 g l^{-1} of L-phenylalanine can be obtained. This is more realistic, based on patent literature, than the low yields in the example considered previously (section 8.6). This automatically means that more glucose will be needed. Let us again presume that instead of 35 g l^{-1} glucose we now need 150 g l^{-1} to achieve this overall yield. The concentrations of the mineral salt are kept the same to maintain good buffering capacity, whilst the concentration of tyrosine and tryptophan are also increased by a factor 4.3 (tyrosine 0.21 g l^{-1} and tryptophan 0.11 g l^{-1})

size of bioreactor

A reasonable size of bioreactor, based on transport and handling considerations, is 200 m^3, with a working volume of 150 m^3. If the fermentation time is 48 hours and down time for reuse about 24 hours, then the total batch time is 72 hours.

As a rule of thumb, total investment of the fermentation plant is 50 dollars per litre of bioreactor. This means the total costs of investment are 200,000 x 50\$ = 10 million dollars. This figure includes all costs concerning engineering, fermentation equipment, recovery equipment, buildings land, etc.

The second calculation is on the production costs themselves. We can discriminate three major parts:

- variable costs;
- total direct costs;
- plant gate costs.

variable costs

Variable costs are costs for raw material used in the fermentation and for recovery (for example if flocculants are used) and utilities like steam (sterilisation, water evaporation), water (medium make up, washing, cleaning, cooling etc) electric power etc.

Let us now reconsider the medium described previously in section 8.6 but adapted for an industrial fermentation.

Major costs for raw materials are those for glucose, tyrosine, tryptophan and salts. The figures are given in Table 8.5.

	kg kg^{-1} phenylalanine	price \$ kg^{-1}	\$ kg^{-1} phenylalanine
glucose	7.5	0.40	3
salts	0.53	1.13	0.6
tyrosine	0.011	30	0.32
tryptophan	0.006	100	0.6
Total			4.52 \$ kg^{-1}

Table 8.5 Raw materials costs for industrial scale fermentation of L-phenylalanine.

As mentioned previously, utilities are steam, water, electricity and gas necessary to run the plant. It is extremely difficult to obtain proper figures from the literature. In addition, costs of utilities differ from plant to plant and from country to country. Not knowing how much of each of these utilities are necessary to run this fictitious plant we have to use "fake" figures.

Based on information in the literature, the costs for utilities are about 10% of the total variable costs, in our case being 0.50 \$ kg^{-1} phenylalanine produced. Variable costs are summarised in Table 8.6.

direct costs

Direct costs are costs for maintenance of materials, operating of supplies labour, control laboratory etc.

If the minimal number of operators necessary to run the plant is for example 10 in a 4 shift system all the year through, than there is a total of 40 operators on the payroll. At an average salary of \$20,000 a year, the total direct costs for labour are \$800,000.

Indirect labour accounts for 40% of the direct labour, being \$320,000, and 85% of the direct labour is used for salaried payroll (\$680,000). Associated payroll costs are assumed to be about 30% of the total payroll costs (indirect+direct+salaried payroll), ie \$540,000.

Maintenance salaries and costs for laboratories are estimated as 3% of the total capital of investment, ie 3% of 30 million dollars is $900,000. The figures for payroll charges calculated back to one kg of L-phenylalanine are given in Table 8.6.

INVESTMENT ($, million)	
Total fixed capital	30
PRODUCTION COSTS ($ kg^{-1} L-phenylalanine)	
Variable costs	
raw materials	4.52
utilities	0.50
total variable costs	5.02
Direct costs	
direct labour	0.80
indirect labour	0.32
salaried payroll	0.68
associated payroll	0.54
maintenance salaries, laboratories etc	0.90
Supplies and expenses:	
maintenance	0.90
supplies	0.20
total direct costs	4.34
Plant gate costs	
depreciation (15 years)	2.00
taxes and insurance	1.20
plant overhead	0.48
total plant gate costs	3.68
total production costs (fixed + variable are thus $13.04 kg^{-1})	

Table 8.6 Cost price evaluation of L-phenylalanine fermentation based on a production of 1,000 tonnes.

As part of the total direct costs we also have to consider costs made for supplies and expenses: maintenance, 3% of total capital; supplies, 25% of direct labour. These figures, calculated back to one kg of L-phenylalanine are given in Table 8.6.

plant gate costs

Plant gate costs are other costs concerning the fermentation plant; for example depreciation, taxes and insurance and other plant overhead costs. Fifteen years depreciation is not abnormal. This brings the depreciation costs to $2 million a year.

Taxes and insurance are estimated to be 4% of the total capital being $1.2 million a year.

Plant overhead accounts for about 60% of total direct labour and is thus calculated to be $480,000 (see Table 8.6).

return on investment criterion

Now that we have dealt with most cost factors, lets see whether the process is profitable or not. A key factor in industry is profitability regardless of technical achievement. Objective procedures to aid such assessment are based on the return on investment (ROI) as the criterion.

This is calculated from the following assumption:

$$ROI = \frac{(A + 9B)/10}{C + D} \times 100 \text{ (expressed as a percentage)} \qquad \text{(Equation 8.4)}$$

where:

ROI = return on investment;

A = net earnings after taxes (50%) in the first year;

B = net earnings after taxes (50%) per annum for the coming period in which the investment should be payed back;

C = original fixed capital investment;

D = working capital (25% of net sales).

10 year projection

To calculate the ROI, again we have to make some more assumptions. Let us assume that we would like to have our costs of investment back in say 10 years, so we can start making real money after that. Let us in addition assume that the selling price for L-phenylalanine is 30 to 40 $ kg^{-1}.

At a production of 1,000 tonnes, this means a total sales of $30 to $40 million a year.

After extraction of costs such as dealer discount, distribution and freight (total about 30%) the net sales are $21 to 28 million a year. Since we calculated a cost price of about 13 dollars per kg phenylalanine (see Table 8.6; $13 million per 1,000 tonnes), the gross profit will be 8 to 15 million dollars.

Assume 12.5% of the net sales for cost of sales and adminstration plus research and development (ie 2.6 to 3.5 million dollars) the profit from operations in the coming period will be 5.4 to 11.5 million dollars.

In the first year, the startup costs are substantial (10% of the capital; in our case 3 million dollars). The profit for the first year is, therefore, not 5.4 to 11.5 million dollars but 2.4 to 8.5 million dollars.

The net earnings after taxes (50%) in the first year are then $1.2 to $4.25 million (A in equation 8.4) and the net earnings per annum for the period to come are $2.7-$5.75 million (B in equation 8.4). The working capital is normally about 25% of the net sales ($5.25 to $7 million; D in equation 8.4) whilst the total fixed capital was assumed to be $30 million (see Table 8.6; C in equation 8.4).

The ROI calculated from equation 8.4 will then be 7 to 15%. Usual figures for ROI in the literature are 15 to 20%.

The total production costs will be increased with sales and administration plus research and development, so the final production costs will be 13.04$ kg^{-1} + ($2.6 to $3.50) giving a total production cost at 100% capacity of $15.64 to $16.54 per kg of phenylalanine.

The product value at 100% capacity will now be (total cost of production + 7 to 15% ROI), ie \$16.04 to \$16.54 + \$1.12 to \$2.48. So the minimum product value will be \$17.16 per kg of L-phenylalanine and the maximum product value \$19.02 per kg of L-phenylalanine.

It is rather difficult to say whether this fictitious process would survive or could compete. Actual data are absolutely necessary. On the other hand this exercise gives us a better understanding of process economics and can also be used to compare a fermentative process for the production of amino acids with, for example, a chemo-enzymatic process.

SAQ 8.7

Calculate the return on investment over a 15 year period for an amino acid fermentation, based on the following data and assumptions.

Production capacity = 500 tonnes per annum
Selling price of product = 50 \$ kg^{-1}
Cost price of product = 24.5 \$ kg^{-1}
Capital = \$40 million
Taxes = 50%.

Assumptions

Cost of dealer discount, distribution and freight = 20% total sales
Startup costs = 10% of capital
Working capital = 25% of net sales
Administration plus R and D costs = 12.5% of net sales.

8.7 Case study: The production of L-phenylalanine by enzymatic methods

Before we deal with the use of enzymes for the production of amino acids, we suggest that you may wish to refresh your memory of the material covered in Chapter 2.

To illustrate a few aspects of amino acid production by enzymatic methods, the production of L-phenylalanine will be considered in some detail. L-Phenylalanine is important as an essential amino acid for human nutrition and is used as an intermediate for the synthesis of the artificial sweetener, aspartame. Other examples of the industrial production of amino acids by enzymatic methods are described briefly in Appendix 2.

In section 8.6 we saw that L-phenylalanine can be produced from glucose by fermentation. In this production method mutants are required to achieve high yields of L-phenylalanine.

Reaction schemes for the production of L-phenylalanine by enzymatic methods are shown in Figure 8.6.

unnatural substrates

Although the term bioconversion can be used to describe all types of productions discussed in this section, it generally refers to the process conducted by micro-organisms and/or enzymes with unnatural substrates (which are not present in normal biochemical pathways). The enzymes available in or from the biomass are used for conversion of the unnatural substrate to obtain the desired product.

Two of the main raw materials used for bioconversion to L-phenylalanine are *trans*-cinnamic acid and acetamido cinnamic acid (reactions 1 and 2 in Figure 8.6.)

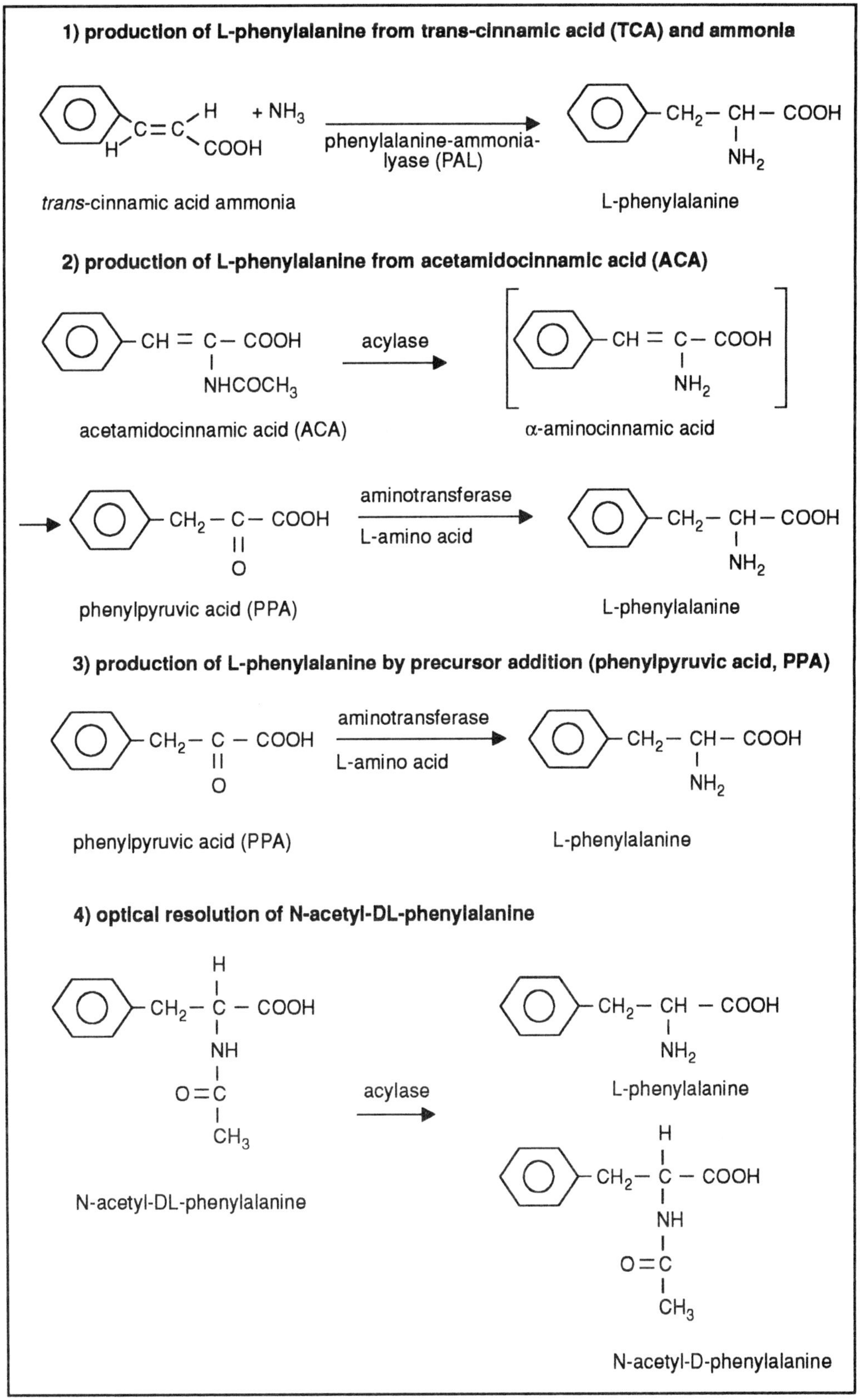

Figure 8.6 Reaction schemes for the production of L-phenylalanine by enzymatic methods.

5) enzymatic reductive amination of α-keto acids

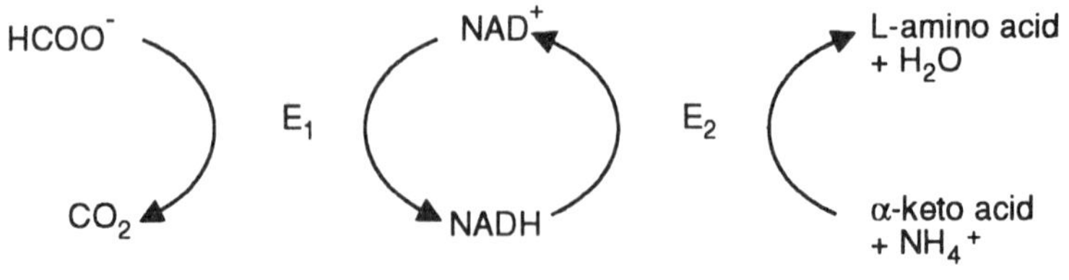

E1 = formate dehydrogenase; E2 = L-amino acid dehydrogenase

for L-phenylalanine production; α-keto acid = phenylpyruvic acid

6) enzymatic reductive amination of α-hydroxy acids

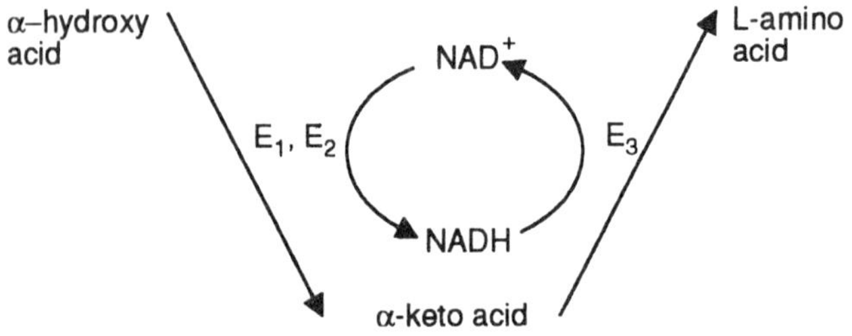

E1 = L-hydroxy acid dehydrogenase E2 = D-hydroxy acid dehydrogenase

E3 = L-amino acid dehydrogenase

for L-phenylalanine production: α-keto acid = phenylpyruvic acid
α-hydroxy acid = phenyllactic acid

Figure 8.6 Continued.

trans-cinnamic acid

Reaction 1 is governed by the enzyme phenylalanine ammonia lyase. This enzyme normally conducts the breakdown of L-phenylalanine to *trans*-cinnamic acid and ammonia. However, the reaction can be reversed leading to the production of L-phenylalanine from *trans*-cinnamic acid by using excess ammonia.

acetamido cinnamic acid

Much research deals with the production of L-phenylalanine from acetamidocinnamic acid (ACA) because of the low price of ACA. Different pathways have been reported but the principles of the main reaction pattern are given in reaction number 2 (Figure 8.6). At least two enzymes, an acylase and an amino transferase, are necessary for the bioconversion of ACA to L-phenylalanine. The amino source usually is an amino acid, L-aspartic acid is often used.

precursor addition

The production of L-phenylalanine by precursor addition is given in reaction 3 (Figure 8.6). The use of intermediates as substrates in L-phenylalanine synthesis avoids inhibition by metabolites. Phenylpyruvic acid, an intermediate precursor in the biosynthesis of L-phenylalanine, can be converted to L-phenylalanine. L-aspartic acid is often used as an amino donor. The amino group can only be transferred from an organic nitrogen source and needs a cofactor (pyridoxalphosphate), which is often present in whole cells. So ammonia cannot be used for amino group addition.

optical resolution

The optical resolution of the chemically synthesised N-acetyl-DL-phenylalanine by an acylase enzyme is given in reaction 4 (Figure 8.6). A selective hydrolysis of N-acetyl-L-phenylalanine is performed.

reductive diamination

Two reactions for the production of L-phenylalanine that can be performed particularly well in an enzyme membrane reactor (EMR) are shown in reaction 5 and 6. The recently discovered enzyme phenylalanine dehydrogenase plays an important role. As can be seen, the reactions are coenzyme dependent and the production of L-phenylalanine is by reductive amination of phenylpyruvic acid. Electrons can be transported from formic acid to phenylpyruvic acid so that two substrates have to be used: formic acid and an α-keto acid phenylpyruvic acid (reaction 5). Also electrons can be transported from an α-hydroxy acid to form phenylpyruvic acid which can be aminated so that only one substrate has to be used: α-hydroxy acid phenyllactic acid (reaction 6).

We can summarise the enzymatic methods of L-phenylalanine production as follows:

- bioconversion (*trans*-cinnamic acid or acetamidocinnamic acid as bioconversion substrate);
- precursor addition (phenylalanine as precursor);
- optical resolution (N-acetyl-DL-phenylalanine addition);
- reductive deamination involving coenzyme regeneration (α-keto acids or α-hydroxy acids addition).

SAQ 8.8

Consider reaction schemes for the production of L-phenylalanine by enzymatic methods. Now match each of the following substrates with the enzyme(s) responsible for L-phenylalanine formation.

Substrates

1) Phenylpyruvic acid (as precursor)

2) N-acetyl-DL-phenylalanine

3) Acetamidocinnamic acid

4) Phenylpyruvic acid (reductive deamination of α-keto acid)

5) *trans*-Cinnamic acid

6) Phenyllactic acid (α-hydroxy acid).

Enzymes

L-hydroxy acid dehydrogenase
acylase
phenylalanine-ammonia-lyase
L-amino acid dehydrogenase
L-amino acid amino transferase

Now that we have seen the possible enzymatic routes to overproduction of L-phenylalanine, in the next section we will consider two processes for production of this amino acid in more detail. These are precursor addition using phenylpyruvic acid and bioconversion using acetamidocinnamic acid. These essentially enzymatic processes are carried out using whole cells in a kind of semi-fermentation. We will see that, in comparison with direct fermentation using glucose (section 8.6), productivity for the semi-fermentations are good although economic feasibility is strongly influenced by the additional cost of the substrate.

8.7.1 Production of L-phenylalanine by precursor addition

transamination of phenylpyruvic acid

The production of L-phenylalanine from the precursor phenylpyruvic acid by transamination is a process which requires two steps:

1) cell production;

2) L-phenylalanine production after addition of phenylpyruvic acid.

fed-batch fermentation

Cell production can be carried out by a normal fed-batch type of fermentation. The feed rate of glucose is increased during the fermentation and the cells grow exponentially.

oxygen limitation

Biomass formation and transamination activity within the cells develop in a similar manner. Growth usually continues until limited by the availability of dissolved oxygen tension (DOT). After 10-15 hours a dry weight biomass concentration of 10 g l^{-1} is normally reached.

resting cells

The culture can be used directly for the conversion of phenylpyruvic acid to L-phenylalanine. Therefore, a batch process with resting cells can be carried out, with some glucose added for maintenance (fed-batch fermentation). Another approach is to harvest the cells from the fermentation broth and to use them in a separate bioreactor in higher concentrations than the ones obtained in the cell cultivation. An advantage of the last method can be that the concentration of compounds other than L-phenylalanine is lower, so that downstream processing may be cheaper.

Figure 8.7 gives a typical time course for the conversion of pyruvic acid to L-phenylalanine.

product formation

Experiments for optimisation of the production of L-phenylalanine are usually carried out at pH = 7.5, a temperature of 37°C, 50% DOT and 10 gram dry weight biomass per litre medium. Maximum productivity is reported to vary between 3 and 6 g l^{-1} h^{-1} and product concentrations of between 11-28 g l^{-1} have been reported. The time necessary for completion of the reaction is about 8 hours (see Figure 8.7).

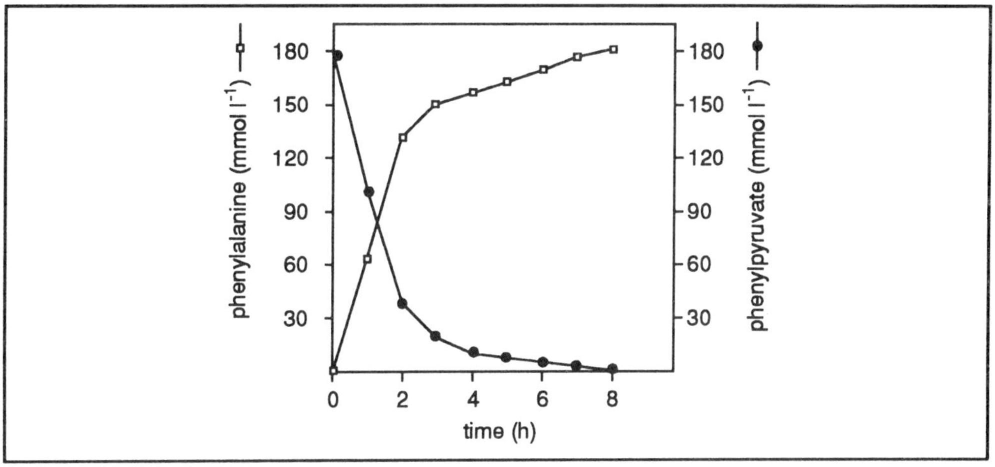

Figure 8.7 Fed-batch fermentation of phenylpyruvic acid to L-phenylalanine.

Specific information about the optimum conditions for the synthesis and the activity of the enzyme has been reported for *Pseudomonas fluorescens*: screening of various micro-organisms resulted in the selection of a *P. fluorescens* strain with an initial rate of conversion of 3 g l^{-1} h^{-1} in an unoptimised state. The following conclusions could be made concerning the production of L-phenylalanine by *P. fluorescens*:

constitutive amino-transferase

- The activity was generally higher in cells grown in a cultivation medium containing D- or L-phenylalanine, this confirmed the inducible nature of the phenylalanine aminotransferase.

influence of amino donor

- The elimination of the amino donor, L-aspartic acid, resulted in an almost complete reduction of activity. Neither cell permeabilisation nor cofactor (pyridoxalphosphate) addition were essential for L-phenylalanine production. Maximum conversion yield occurred (100%, 22 g l^{-1}) when the amino donor concentration was increased. Aspartic acid was a superior amino donor to glutamic acid; 35 g l^{-1} was used.

influence of pH

- Maximal conversion was observed at 30-40°C and at alkaline pH. pH 11 resulted in very rapid initial conversions (12 g l^{-1} h^{-1}), which decreased to a rate of approximately 3 g l^{-1} h^{-1}. The high conversion rate at high alkaline pH illustrates the potential for biochemical manipulation of this strain. The solubility of L-phenylalanine is greatly increased at such high pH's and enzymes catalysing the degradation of L-phenylalanine and the formation of byproducts may be significantly inhibited at such high pH's. Furthermore, at this high pH the phenylpyruvic acid substrate will rapidly equilibrate between the keto and enol forms, so facilitating the future use of high concentrations of substrate with little inhibition of the reaction rate by the keto acid itself.

end product inhibition

- The final conversion yield decreased when substrate concentration was increased from 2% to 4%. This was attributed to end product inhibition by the L-phenylalanine produced. Thus although faster conversion rates were observed with addition of high substrate concentrations, the product titres never exceeded 16 g l^{-1}. As already discussed the rate of yield of the conversion was proportional to the concentration of amino donor employed. Using a ratio of 1:3 substrate to amino donor, almost a 90% conversion was achieved in 3 hours.

cell concentration

- The optimum cell concentration was between 10 and 20 g l^{-1}, which is not different from the concentrations employed in direct fermentation.

cell immobilisation

- A rapid activity loss (a few days) was observed with whole cells. Immobilisation increased the stability and continuous production of L-phenylalanine was possible using alginate bead immobilised cells of *P. fluorescens* for 60 days. However, to achieve this the cofactor pyridoxalphosphate had to be continuously added to the beads to correct for the dissociation of the cofactor and loss from the cells.

It would seem that immobilisation looks very promising, but remember that the costs of immobilisation and the addition of cofactor have to be compared with the cost of production by free cells.

The results obtained with the *P. fluorescens* strain without biochemical manipulation compare well with those reported for a *E. coli* strain. Both achieve volumetric rates of 3 g l^{-1} h^{-1} under normal conditions. So it appears that the efficiency of the process can be increased by a few simple operations: increasing the pH and amino donor concentration (aspartic acid).

mutant strains of *E. coli*

aspartate phenylalanine transaminase

A possible explanation for the superiority of the amino donor, L-aspartic acid, has come from studies carried out on mutants of *E. coli*, in which only one of the three transaminases that are found in *E. coli* are present. It is believed that a branched chain transaminase, an aromatic amino acid transaminase and an aspartate phenylalanine aspartase can be present in *E. coli*. The reaction of each of these mutants with different amino donors gave results which indicated that branched chain transminase and aromatic amino acid transminase containing mutants were not able to proceed to high levels of conversion of phenylpyruvic acid to L-phenylalanine. However, aspartate phenylalanine transaminase containing mutants were able to yield 98% conversion on 100 mmol l^{-1} phenylpyruvic acid. The explanation for this is probably that both branched chain transaminase and aromatic amino acid transminase are feedback inhibited by L-phenylalanine, whereas aspartate phenylalanine transaminase is not inhibited by L-phenylalanine. In addition, since oxaloacetate, which is produced when aspartic acid is used as the amino donor, is readily converted to pyruvic acid, no feedback inhibition involving oxaloacetate occurs. The reason for low conversion yield of some *E. coli* strains might be that these *E. coli* strains are deficient in the aspartate phenylalanine transaminase.

Application of the desired biotransformation to give a practical and economical process required high molar conversion yields, high amino transferase activities, highly efficient product recovery and an inexpensive source of phenylpyruvic acid. With genetic and/or biochemical manipulation considerable progress can be made towards meeting some of these requirements.

SAQ 8.9

Consider the production of L-phenylalanine using *P. fluorescens* by precursor (phenylpyruvic acid) addition. Explain briefly why:

1) Growth of cells in presence of D- or L-phenylalanine is desirable.

2) An alkaline pH is preferred.

3) Precursor concentration in the medium should be less than 4%.

4) Cell immobilisation is costly in comparison to the production of free cells.

8.7.2 Bioconversion of acetamidocinnamic acid

inducible enzymes

Recent progress published on the production of L-phenylalanine by bioconversion of acetamidocinnamic acid (ACA) mentions selection of special highly productive strains. Organisms were obtained by screening for growth on ACA as C and N source (which requires acylase activity) and testing for L-phenylalanine accumulation. The enzymatic activity for forming L-phenylalanine from ACA was considered to be inducible since this activity was not detected when ACA was omitted.

acylase

It was proven that the pathway of L-phenylalanine formation involved phenylpyruvic acid as intermediate and two steps could be distinguished (see Figure 8.6; section 8.7):

1) an 'acylase' enzyme that splits the acetyl group and gives rise to phenylpyruvic acid;

2) a reaction catalysed by a pyridoxal-dependent enzyme that binds the amino group and gives rise to L-phenylalanine formation (transaminase reaction).

aminotransferase

Two strains producing high L-phenylalanine levels were chosen for experiments: *Alcaligenes faecalis* and *Bacillus sphaericus*. Both strains showed equal production rates of 3.3 g l^{-1} d^{-1} at 30°C, pH 7 and 40 g l^{-1} biomass. It was observed, however, that it was difficult to elevate aminotransferase activity together with acylase activity. Further research was therefore conducted on two special strains, each performing a specific reaction step in the formation of L-phenylalanine from ACA. One strain with a high acylase activity and the other with a high aminotransferase activity. These strains were obtained by screening for aminotransferase and acylase activity, resulting in *B. sphaericus* as an acylase specialist and *Paracoccus denitrificans* as an aminotransferase specialist. Acylase activity in the cells of *B. sphaericus* was only induced in a medium with acetamidocinnamic acid.

specialist micro-organisms

To establish the most advantageous conditions for production of L-phenylalanine from acetamidocinnamic acid using two micro-organisms the following factors were investigated: pH, amino donor and ratio of two enzyme activities.

influence of pH

The optimum pH for both enzymes was 7.5-8.0. For practical production it is desirable to carry out a reaction in a single reactor at optimum pH. Therefore, it is most advantageous to carry our the two step enzyme reaction at pH 8.0.

amino donor

The best results were obtained with L-aspartic acid as the amino donor for *P. denitrificans* and phenylpyruvic acid as the amino acceptor. With L-aspartic acid, conversion of phenylpyruvic acid exceeded 90%. This may be attributed to absence of feedback inhibition of the reaction due to metabolism of the reaction product, oxaloacetic acid. When using glutamic acid the conversion of phenylpyruvic acid did not exceed 60%.

cell concentrations

Practical conversion from ACA to L-phenylalanine was best achieved when acylase and aminotransferase activities were equal. This could be achieved by using 10 gram *B. sphaericus* and 2.5 gram *P. denitrificans* per litre medium.

Under optimal conditions (pH = 8.0, 67 g l^{-1} L-aspartic acid, 30°C, 1:1 ratio of enzyme activities) after addition of pyridoxal phosphate, 76 g l^{-1} L-phenylalanine could be produced within 72 hours (92% conversion). This illustrates how simple biochemical manipulation can increase productivity dramatically.

coimmobilised cells

A cell immobilisation method for production of L-phenylalanine from ACA has also been investigated. Co-immobilised cells showed highest activity. This might be explained by the fact that phenylpyruvic acid produced by the acylase reaction was immediately converted to L-phenylalanine by the aminotransferase reaction without diffusion resistance of phenylpyruvic acid. A ratio of 2:1 acylase:aminotransferase was better for a stable L-phenylalanine forming activity as compared to a 1:1 ratio. This might be caused by the difference of stability between the two enzymes. The operational stability of L-phenylalanine forming activity increased with decreasing temperature, so the half-life time increased with decreasing temperature. The activity increased with increasing temperature. The productivity at temperatures higher than 30°C decreased because of low stability of L-phenylalanine forming activity. When the space velocity was 0.06 h^{-1} at 30°C and pH 8.0, 24 g l^{-1} L-phenylalanine was produced with a 98% conversion rate from ACA. A volumetric productivity of 1.5 g l^{-1} h^{-1} was achieved, 17% higher than with whole cells. The half life under these conditions of the L-phenylalanine activity of the co-immobilised cells was calculated to be 14 days.

8.7.3 Summary of process conditions for L-phenylalanine production

The most important process conditions for the production of L-phenylalanine by the three methods discussed in this chapter are summarised in Table 8.7.

	Ferm	PPA	ACA	Units
maximum productivity	0.6	3.5	1.5	g l^{-1} h^{-1}
time	24	8	60	h
pH	7	7.5	8	
temp	35	35	35	°C
biomass	20	10	40	g l^{-1}

Table 8.7 The most important process conditions for the production of L-phenylalanine by direct fermentation; precursor addition, phenylpyruvic acid (PPA); bioconversion, acetamidocinnamic acid (ACA).

As can be seen from Table 8.7 productivity (expressed in g l^{-1} h^{-1}) is highest for precursor addition. The production of L-phenylalanine from phenylpyruvic acid also has the shortest reaction time to obtain high conversions. The pH commonly used is around 7.5, quite normal for biological processes. Only the enzyme phenylalanine ammonia lyase shows an optimum pH of 10.The process temperature varies between 30 and 40°C with an average of 35°C. No extreme temperatures have been reported due to the fact that denaturation occurs at high temperatures. The optimal concentration for cells frequently used is 10-20 g l^{-1}. However, conversion of ACA is done with high cell mass concentrations in recent studies; possibly to compensate for substrate inhibition and thus to maintain high product concentration. The processes using PPA and ACA need an amino acid as amino donor, usually L-aspartic acid is used.

8.7.4 Cost of production by precursor addition and by bioconversion

Extra cost compared to direct fermentation are mainly concerned with addition of PPA (precursor addition) or ACA (bioconversion). Table 8.9 compares estimated costs for the PPA and ACA process based on the data considered previously (section 8.6.1 and 8.6.2). Costs are estimated for production of 100 tonnes per year.

	PPA $,thousands	ACA $, thousands
Bioreactor	38 ($3m^3$)	88
Personnel	225	225
Cultivation	328	102
Recovery	1000	1000
Utilities	59	41
Substrate - PPA	1050	- ACA 1087
Substrate - L-ASP	805	- L-ASP 479
Total	3505	3022
Cost price per kg		
L-phenylalanine	0.035	0.030

Table 8.8 Estimated costs of production of L-phenylalanine by enzymatic methods (100 tonnes per year). PPA = phenylpyruvic acid, ACA = acetamidocinnamic acid, L-ASP = L-aspartic acid.

substrate costs

We can see from Table 8.8 that substrate costs represent about half the total costs of the production. It follows that the best way of reducing costs is to use cheaper substrates. One possibility is to develop your own production processes for the chemicals. It is not possible to directly compare overall cost of fermentation with the PPA and ACA processes because published information on the processes are for different scales of production (1,000 tonnes per year for direct fermentation and 100 tonnes per year for PPA and ACA processes). However, we can say that cost savings on direct fermentation is possible when higher product yields are obtained. This will result in a lower substrate consumption and a smaller reactor volume, resulting in lower direct costs. Strain improvement is necessary to achieve significantly higher product yields in direct fermentation.

SAQ 8.10

Assume that the cost price of L-phenylalanine produced by direct fermentation is 28.5 $ kg^{-1} (100 tonnes per annum capacity). What percentage reduction in substrate costs are required for 1) precursor feeding and 2) biotransformation to be competitive on a cost price basis with direct fermentation?

Summary and objectives

In this chapter we have seen that amino acids can be produced as pure enantiomers by fermentation and by chemo-enzymatic processes. In direct fermentation of amino acids, mutant strains or DNA-recombinant micro-organisms are used to overcome feedback inhibition and repression by endproducts of metabolic pathways. Fermentation problems include those associated with back mutation of the process micro-organism, loss of genetic material and phage contamination. The problems can be minimised by altering the fermentation conditions. Several methods can be used to recover amino acids from fermentation broths. Production of L-phenylalanine by direct fermentation was considered in detail and we saw that carbon balance analysis can be used to characterise the fermentation. Economic evaluation of a bioprocess involves a consideration of investment costs, direct costs, variable, direct and plant gate costs. A key factor in industry is profitability regardless of technical achievement, which is based on the return on investment criterion. L-Phenylalanine can also be produced by various enzymatic processes, including precursor addition (phenylpyruvic acid) and bioconversion (acetamidocinnamic acid). Productivities of these processes are higher than that of direct fermentation. However, precursor addition and bioconversion processes have substantially greater costs than direct fermentation. Other factors that can substantially increase the cost of chemo-enzymatic processes include the possible requirement for immobilisation of the biocatalyst and for addition of cofactors.

Now that you have complete this chapter you should be able to:

- explain the importance of producing pure enantiomeric compounds;
- explain why auxotrophic mutants and regulatory mutants are often used for production of amino acids by fermentation;
- describe common problems associated with amino acid fermentations and outline strategies that can be used to overcome them;
- describe the production of L-phenylalanine by direct fermentation from glucose, and explain how carbon balance analysis can be used to characterise the process;
- calculate the return on investment for a process, based on cost analysis data;
- compare and contrast the production of L-phenylalanine by precursor feeding and by bioconversion.

Appendix 8.1: Representation of and the nomenclature for stereo-isomers

It is rather difficult to represent a tetrahedron which is three-dimensional with a drawing of a formula that is two-dimensional. Two types of representations are used, perspective and projection formulae. Fischer projections are widely used because of their simplicity.

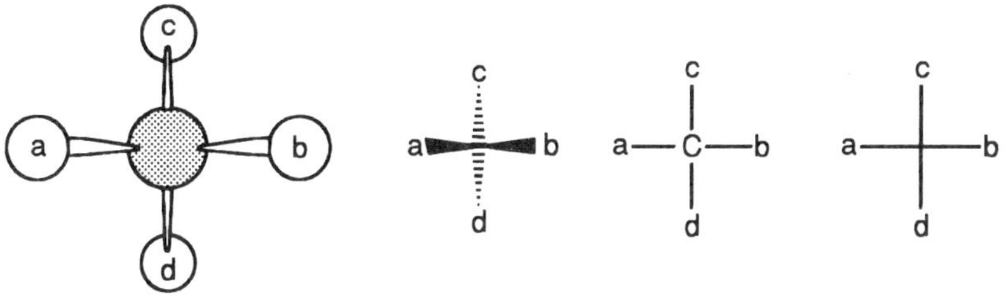

perspective formulae   projection formulae   Fischer projection

You should note that on a Fischer projection the two formulae

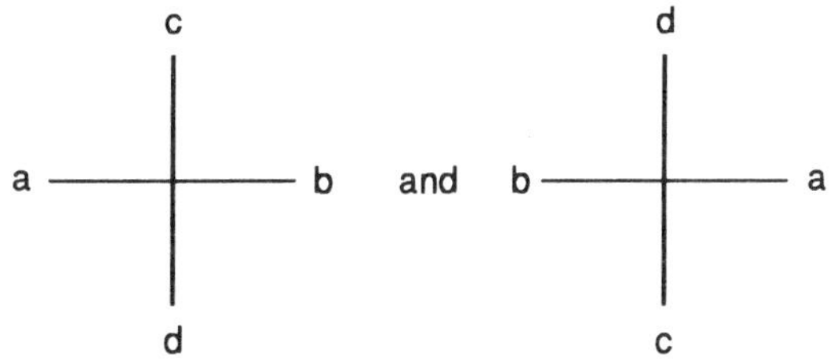

refer to the same molecule

Nomenclature for stereo-isomers

A system of nomenclature has been devised to describe optical isomers conveniently. These isomers differ in configuration and we have to be able to specify the configuration at the asymmetric atom unambiguously.

The original method had as a starting point the enantiomers of a standard compound, glyceraldehyde.

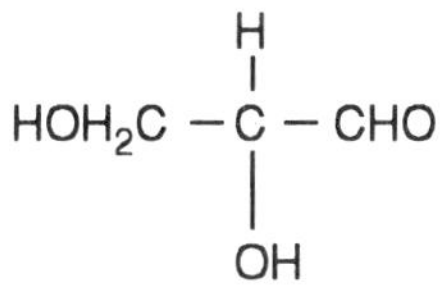

One enantiomer rotated polarised light to the right or clockwise, and it was referred to as (+)-glyceraldehyde. The other rotated polarised light to the left or counterclockwise and was referred to as (-)-glyceraldehyde. The two enantiomers of glyceraldehyde were, according to their Fischer projections called D and L, from the Latin word dexter (right) and laevus (left).

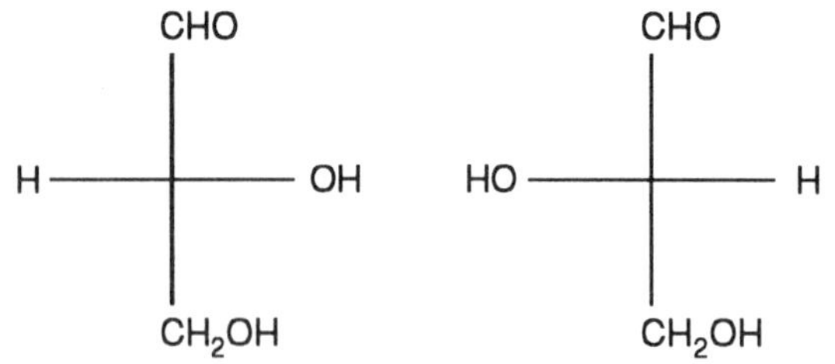

D-glyceraldehyde [(+)-glyceraldehyde]   L-glyceraldehyde [(-)-glyceraldehyde)]

The D-configuration is assigned to (+)-glyceraldehyde and the L-configuration is assigned to (-)-glyceraldehyde.

Compounds related to D-glyceraldehyde were said to have a D-configuration while those related to L-glyceraldehyde were said to have a L-configuration (as is the case with naturally occurring amino acids and carbohydrates). However, it is not always clear to which glyceraldehyde a given compound is 'related'. For example, although it is clear enough that the compounds

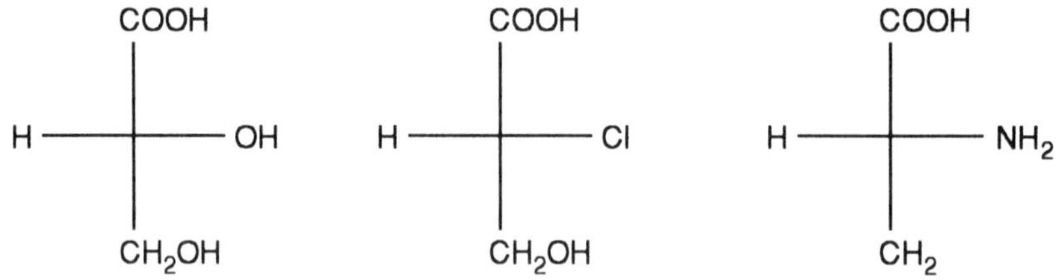

are D rather than L, it is not clear whether

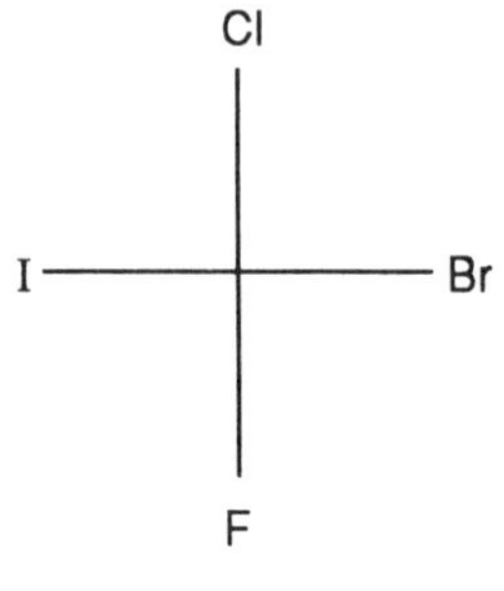

is D or L.

Because of the somewhat ambiguous definitions of D and L, another system of nomenclature was devised for asymmetric compounds, and it has largely replaced the old D, L-system since the 1960s. However, the D, L-nomenclature is still used for amino acids and sugars.

The RS designation of chirality

The absolute configuration of any chiral centre can be unambiguously specified using the RS notation.

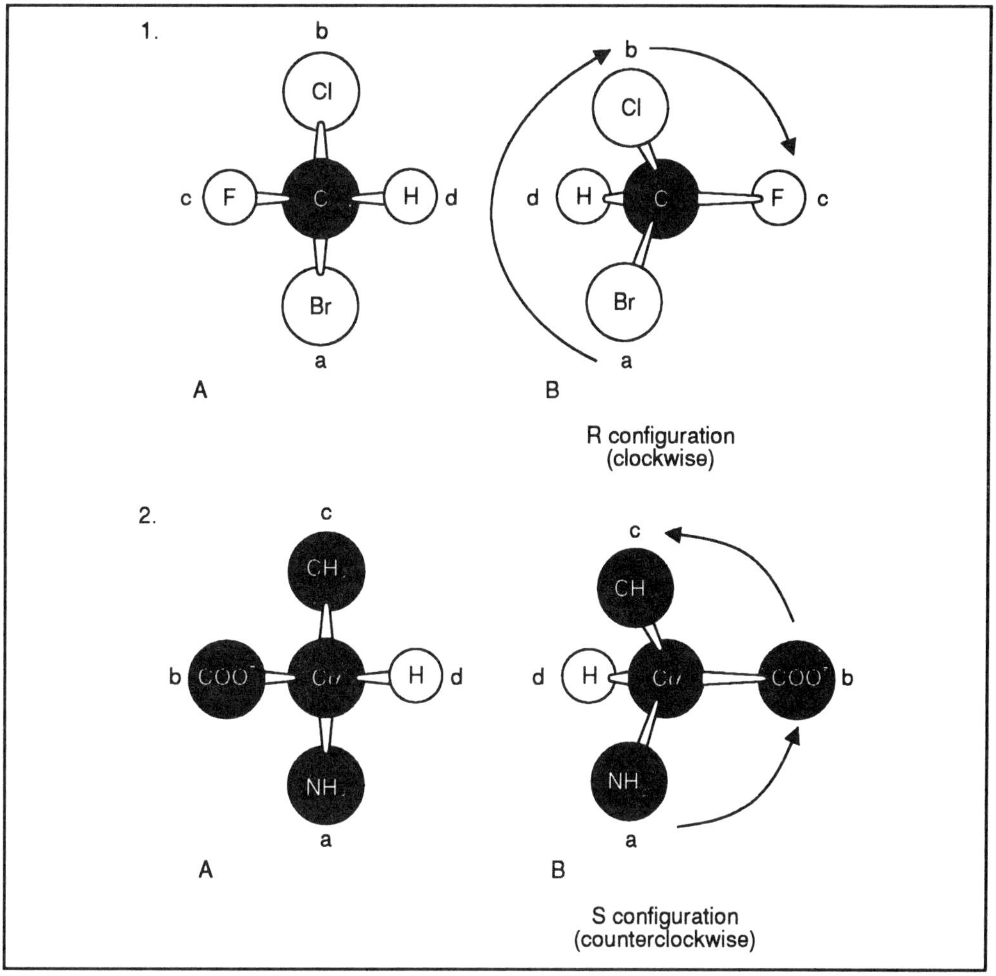

Figure A8.1 RS designation of chirality. 1) CHFClBr and 2) alanine.

The stereoisomer of CHFClBr shown in Figure A8.1 has the R configuration, whereas that of alanine has the S configuration.

The first step in using the RS designation is to assign a priority sequence to the four substituents by applying the rule that an atom with a higher atomic number has a higher priority than an atom with a lower atomic number. Hence, the priority sequence of these four substituents is a) Br, b) Cl, c) F, and d) H, with Br having the highest priority. The next step is to orient the molecule so that the group of lowest priority d) points away from the viewer. For CHFClBr, this means that the molecule should be oriented so that H is away from us (behind the plane of the page). We then ask whether the path from a to b to c under these conditions is clockwise or counterclockwise. If it is clockwise (right-handed), the configuration is R (from the Latin *rectus*, right). If it is counterclockwise (left-handed), the configuration is S (from the Latin *sinister*, left).

Now let us consider the RS designation of alanine (Table 8.2). The four atoms bonded to the α-carbon are N, C, C and H. The priority sequence of the methyl carbon and the carboxyl carbon is determined by going outward to the next set of atoms. It is useful to note the following priority sequence for biochemically important groups: -SH (highest), -OR, -OH, -NHR, $-NH_2$, -COOR, -COOH, -CHO, $-CH_2OH$, C_6H_5, CH_3, -T, -D, -H (lowest). Thus, the priority sequence of the four groups attached to the α-carbon of alanine is a) $-NH_3+$, b) $-COO^-$, c) $-CH_3$ and d) -H. The next step is to orient L-alanine so that its lowest priority group (-H) is behind the plane of the page. The path from a) $-NH_3+$ to b) $-COO^-$ to c) $-CH_3$ is then counterclockwise (left-handed) and so L-alanine has an S configuration.

Appendix 8.2 Examples of industrial production of amino acids by enzymatic methods

A8.2.1 Enzymatic resolution of racemates

Several classes of enzymes have been used to separate stereoisomers of α-H-and α-disubstituted amino acids, eg amidases, nitrilases, hydantoinases, acylases and esterases.

A8.2.1.1 Amidases

A very efficient and universal method has been developed for the production of optically pue L- and D-amino acids. The principle is based on the enantioselective hydrolysis of D,L-amino acid amides. The stable D,L-amino acid amides are efficiently prepared under mild reaction conditions starting from simple raw materials (Figure A8.2). Thus reaction of an aldehyde with hydrogen cyanide in ammonia (Strecker reaction) gives rise to the formation of the amino nitrile. The aminonitrile is converted in a high yield to the D,L-amino acid amide under alkaline conditions in the presence of a catalytic amount of acetone. The resolution step is accomplished with permeabilised whole cells of *Pseudomonas putida* ATCC 12633. A nearly 100% stereoselectivity in hydrolysing only the L-amino acid amide is combined with a very broad substrate specificity.

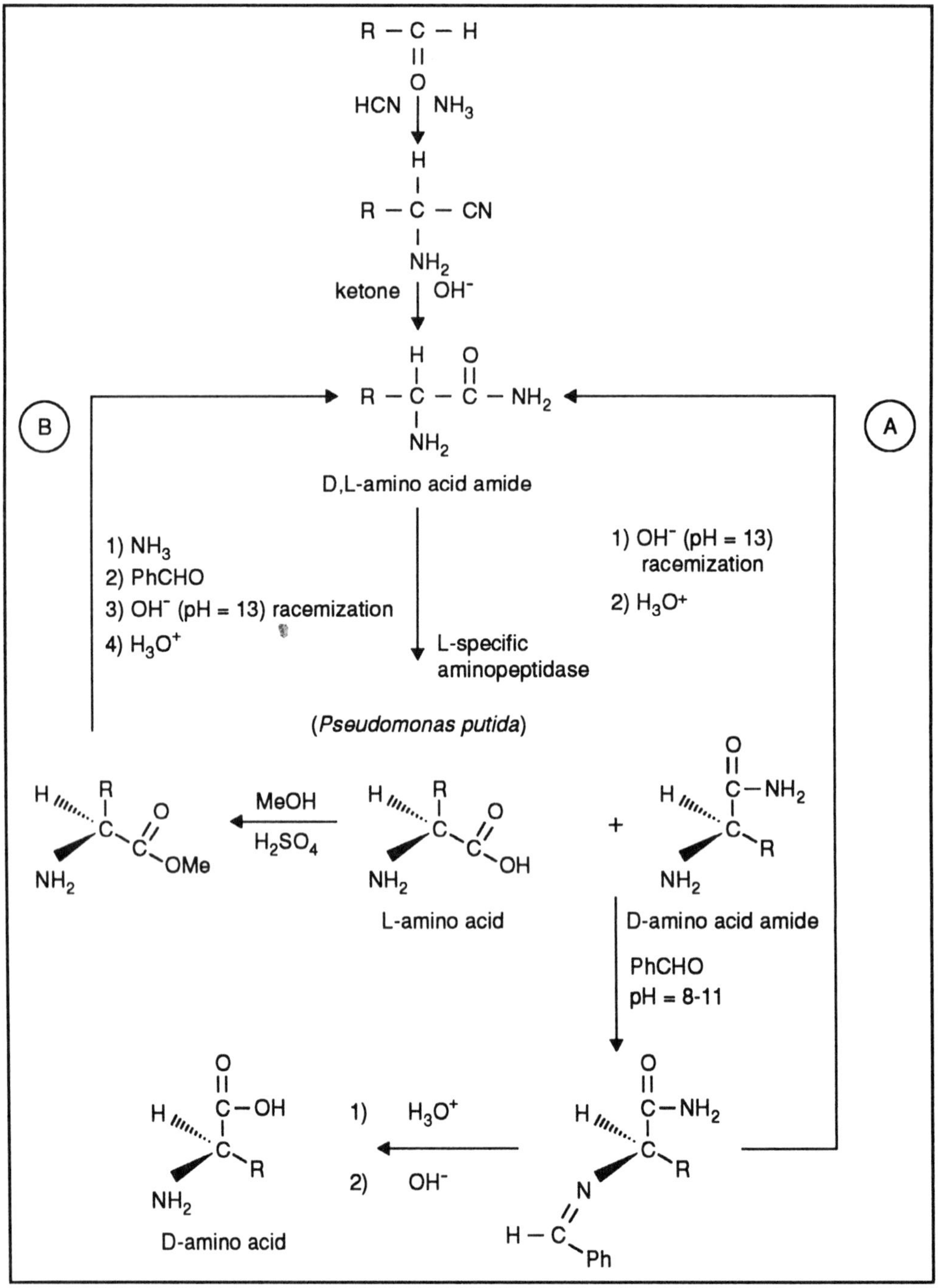

Figure A8.2 The production of optically pure L- and D-amino acid. (See text for further explanation).

Not only the smallest optically active amino acid (alanine), but also leucine, several (substituted) aromatic amino acids, heterosubstituted amino acids (methionine, homomethionine and thienylglycine) and even an iminoacid, proline, are obtainable in both the L- and D-form.

No enzymatic side effects are observed and substrate concentrations up to 20% by weight can be used without affecting the enzyme activity. The biocatalyst is used in soluble form in a batchwise process, thus poorly soluble amino acids can be resolved without technical difficulties. Re-use of the biocatalyst is in principle possible.

A very simple and elegant alternative to the use of ion-exchange columns or extraction to separate the mixture of D-amino acid amide and the L-amino acid has been elaborated. Addition of one equivalent of benzaldehyde (with respect to the D-amino acid amide) to the enzymic hydrolysate results in the formation of a Schiff base with the D-amino acid amide, which is insoluble in water and, therefore, can be easily separated. Acid hydrolysis (H_2SO_4, HX, HNO_3, etc.) results in the formation of the D-amino acid (without racemization). Alternatively the D-amino acid amide can be hydrolysed by cell-preparations of *Rhodococcus erythropolis*. This biocatalyst lacks stereoselectivity. This option is very useful for amino acids which are highly soluble in the neutralised reaction mixture obtained after acid hydrolysis of the amide.

Process economics dictate the recycling of the unwanted isomer. Path A in Figure A8.2 illustrates that racemisation of the D-N-benzylidene amino acid amide is facile and can be carried out under very mild reaction conditions. After removal of the benzaldehyde the D,L-amino acid amide can be recycled; 100% conversion to the L-amino acid is theoretically possible. Another method for racemisation and recycling of the L-amino acid (path B, Figure A8.2) comprises the conversion of the L-amino acid into the ester in the presence of concentrated acid, followed by addition of ammonia, resulting in the formation of the amide. Addition of benzaldehyde and racemisation by OH- (pH = 13) gives the D,L-amino acid amide. In this way 100% conversion to the D-amino acid is possible.

A8.2.1.2 Nitrilases

The bioconversion of α-aminonitriles, although up until now not used on an industrial scale, is of practical interest in the production of optical active α-amino acids. This, however, will only be the case if one can select a nitrilase that enantioselectively hydrolyses the aminonitrile.

As illustrated in Figure A8.3 nitrilases catalyse conversions of nitriles directly into the corresponding carboxylic acids (route A), while other nitrile converting enzymes, the nitrile hydratases, catalyse the conversion of nitriles into amides (route B) which, by the action of amidases usually present in the whole cell preparations, are readily transformed into carboxylic acids (route C).

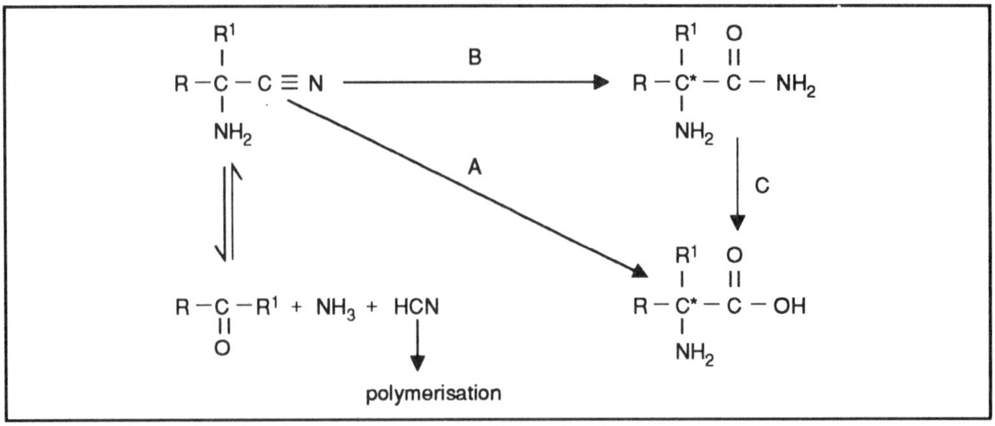

Figure A8.3 Preparation of optically active α-amino acids via bioconversion of the corresponding α-aminonitriles.

L-Amino acids could be produced from D,L-aminonitriles with 50% conversion using *Pseudomonas putida* and *Brevibacterium* sp respectively, the remainder being the corresponding D-amino acid amide. However, this does not prove the presence of a stereoselective nitrilase. It is more likely that the nitrile hydratase converts the D,L-nitrile into the D,L-amino acid amide, where upon a L-specific amidase converts the amide further into 50% L-amino acid and 50% D-amino acid amide. In this respect the method has no real advantage over the process of using a stereospecific L-aminopeptidase (*vide supra*).

A bacterial isolate APN has been shown to convert α-aminopropionitril enantioselectively to L-alanine (94% yield, 75% e e). However, the major disadvantage of this approach, is the low stability of most aminonitriles in water (for example α-aminophenylacetonitrile in water of pH 7, degrades completely within 48 hours). The aminonitriles are always in equilibrium with the aldehyde or ketone and ammonia/HCN. Polymerisation of hydrogen cyanide gives an equilibrium shift resulting in the loss of the aminonitrile. Therefore, a low yield in amino acids is to be expected, which makes this method less attractive for the industrial synthesis of optically active amino acids.

A8.2.1.3 Acylases

General

Several L-amino acids are produced on a large scale by enzymatic resolution of N-acetyl-D,L-amino acids (Figure A8.4). Acylase immobilised on DEAE-Sephadex is for example employed in a continuous process while Degussa uses the free acylase retained in a membrane reactor. In the latter process the advantage of reuse of the enzyme and homogeneous catalysis are combined.

However, the products are separated using ion-exchange columns and the starting material is a derivative rather than a precursor of the racemic amino acid, thus making the total process circuitous since it involves several chemical steps in addition to the enzymatic resolution step. Furthermore, racemisation of the unwanted isomer is not easily accomplished.

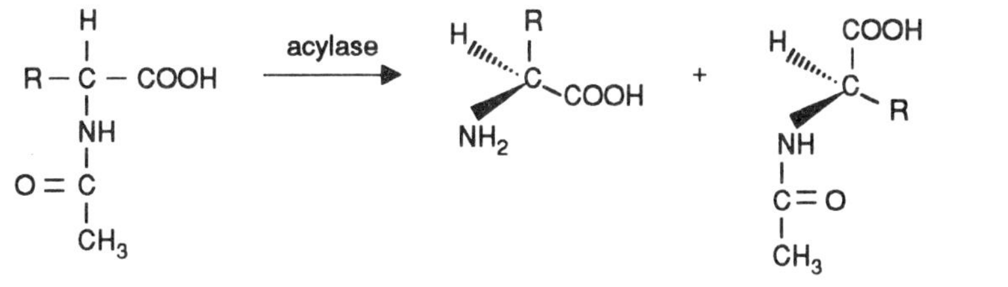

Figure A8.4 Commercial process for the enzymatic production of L- and D-amino acids from N-acetyl-D,L-amino acids.

Also the scope of the acylase method is limited in practice because chemical acylation of amino acids is difficult, and some N-acylamino acids are unsusceptible to the enzyme.

Continuous process of amino acid resolution using immobilised acylase

The Japanese firm Tanabe Inc Ltd has been operating, since 1969, the optical resolution of DL-amino acids using aminoacylase. The principle is based on the asymmetrical hydrolysis of N-acyl-DL-amino acid by amino acylase which gives the L-amino acid and the unhydrolysed acyl-D-amino acid.

To develop a continuous process, the immobilisation of aminoacylase of *Aspergillus oryzae* by a variety of methods was studied, for example ionic binding to DEAE-Sephadex, covalent binding to iodo-acetyl cellulose and entrapment in polyacrylamide gel. Ionic binding to DEAE-Sephadex was chosen because the method of preparation was easy, activity was high and stable, and regeneration was possible.

The flow diagram of the enzyme reactor for continuous production of the L-amino acid is given in Figure A8.5. The acetyl amino acid is continuously charged into the enzyme column through a filter and a heat exchanger. The effluent is concentrated and the L-amino acid is crystallised. The acyl-D-amino acid contained in the mother liquor is racemised by heating in a racemisation tank, and reused.

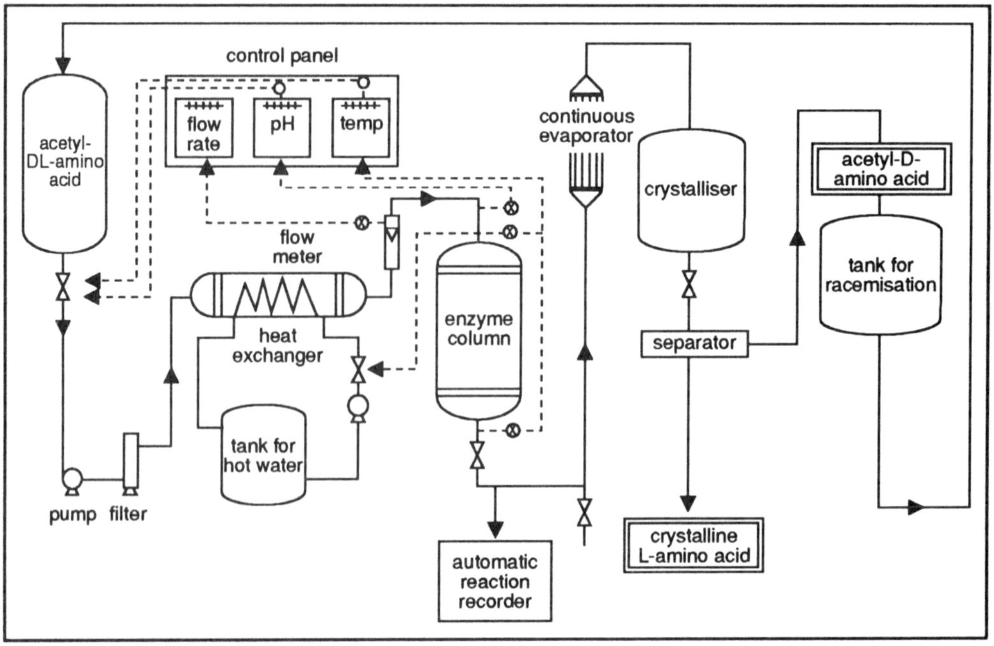

Figure A8.5 Flow diagram for the continuous production of L-amino acids by immobilised aminoacylase.

The advantages of the continuous process over the batch process are:

- saving of enzymes;
- less human labour due to automation;
- increase in reaction yield, due to the easy isolation of L-amino acid from the reaction mixture.

Due to these advantages the overall production costs for the immobilised continuous process were found to be 40% lower than that of the batch process. In Figure A8.6 a comparison is given between the batch process costs and the continuous production costs.

The plant is mainly used for L-methionine, L-valine, L-phenylalanine. The German firm Degussa AG

Degussa AG uses immobilised acylase to produce a variety of L-amino acids, for example L-methionine (80, 000 tonnes per annum). The principles of the process are the same as those of the Tanabe-process, described above. Degussa uses a new type of reactor, an enzyme membrane reactor, on a pilot plant scale to produce L-methionine, L-phenylalanine and L-valine in an amount of 200 tonnes per annum.

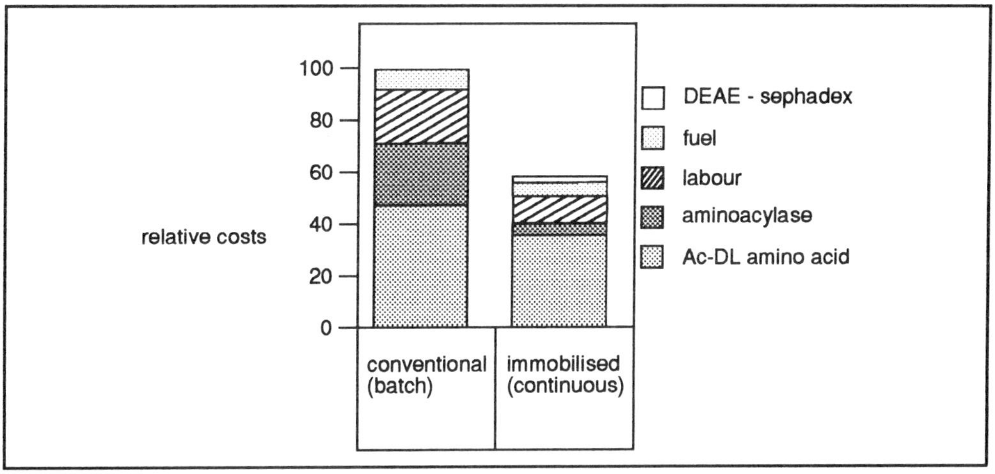

Figure A8.6 Economic comparison of batch and continuous production of L-amino acids using amino acylases.

α-Disubsituted amino acids are not easily prepared using acylases. Hog renal acylases I is unable to catalyse the hydrolysis of N-acetyl-α-methylphenylalanine. In contrast, carboxypeptidase A, is able to hydrolyse stereoselectively N-trifluoroacetyl-α-methylamino acids (Figure A8.7). Optically pure L-α-methylvaline and L-α-methylphenylalanine can be obtained using this method, but the use of an expensive trifluoroacetyl group makes it rather unattractive.

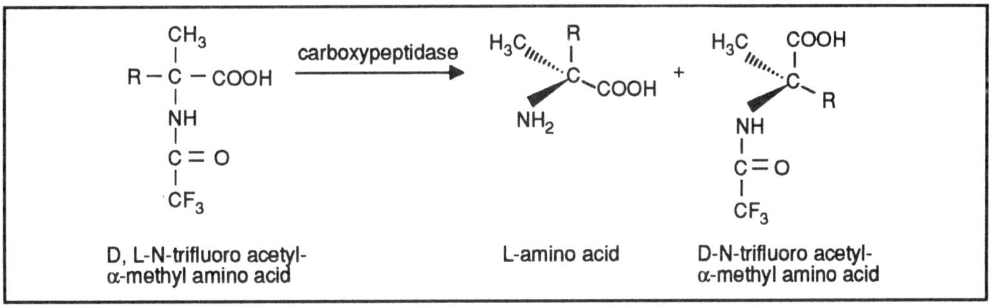

Figure A8.7 Preparation of optically pure α-methyl amino acids using carboxypeptidase A.

A8.2.1.4 Hydantoinases

Another option for resolution of optically active amino acids is illustrated in Figure A8.8.

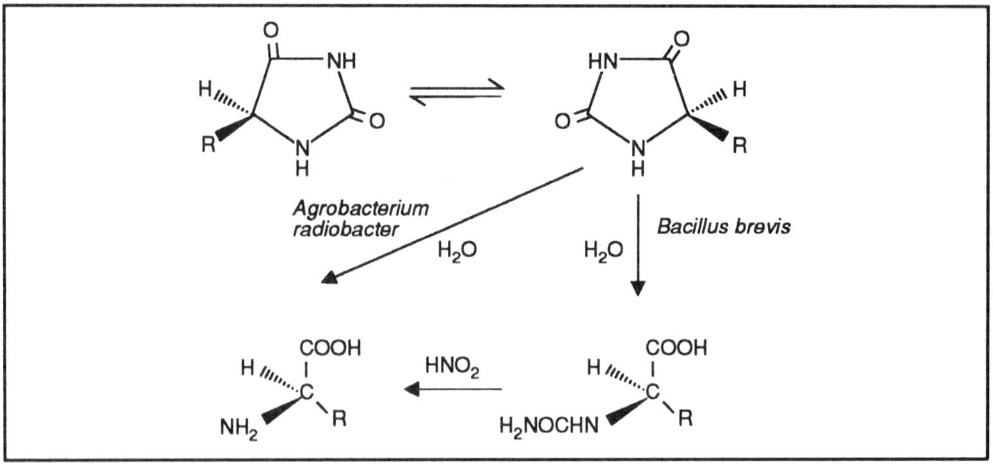

Figure A8.8 Enzymatic processes for the production of optically active α-amino acids via resolution of the racemic hydantoins.

Chemically synthesised D,L-hydantoins prepared from the corresponding aldehydes via the Bucherer Berg reaction are converted by the bacterial cells (*Bacillus brevis*), containing a D-specific hydantoinase, to a mixture of D-N-carbamoyl amino acid and L-hydantoin. The latter compound undergoes rapid and spontaneous racemisation under the conditions of the reaction, therefore, in principle 100% of the hydantoin is converted into the D-N-carbamoyl compound. The D-amino acid is obtained after treatment of the D-N-carbamoyl compound with nitrous acid. This process is operated on an industrial scale by the Japanese firm Kanegafuchi.

An even more elegant approach for the production of D-phydroxyphenylglycine on an industrial scale uses the bacterium. *Agrobacterium radiobacter* (Figure A8.8). The organism is able to produce both D-hydantoinase and a second enzyme, N-carbamoyl-D-amino acid aminohydrolase, which catalyse the hydrolysis of N-carbamoyl-D-amino acid.

The stereoselective hydrolysis of D,L-N-carbamoyl-methionine by the use of micro-organisms of the type *Agrobacterium rhizogenes* IFO 13259, *Corynebacterium sepedonicum* IFO 3306, and *Mycobacterium smegmatics* ATCC 607, producing L-methionine and D-N-carbamoyl-methionine, has also been described.

The stereoselective cyclisation of N-carbamoyl-α-disubstituted amino acids into the corresponding D-hydantoins is possible using bacteria belonging to the genera *Aerobacter*, *Agrobacterium*, *Bacillus* etc. D- and L-α-disubsituted amino acids can be prepared from the D-hydantoins and the remaining L-N-carbamoyl compounds, respectively (Figure A8.9).

Although this route seems quite elegant, there are some disadvantages. From the point of economic feasibility, it would be more attractive to start with a substrate that is a precursor of the optically pure α-disubstituted amino acid. In the method described, first the D,L-amino acid has to be prepared and protected. After the enzymatic step deprotection is necessary to obtain the D-amino acid and the total process is circuitous. Furthermore, the unwanted isomer cannot be racemised.

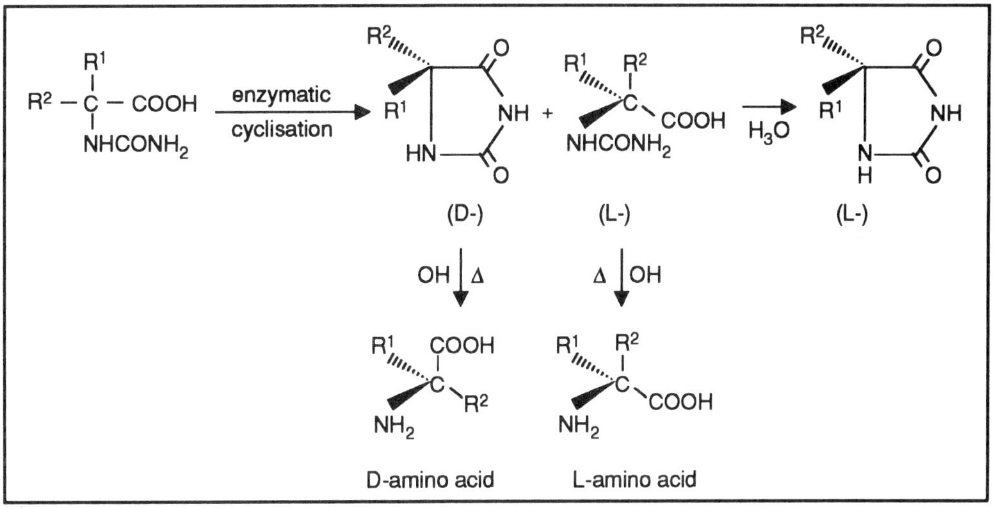

Figure A8.9 Procedure for the preparation of optically active α-disubstituted amino acids through stereoselective enzymatic cyclisation of the N-carbamoyl derivatives.

A8.2.1.5 Esterases

Alcalase selectively catalyses the hydrolysis of D,L-amino acid methyl and benzyl esters to provide L-amino acids and D-amino acid esters with high optical purity (Figure A8.10).

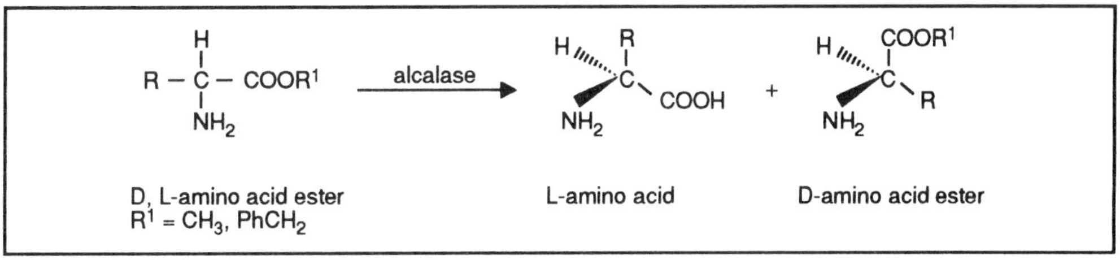

Figure A8.10 Enzymatic resolution of D,L-amino acid esters.

Microbial serine proteases, such as chymotrypsin, catalyse the hydrolysis of N-acetyl-L-amino acid esters (Figure A8.11).

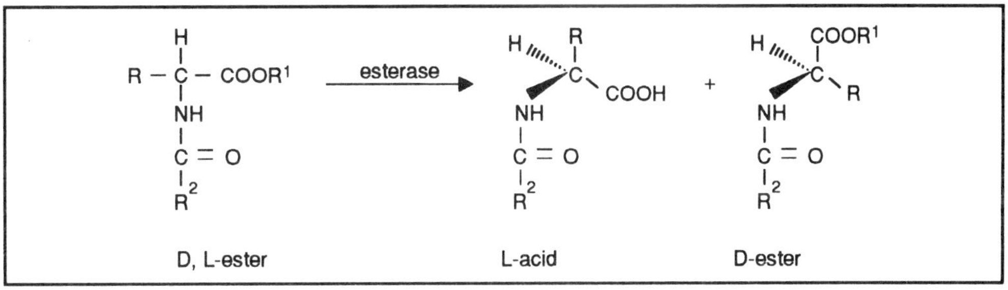

Figure A8.11 Enzymatic resolution of N-acetyl-D,L-amino acid esters.

Numerous other examples of enantioselective hydrolysis of esters have been reported. For example chymotrypsin, immobilised in a liquid membrane of kerosene or cyclohexane, can be used for resolution of D,L-amino acid esters in an emulsion type reactor. This emulsion-type enzyme membrane reactor is used principally for continuous racemate resolution.

These are major disadvantage of the esterase resolution process. Since the optimum pH of the enzymic reaction is generally on the alkaline side, the esters used as substrates are non-enzymatically hydrolysed and the optical purity of the L-amino acids obtained is generally low. Also the substrate has to be protected at the amino group in most cases in order to prevent formation of diketopiperasines. The esterase method is not attractive in practice and to the best of our knowledge is not used on an industrial scale.

A8.2.2 Enzymatic asymmetric synthesis

A8.2.2.1 Ammonialyases

L-Phenylalanine can be synthesised from *trans*-cinnamic acid (Figure A8.12) catalysed by a L-phenylalanine ammonia-lyase from *Rhodococcus glutinis*. The commercialisation of the process was limited by the low conversion (70%), low stability of the biocatalyst and the severe inhibition exerted by *trans*-cinnamic acid. These problems were largely overcome by researchers at Genex. The process, commercialised for a short period by Genex, involves a cell-free preparation of phenylalanine-ammonia-lyase activity from *Rhodotorula rubra*.

Of industrial importance at present is the biotransformation of fumarate to L-aspartic acid by *Escherichia coli* aspartase. Modified versions have been developed, such as the continuous production of L-aspartic acid using duolite-ADS-aspartase. A conversion higher than 99% during 3 months on a production scale has been achieved.

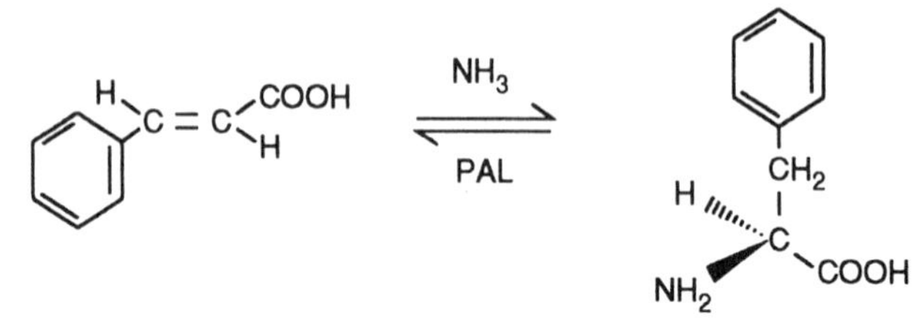

Figure A8.12 Enzymatic conversion of *trans*-cinnamic acid to L-phenylalanine.

L-alanine can be prepared from aspartic acid (Figure A8.13). L-Aspartate-β-decarboxylase produced by *Xanthomonas oryzae* No 531 has been used to prepared L-alanine in 95% yield from 15% L-aspartic acid solution. Other strains, ie *Pseudomonas dacunhae* or *Achromobacter pestifer*, give comparable yields of L-alanine. The process has been commercialised by Tanabe.

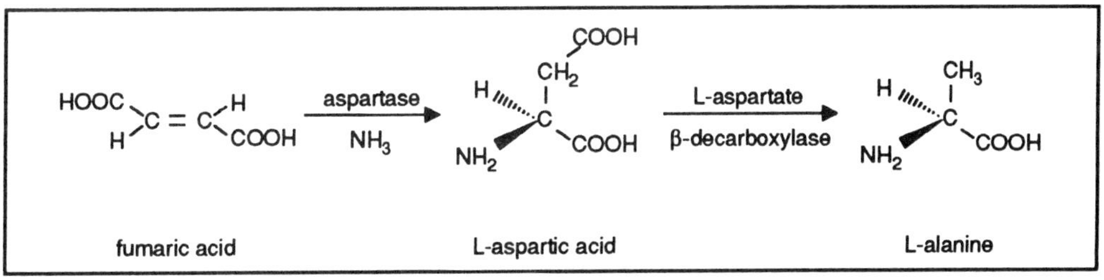

Figure A8.13 Two step enzymatic synthesis of L-alanine from fumaric acid.

A8.2.2.2 Industrially operated enzymatic synthesis

Enzymatic synthesis of L-aspartic acid and L-alanine

L-aspartic acid has been produced on an industrial scale by the Tanabe Seiyaku Co Ltd, Japan, in a batchwise process using whole cells of *Escherichia coli* with high aspartase activity. In this process, L-aspartic acid is produced from fumaric acid and ammonia using aspartase, as described in Figure A8.13.

To develop a continuous process, the immobilisation of the aspartase was studied, but the operational stability of the preparations was not sufficient for industrially production. To overcome these disadvantages the *E. coli* cells were directly immobilised by entrapping them in polyacrylamide gel. For the industrial application, a continuous aspartase reactor system, using a column packed with the immobilised *E. coli* cells, was designed. The column used for the industrial production was designed as a multi-stage system with heat exchanger, because the reaction is exothermic. The half life of the conventional polyacrylamide gel immobilised *E. coli* was 120 days at 37°C. *E. coli* cells immobilised with X-carrageenan showed a half life time of 680 days at 37°C. From 1978 the carrageenan method has been used to immobilise the *E. coli.*

L-alanine can be produced from L-aspartic acid using L-aspartate-4-decarboxylase. This reaction has been described in (Figure A8.13). L-alanine has been produced since 1965 by a batchwise enzyme reaction, using the L-aspartase-β-decarboxylase of *Pseudomonas dacunhae*. Again a continuous process was developed using *Pseudomonas dacunhae* immobilised with x-carrageenan. One of the problems in the continuous production is the evolution of carbon dioxide during the reaction. This carbon dioxide gas makes it difficult to obtain plug flow conditions in the reactor. Therefore, a closed column was designed, which allows the enzyme reaction to perform at high pressure (about 4 bar). The flow diagram of the closed column reactor is shown in Figure A8.14. The carbon dioxide remains dissolved in the reaction mixture, obtaining almost plug flow conditions. The efficiency of the immobilised cells under high pressure is 42% higher than the conventional column system.

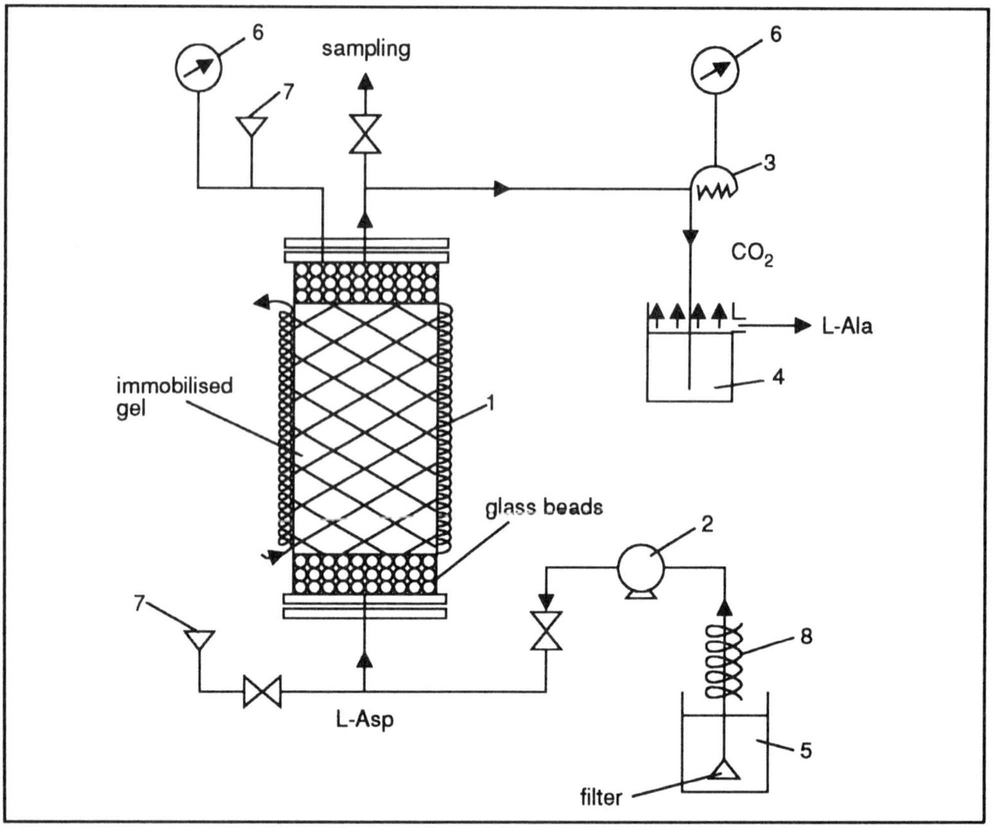

Figure A8.14 Closed column reactor for the production of L-alanine. 1) reactor; 2) plunger pump; 3) pressure control valve; 4) reservoir; 5) substrate tank; 6) pressure gauge; 7) safety valve; 8) heat exchanger.

Aspartic acid may be produced continuously from ammonium fumarate using immobilised *E. coli*. L-alanine is continuously produced using immobilised *Ps. dacunhae*. If both *E. coli* and *P. dacunhae* are employed simultaneously, L-alanine could be produced more efficiently from ammonium fumarate according to the reaction shown in Figure A8.13.

To prevent the formation of byproducts like L-malic acid and D-alanine, the cells undergo a pH-treatment to inactive fumarase and alanine racemase. Several reactor conformations have been investigated, but a two reactor system was found to be the most effective. The flow sheet of this two reactor system is given in Figure A8.15. In the first reactor L-aspartic acid is formed, which reacts in the second reactor to L-alanine.

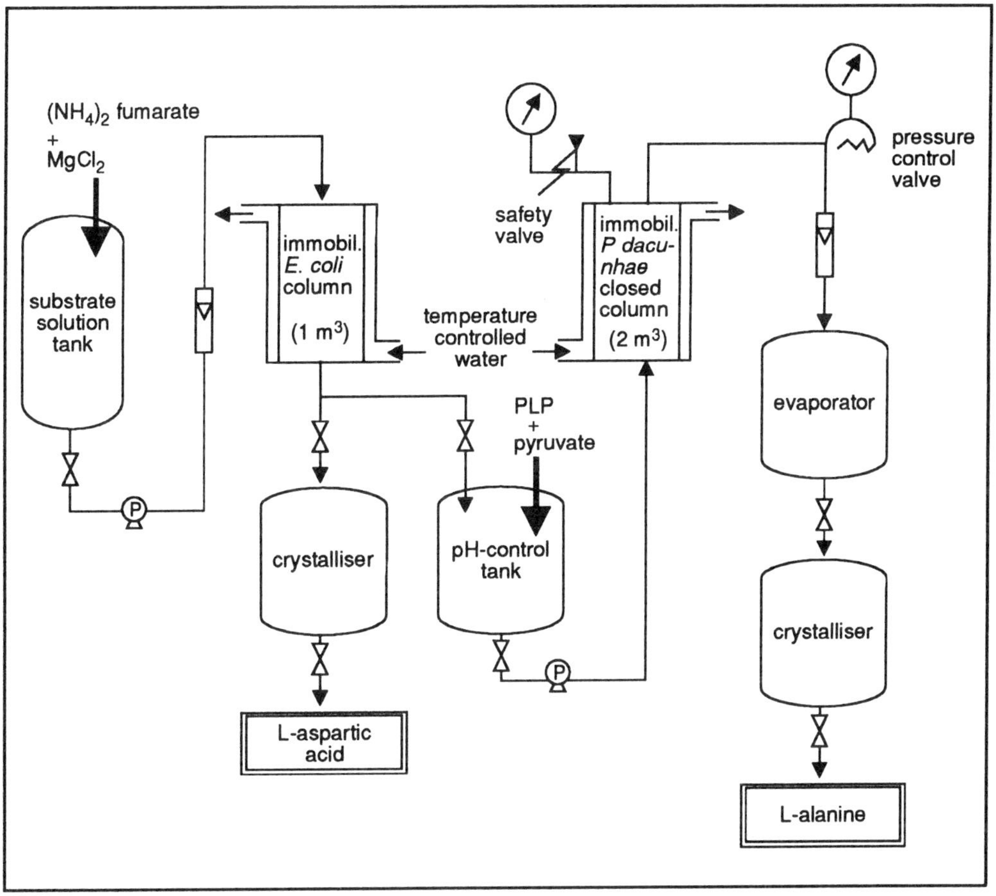

Figure A8.15 Flow diagram for the continuous production of L-aspartic acid and L-alanine.

A8.2.3 Oxidation - Reduction reactions

A8.2.3.1 Aminotransferases

A new development is the industrial production of L-phenylalanine by converting phenylpyruvic acid with pyridoxalphosphate-dependent phenylalanine transaminase (see Figure A8.16). The biotransformation step is complicated by an unfavourable equilibrium and the need for an amino-donor (aspartic acid). For a complete conversion of phenylpyruvic acid, oxaloacetic acid (deamination product of aspartic acid) is decarboxylated enzymatically or chemically to pyruvic acid. The use of immobilised *E. coli* (covalent attachment and entrapment of whole cells with polyazetidine) is preferred in this process (Figure A8.17).

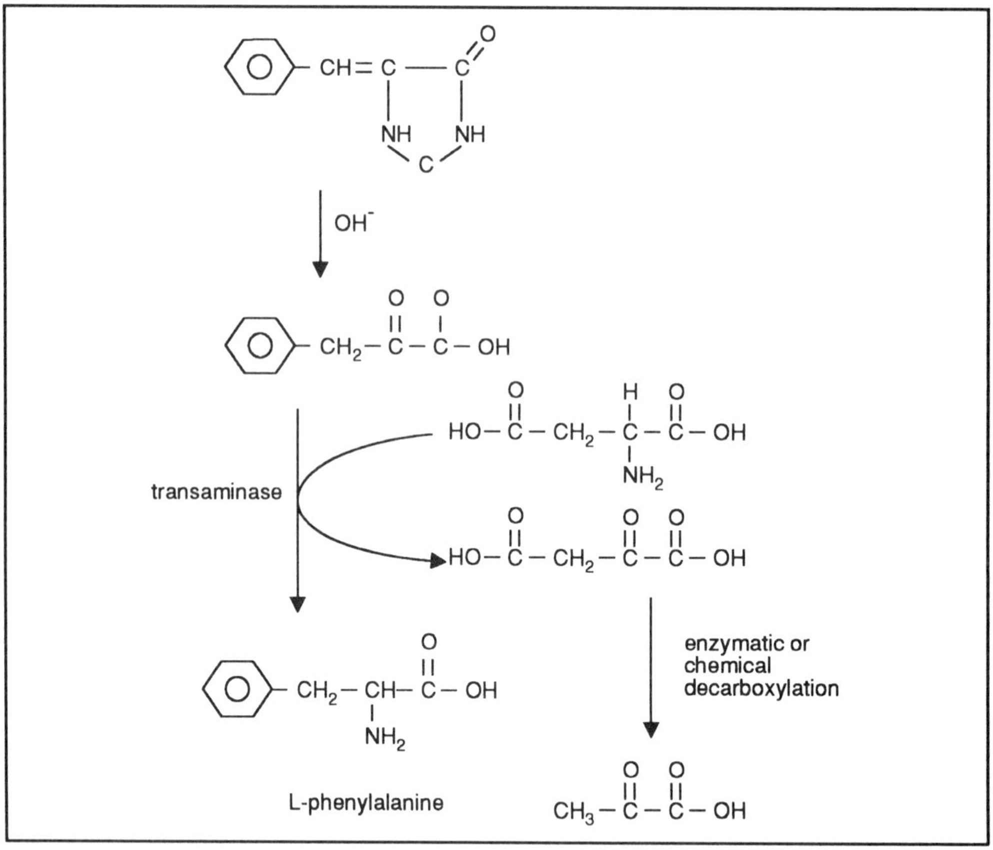

Figure A8.16 Production of L-phenylalanine via phenylalanine transaminase.

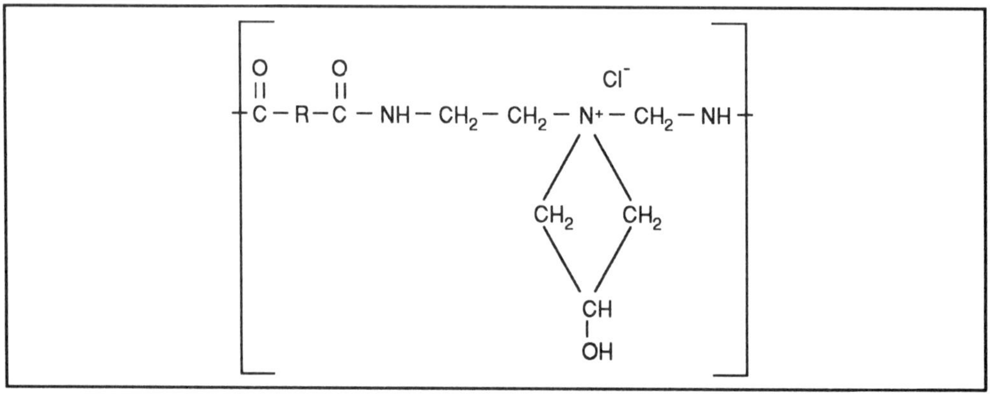

Figure A8.17 Idealised structure of a representative polyazetidine used for the immobilisation of whole cells.

Polyazetidine prepolymer may be cross-linked in aqueous solution by reaction with amine, thiol, hydroxyl, carboxylic acid or other polyazetidine groups. Cross-linking occurs upon water removal, heating or by changing to a basic pH. The immobilised cell/polymer composition may be prepared in the form of membranes, fibres, tubes or beads.

The industrial development of biotransformations is hampered currently by lack of commercial availability of biocatalysts at a reasonable price, insufficient operational stability of most biocatalysts and the practical problems associated with the exploitation of cofactor-dependent biocatalysts.

Many procedures have been suggested to achieve efficient cofactor recycling, including enzymatic and non-enzymatic methods. However, the practical problems associated with the commercial application of coenzyme dependent biocatalysts have not yet been generally solved. Figure A8.18 illustrates the continuous production of L-amino acids in a multi-enzyme-membrane-reactor, where the enzymes together with NAD^+ covalently bound to water soluble polyethylene glycol 20,000 (PEG-20,000-NAD) are retained by means of an ultrafiltration membrane.

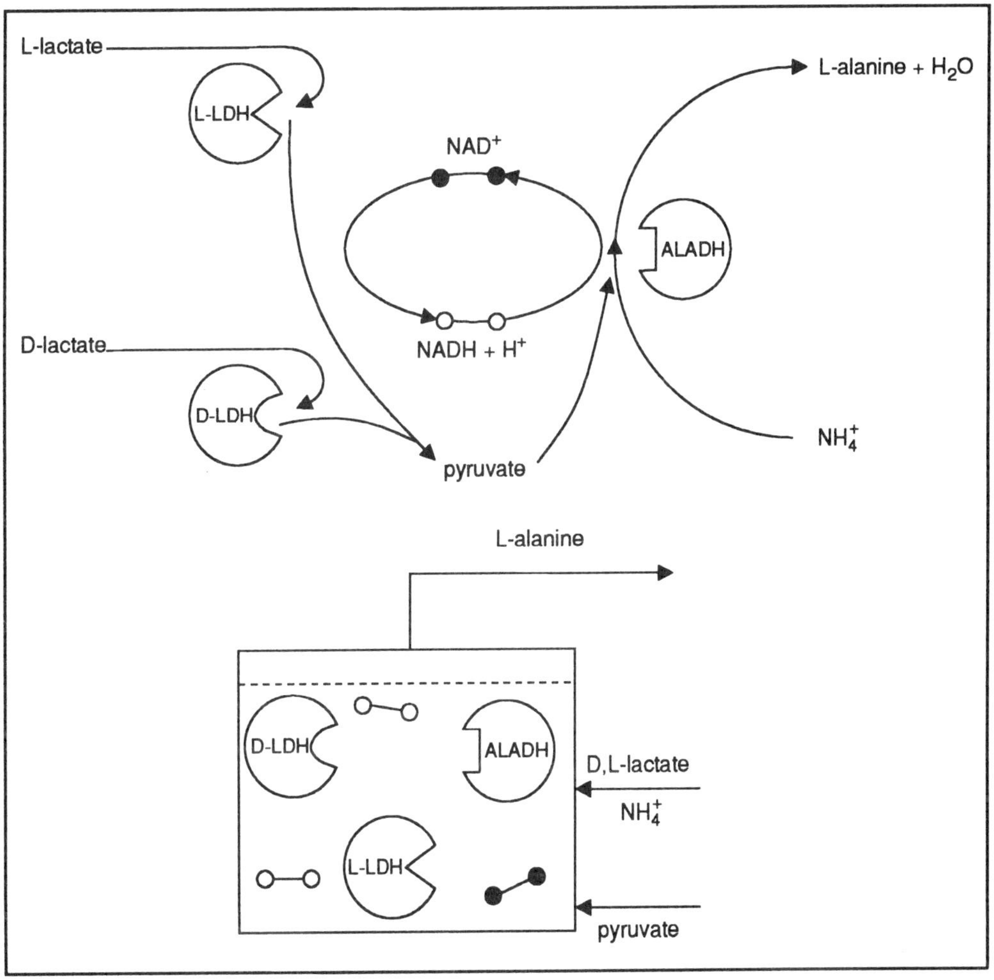

Figure A8.18 A racemic mixture of α-hydroxyacids (like L, D-lactate) can be transformed via the corresponding α-ketoacid (pyruvate) to the desired L-amino acid (L-alanine) with cofactor recycling.

So far the economic feasibility of co-enzyme dependent biocatalyses is confined to relatively small market niches comprising products with high added value.

Biotransformation of lipids

Biotransformation of lipids

9.1 Introduction

lipids include a wide variety of molecular types

In this chapter we will examine how cells and enzymes are used in the transformation of lipids. The lipids are, of course, a very diverse and complex series of molecular entities including fatty acids, triglycerides, phospholipids, glycolipids, aliphatic alcohols, waxes, terpenes and steroids. It is usual to teach about these molecules, in a biochemical context, in more or less the order given above, since this represents a logical sequence leading from simple molecules to the more complex. Here, however, we have adopted a different strategy.

Clearly, the technical and commercial aspects of industrial lipid transformation are both diverse and complex. We have, therefore, to be selective in what we can include in a single chapter. We have decided to begin the chapter by discussing the transformation of sterols and steroids since these transformations are illustrative of the potency of biocatalysts in bringing about selective and stereospecific chemical transformation of quite complex molecules. This part of this chapter is a logical extension of the issues discussed in the previous chapter. We have also elected to focus on the technical rather than the commercial issues although attention is drawn to the importance of commercial criteria in the selection of strategies for production. The reader should be aware, however, that the production of steroids for pharmaceutical use and as contraceptives is a large market. Estimates of annual sales of these materials vary widely (WHO 1990 Year Book $\$3x10^8$ annum^{-1} - $\$10^{10}$ annum^{-1}) reflecting the difficulty of accessing details on such a diverse group of compounds.

After discussing the biological capability to transform steroids, we briefly examine the biotransformation of other terpenoids to ensure that the reader develops an awareness of the potential of biotechnology to modify or produce derivatives of a wide range of natural materials that are of tremendous potential, commercial value in the food and health care sectors. We also include a brief consideration of the use of biocatalysts to transform a range of other hydrocarbon compounds.

The bulk of the chapter is therefore concerned with highly specific reactions arising to produce molecules of known structure. However, in a chapter on lipid transformation, we should not miss the use of general lipases to change the composition of triglycerides. Although the replacement of one set of fatty acids in trig;ycerides is, from a chemical/biochemical point of view, not as stimulating as the biotransformation of steroids, it is of major commercial value in the food industry. The biotransformation of triglycerides (and phospholipids) to produce food materials which have desirable organoleptic properties (eg melt in the mouth feel) potentially dwarfs the steroid market in terms of volume and turnover. The placing of this topic towards the end of the chapter does not imply that it is of limited commercial value.

The final part of the chapter briefly explores the potential of using non-aqueous solvents for lipid transformation.

Despite the technical emphasis of this chapter, we have included some examples of industrial processes.

9.2 The structure, roles and abundance of sterols and steroids

steroidal ring structure

Some examples of sterols and steroids are given in Figure 9.1. Also included in this Figure are some examples of bile salts. You should realise that the structures shown are only a few of the many hundreds of compounds which occur in nature. All of these compounds include the steroidal ring structure which is numbered as shown below.

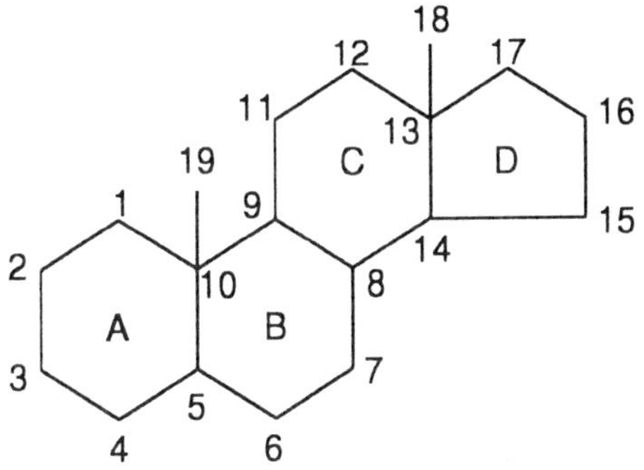

Substituents are designated as in the α configuration if they are below the plane of the steroidal nucleus, and as β if above the plane:

Thus

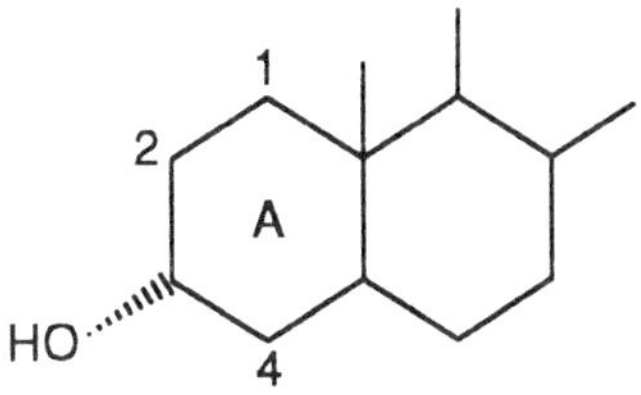

is 3α hydroxy -

whilst

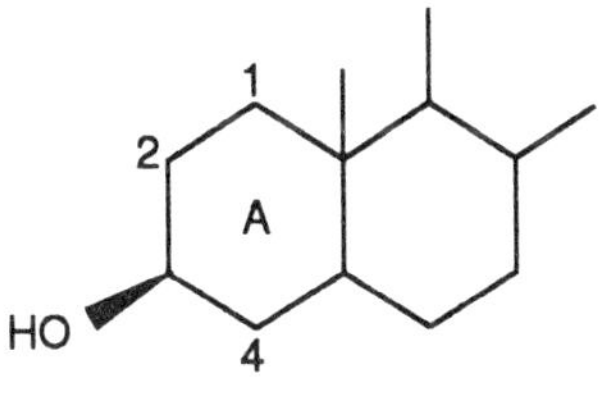

is 3β hydroxy -

Π Examine Figure 9.1 and sec if you can distinguish between the roles of the sterols (at the top of the figure) and the C_{18}, C_{19}, C_{21} and C_{24} steroidal compounds.

In general, the sterols perform a structural function, for example as components of the lipid layers of membranes. The C_{18}, C_{19} and C_{21} steroids mainly perform an endocrine function. In other words they are hormones. The bile salts (C_{24}-steroids) fulfil a functional role in digestion in animals.

Π Bearing this in mind, which of the three groups are likely to occur a) in greatest amounts, b) in lowest concentration in biological systems?

You should have predicted that the sterols are present in greatest quantity in biological systems. Your knowledge of biology should have enabled you to identify the steroid hormones as being present in lowest concentrations since hormone, in general, are effective at very low concentrations.

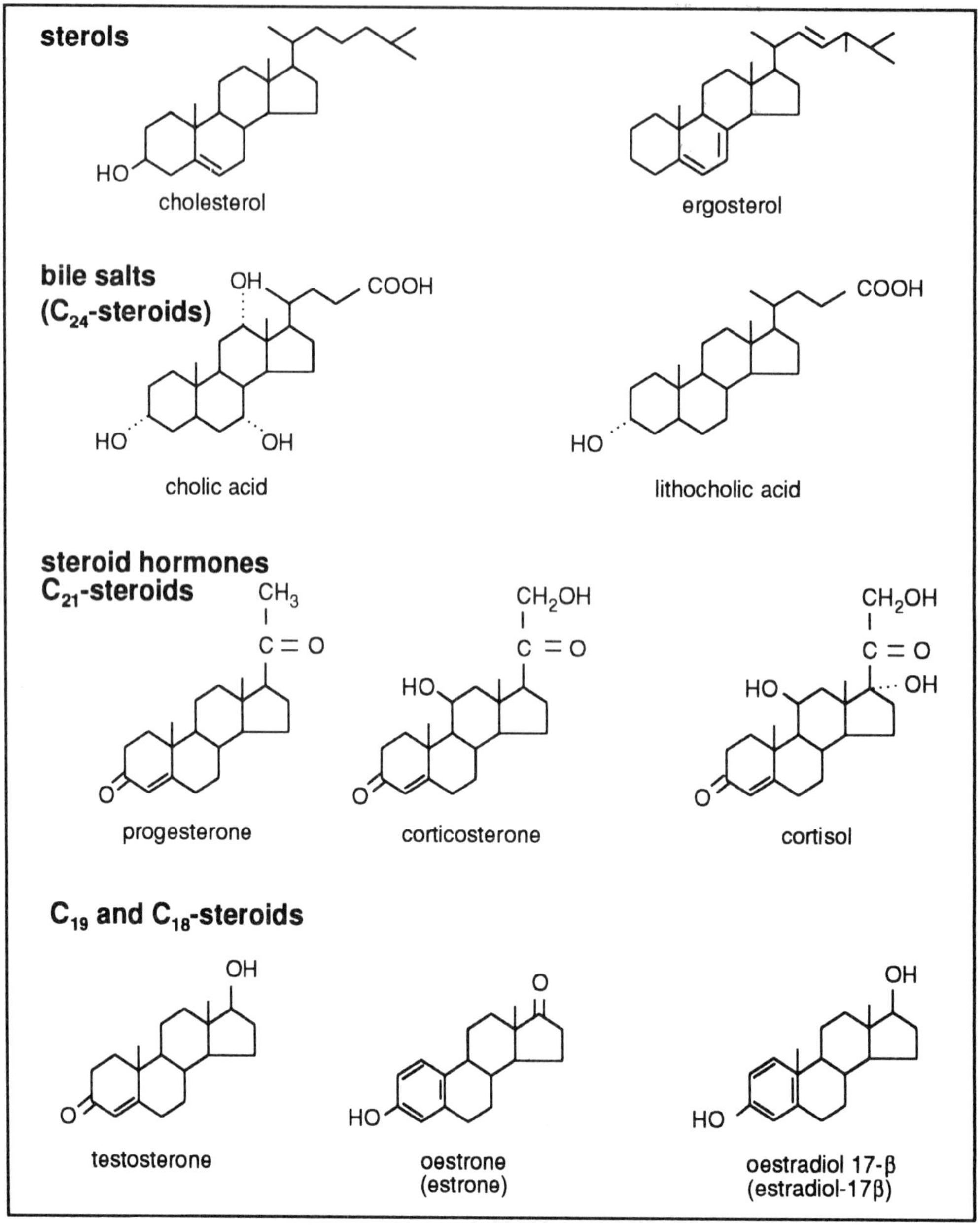

Figure 9.1 Some examples of sterols, bile salts and steroids.

steroids have pharmaceutical value

Although representatives of all of the classes of sterols and steroids are essential to humans, the biological (pharmacological) activities of the C_{18}, C_{19} and C_{21} steroids make these potentially very useful as therapeutic agents. It has long been realised that variations on the structures of naturally occurring steroids lead to products with greatly modified biological activities. Thus we can visualise the situation in which steroids may be modified to produce substances which have enhanced or reduced activities. This has far reaching implications in the healthcare sector. For example, natural and modified corticosteroids have applications as anti-inflammatory agents and may be used where the immune response needs to be moderated. Similarly the ability of C_{19} and C_{18} steroids to modulate reproductive capabilities makes them useful as fertility agents and as contraceptives.

Π Which of the groups of compounds shown in Figure 9.1 is (are) likely to be of greatest commercial (and social) value?

steroids are of great commercial value

Again, the answer should be fairly obvious. The potential therapeutic value of the steroid hormones makes these of tremendous commercial value. The commercial market for these is of the order of hundreds of millions of dollars per year. There is no comparable market for sterols and bile salts. We are faced with the interesting situation, therefore, that sterols are relatively abundant in natural sources but of relatively low commercial value, whilst steroids occur naturally at very low concentrations but are of great commercial value. Although there are tremendous variations amongst different products, steroids with desirable properties command market prices that are (ten to one thousand fold) greater than their sterol counterparts.

Π Bearing in mind the relative abundance of sterols and steroids and their chemical structures, which of the following strategies for producing steroids is most likely to be commercially successful?

1) Extraction from animals.

2) Total chemical synthesis.

3) Partial chemical synthesis starting from a natural product.

4) Total biosynthesis.

5) Enzymatic transformation of natural products.

Below we have considered each of these strategies in turn.

1) Although animals produce steroids, the low concentrations of these compounds does not make these commercially (nor ethically) attractive sources of these substances. Furthermore, they could only serve as sources of naturally occurring steroids. Thus we would not have selected this option.

2) The steroid ring structure is complex and contains many chiral carbons (for example at positions 5, 8, 9, 10, 13, 14 and 17) thus many optical isomers are possible. (The actual number of optical isomers is given by 2^n where n = the number of chiral carbons). From your knowledge of biochemistry you should have realised that only one of these optical isomers is likely to be biologically active. Synthesis of such a complex chemical structure to produce a single isomeric form is extremely difficult, especially when it is realised that many chemical reactions lead to the formation of racemic mixtures. Thus, for complete chemical synthesis, we must anticipate that

we would need a multistage reaction and that the desired isomer would only be a small fraction of the final product. We would also be presented with the difficulty of isolating the desired isomer from its isomeric partners. We, therefore, conclude that, although technically feasible, this approach is not a realistic commercial option.

3) Partial chemical synthesis is perhaps more realistic. If, for example, we wish to slightly modify the structure of a naturally occurring substance, then this might be possible using chemical processes. The problem here is to identify reagents and reactions which will be specific, both in terms of the site of attack on the natural product and the stereospecificity of the reaction. We must anticipate, therefore, that chemical reactions may be used in some cases, but this is not a universally applicable strategy.

4) The natural systems that produce steroids do so in quantitatively small amounts. Although in principle the cells producing these might be isolated and cultivated *in vitro,* the quantities produced will still be small and the costs of cultivation are high. This approach is, therefore, not generally commercially viable. You may have considered the option of transferring the genes, which encode for the enzymes involved, into an easy to cultivate system (for example a yeast or bacterium) and to control the expression of these genes using strong promoters. Although this approach is theoretically possible using the techniques of genetic engineering, the difficulties in isolating the necessary genes and the multiplicity of the enzyme steps needed for steroid biosynthesis makes the development costs of this approach extremely high. In the longer term, this may become a realistic option, but is not, currently, commercially viable.

5) The enzymatic transformation of natural products is by far the most attractive option. In this approach, it can be envisaged that sterols, which are relatively abundant, may be selectively modified to produce desired products. The diversity of enzyme activities, their reaction specificity, regiospecificity and stereospecificity are all features which could contribute to carrying out the desired changes. This does not mean, however, that transformations using enzyme systems are simple. Nevertheless, biotransformations have become of vital importance in the production of steroids.

In the following sections we will explain some applications of enzymes (and cells) in the transformation of sterols and steroids. You should realise, however, that for each process a decision has to be made whether to use an enzyme-mediated transformation or to use a chemical reaction. In many instances the biotransformation process is the most attractive but, as we will see later, this is not always the case.

9.3 Selective degradation of the sterol side chain

Π Re-examine the structures shown in Figure 9.1 and see if you can identify the fundamental difference between sterols and the steroid hormones.

Although there are many differences between these two groups of molecules, the fundamental difference between them is that the steroids do not possess the long side chain attached to position 17 that occurs in sterols. Thus, if we are to use sterols as the starting point for producing steroids, then we need to selectively remove this side chain.

micro-organisms may selectively degrade the side chain

Fortunately many micro-organisms can be used to selectively remove the side chain of abundant, naturally occurring sterols such as cholesterol, β-sitosterol and compesterol. These organisms include members of the genera *Nocardia*, *Pseudomonas*, *Mycobacterium*, *Corynebacterium* and *Arthrobacter*. They are capable of using sterols as their sole source of carbon. Unfortunately, the natural occurring organisms catabolise both the side chain and the ring structure of the sterols. The catabolism of these two components may occur simultaneously. Therefore, methods have to be found to prevent ring structure catabolism whilst allowing the degradation of the side chain.

Π See if you can identify two strategies for achieving this objective.

use of mutants

In practice, several strategies have been used. In one, mutants are produced which are defective in the enzymes involved in ring structure catabolism but still retain the enzymes involved in side chain catabolism.

Π Would such mutants grow on a) cholesterol b) testosterone?

a) We would anticipate that such mutants would grow, albeit slowly, on cholesterol as they could still derive carbon and energy from catabolising the side chain.

b) The mutants would probably not grow on testosterone as their is no side chain for them to catabolise.

We could use these differences to identify putative mutants with the desired metabolic block.

inhibition of a specific enzyme

A second strategy is to find a way of inhibiting an enzyme involved early in the catabolism of the ring. One such enzyme is a 9α-hydroxylase (it hydroxylates carbon 9). This enzyme has an absolute requirement for Fe^{2+} ions. By adding chelating agents which complex with these ions, the enzyme can be inhibited.

modification of the substrate

A third option is to modify the ring structure of the sterol so that it no longer serves as a substrate for the ring-catabolising enzyme. In this approach, a chemical reaction is used to modify the ring-structure and the product is subsequently incubated with the catabolising organism. For example, hydroxylation at C-19 prevents ring cleavage. Some examples in which modified sterols have been used for selective side chain degradation are given in Table 9.1. This table also indicates the nature of the products formed and the organisms used. We would not expect you to remember all of the details of these substrates, products and organisms. We will, however, examine some examples in more detail to illustrate the principles involved.

Substrate	Product	Micro-organism
19-hydroxysterols, 19-norsterols, 3-hydroxy-19-nor-$\Delta^{1,3,5}$-sterols	estrone	*Nocardia restrictus* ATCC 14887 *Nocardia* sp ATCC 19170 *Arthrobacter simplex* IAM 1660 *Corynebacterium* sp Mycobacteria
19-hydroxy-$\Delta^{4,7}$-sterols, 3-hydroxy-19-nor-$\Delta^{1,3,5,7}$-sterols	equilin, equilenin, estrone	*Mycobacterium* sp *Corynebacterium simplex* *Nocardia rubra*
6β,19-oxidostenones, 3β-acetoxy-5α-chloro(fluoro)-6β,19-oxidosterols	6β,19-oxido-4-androstene-3,17-dione	*Nocardia* sp ATCC 19170 Mycobacteria
3β-acetoxy-5α-bromo-6β,19-oxidosterols	5α-bromo-6β,19-oxidoandrostane-3,17-dione	*Nocardia* sp ATCC 19170
5α,5α–cyclosterols	3α,5α-cycloandrostane-17-one	*Mycobacterium phlei*
3α,5α–cyclo-6β,19-oxido-sterols	3α,5α-cyclo-6β,19-oxido-androstane-17-one	*Arthrobacter* spp Corynebacteria
sterol-3-oximes	4-androstene-3,17-dione (after hydrolysis)	*Mycobacterium* sp
4-hydroxycholestenone	3β-hydroxy-5α-androstane-4,17-dione 3α-hydroxy-5α-androstane-4,17-dione, 3β,4α–dihydroxy-5α–androstane-17-one	*Mycobacterium phlei*
25D-spirost-4-ene-3-one	1,4-androstadiene-3,16-dione, 20α-hydroxy-4-pregnene-3,16-dione 3α,11β,20α-trihydroxy-5α–pregnane-16-one	*Fusarium solani* *Verticillium theobromae* *Stachylidium bicolor*
ponasterone A	rubrosterone	*Fusarium lini* 9593

Table 9.1 The use of modified sterols to allow selective cleavage of the side chain (based on Martin, CKA Sterols in Biotechnology Volume 6a Edited by Kieslich K 1984 Verlag Chemie, Weinheim).

9.3.1 Use of modified sterols

First, let us briefly examine the route of side chain degradation in micro-organisms. The pathway is illustrated in Figure 9.2.

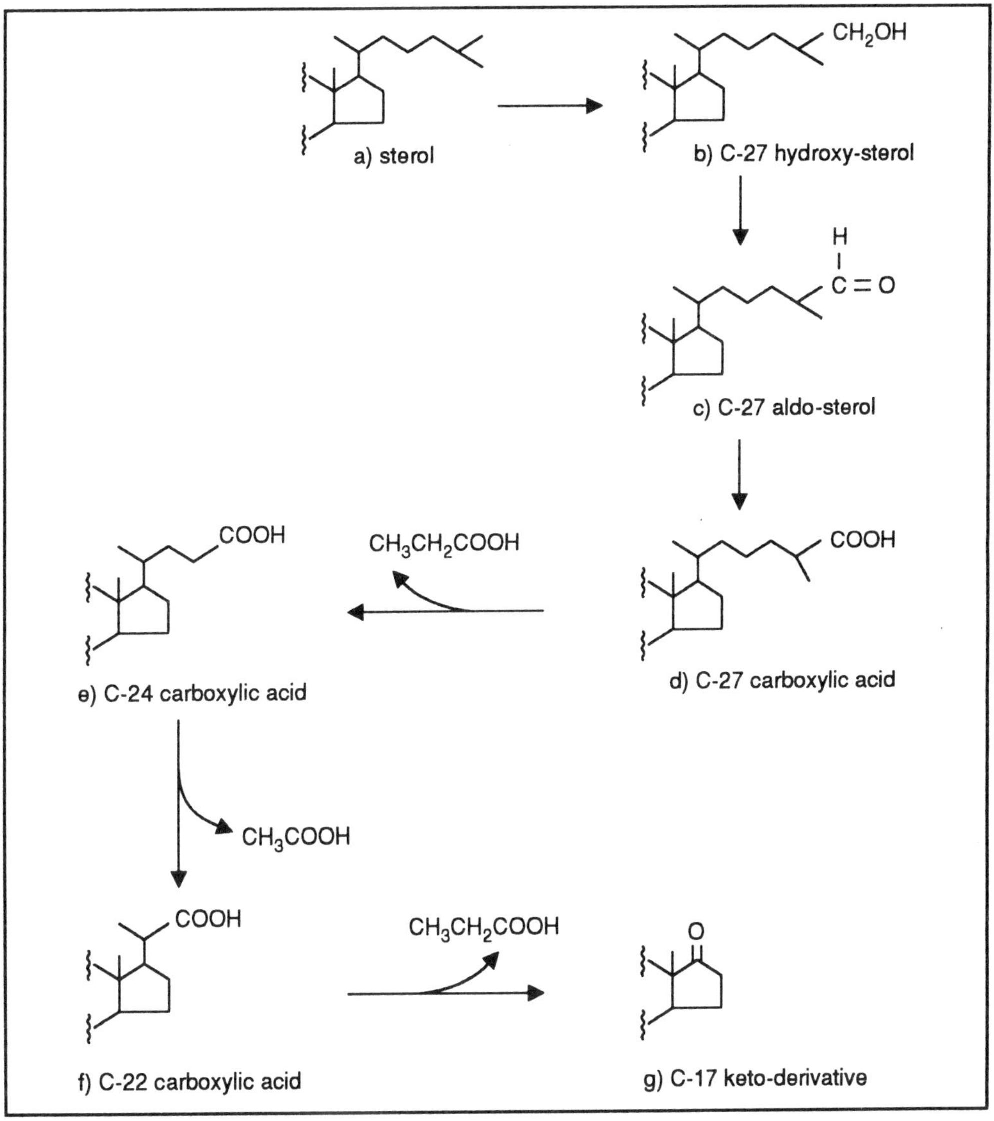

Figure 9.2 Generalised metabolic sequences of sterol side chain degradation by micro-organisms.

side chain degradation

First carbon 27 is hydroxylated and oxidised to a carboxylic acid. The resulting acid is then cleaved to release propionic and acetic acids and a second propionic acid. The final reaction in this sequence results in the formation of a keto group at C-17.

hydroxylation at C-19

If a hydroxyl group is introduced into position C-19 then complete breakdown of the ring structure is prevented, although it may be subjected to some modification. In the examples shown in Figure 9.3, the incubation of 19-hydroxysterols with *Nocardia restrictus* ATCC 14887 or *Nocardia* sp ATCC 19170 leads to the production of estrone. In these cases, you will notice that ring A has become modified but the ring structure is not broken. The yields of estrone using these substrates and organisms are of the order of 30%.

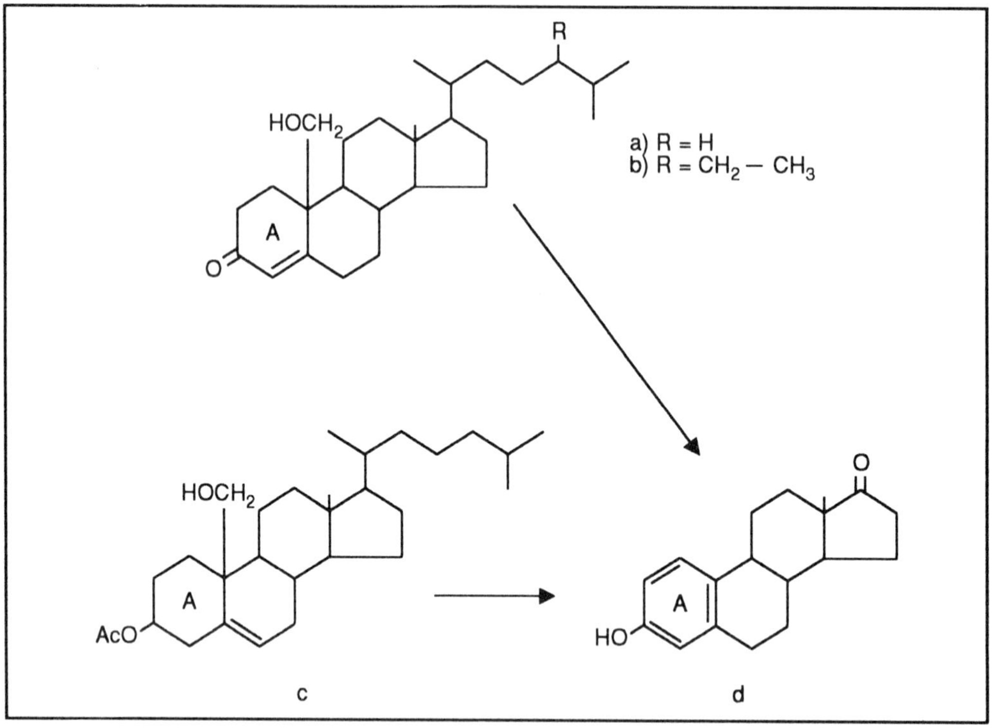

Figure 9.3 The figure shows the degradation of the side chain of sterols which have substitutions at C-19. Removal of the C-19 methyl group (eg 19 norcholesta-1,3,5 (10) triene-3-ol) also prevents ring breakdown. Note, however, hydroxylation of C-19 does not prevent all ring modifications.
a: 19-hydroxycholesterone, b: 19-hydroxysitosterone, c: 3β-acetoxy-19-hydroxy-5-cholestene, d: estrone.

Other modifications which restrict ring cleavage are the formation of 6β,19-oxido derivatives and 3α, 5α cyclo-derivatives. The structures of some of these are given in Figure 9.4.

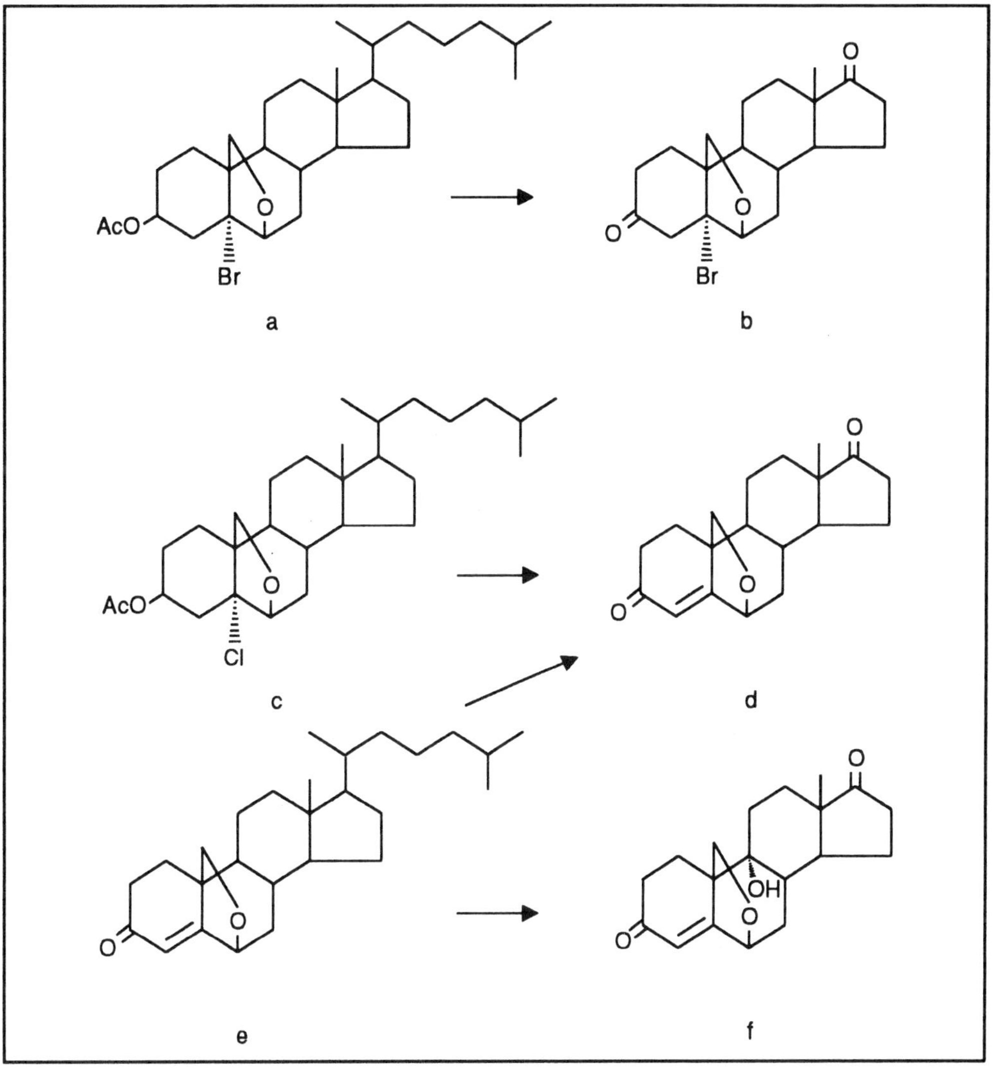

Figure 9.4 Some examples of side chain degradation of 6β and 19 oxidosterols.
a: 3β-acetoxy-5α-bromo-6β, 19-oxidocholestane,
b: 5α-bromo-6β, 19-oxidoandrostane-3,17-dione,
c: 3β-acetoxy-5α-chloro-6β, 19 oxidocholestane,
d: 6β, 19-oxidoandrost-4-ene-3,17-dione,
e: 6β, 19-oxido-4-cholestene-3-one,
f: 6β, 19-oxido-9α-hydroxyandrost-4-ene-3,17-dione.

9.3.2 Use of enzyme inhibitors

Earlier we indicated that an alternative strategy to prevent ring cleavage is to use selective inhibition of the enzymes which catabolise the ring structure.

The inhibitors used include:

- chelating agents;
- metal ions with similar ion radii which will displace Fe^{2+};

- inorganic -SH reagents;
- auto-oxidisable redox dyes.

A list of some examples of these is given in Table 9.2.

Π Are these compounds likely to be toxic to cells?

Since they inhibit metabolism, they are likely to be toxic to cells. It is, therefore, usual to add these compounds after the culture has grown and to subsequently add the sterol to be metabolised.

ring cleavage inhibitors

Of the compounds listed in Table 9.2, 8-hydroxyquinoline, α,α^1-dipyridyl and 1,10-phenanthroline have found most use. They are usually used in the concentration range 0.1 mmol l^{-1}. It is, however, essential to use an optimal concentration of these agents in order to get high yields.

Mechanism of action	Compound
Chelating agents for Fe^{2+}	Cupferron
	Diethyldithiocarbamate
	α,α^1-Dipyridyl
	Diphenylthiocarbazone
	8-Hydroxyquinoline
	Isonicotinic acid hydrazide
	4-Isopropyltropolone
	5-Nitro-1,10-phenanthroline
	1,10-Phenanthroline
	o-Phenylenediamine
	Tetraethylthiuram-disulphide
	Xanthogenic acid
Metal ions replacing iron or blocking	Ni^{2+}, Co^{2+}, Pb^{2+}
SH-functions	SeO_3^{2-}, AsO_2^-
Redox dyes	Methylene blue
	Resazurine

Table 9.2 Compounds used to inhibit steroid ring degradation (based on Martin CKA Sterols in Biotechnology Volume 6a Edited by Kieslich K 1984. Verlag Chemie, Weinheim).

Π What will happen if the concentration of the inhibitor is a) too low b) too high?

a) If it is too low, complete degradation of the substrate may occur.

b) If it is too high, inhibition of other enzymes, including those involved in side chain degradation, may occur.

optimal inhibitory concentrations

Achieving optimal concentrations is difficult because components in the medium and the biomass itself may neutralise the inhibitory effects of these reagents. This is especially true if complex, non-defined media, such as cornsteep liquor, are used.

This approach has been used with a wide variety of organisms, although it is more commonly employed using *Arthrobacter simplex*, *Brevibacterium lipolyticum*, *Corynebacterium* spp and certain strains of *Nocardia*.

Π Because of the difficulties in creating optimal inhibitory concentrations, do you think that incubation of a culture of sterol degrading organisms in the presence of cholesterol and inhibitor will lead to the production of a single steroidal product?

The answer is that a mixture of compounds is usually produced because the inhibition is somewhat imprecise and therefore "leaky". The major product is usually 1,4-androstadiene-3,17-dione.

4-Androstene-3,17-dione and other androstane and testosterone-related steroids are often also produced.

The key to successfully using the inhibitor approach to convert sterols like cholesterol to steroids, is to reduce the further metabolism of the C-17 keto steroid as it accumulates.

reduction of toxicity

In some cases, the chelating agents used to trap Fe^{2+} to reduce 9-hydroxylase activity are toxic. The toxicity may be reduced by using absorbants such as styrene-divenylbenzene copolymers. Although this reduces the toxicity of the chelating agent it does not seem to prevent the chelating agent from trapping Fe^{2+} ions. This increases yields, especially of 1,4-androstadiene-3,17-dione. Yields can be further increased by the addition of oils: linseed and soya oils are most effective.

Typically, sterol concentrations of 3 to 4 g l^{-1} are used and incubation times of about 100h. Yields are dependent upon the species and substrates used. Some data relating to the yields of 1,4-androstadiene-3,17-dione from various sterols and steroids using cultures of *Arthrobacter simplex* are reported in Table 9.3.

Substrate	Yield of 1,4-androstadiene-3,17-dione (as % of substrate)
campesterol	38
cholesterol	58
cholestanol	33
7-dehydrocholestrol	16
ergosterol	5
lithocholic acid	63
β-sitosterol	39

Table 9.3 Yields of 1,4 androstadiene-3,17, dione using a variety of sterols and steroids as substrates and employing cultures of *A. simplex* (data from Martin CKA Sterols in Biotechnology Vol 6a. Edited by Kieslich K 1984 Verlag Chemie, Weinheim).

Π The data shown in Table 9.3 indicate that highest yields of 1,4-androstadiene-3,17-dione are obtained using lithocholic acid as substrate. However, this substrate is not necessarily the one of choice for the commercial production of 1,4-androstadiene-3,17-dione using *A. simplex*. List factors that could determine the choice of substrate.

Lithocholic acid costs 2 or 3 times more than cholesterol. Thus, although the yields are slightly lower with cholesterol, it is cheaper to use it. Furthermore, cholesterol is more widely available and in greater quantities than lithocholic acid. These two factors tend to favour the use of cholesterol. Lithocholic acid does have the advantages, however, of being more water soluble and is, therefore, more easily supplied to cultures in aqueous media. The costs of recovery of the desired product from the reaction brew are also commercially important. The point we are making in this in-text activity is that in selecting a substrate we need to consider more than simply the conversion efficiency and the cost of the substrate.

SAQ 9.1

Assume that you have a culture of a *Mycobacterium* sp which is able to use cholesterol or β-sitosterol as its sole source of carbon and energy.

You incubate aliquots of this culture with samples of media containing one of the following as the sole carbon source.

1)

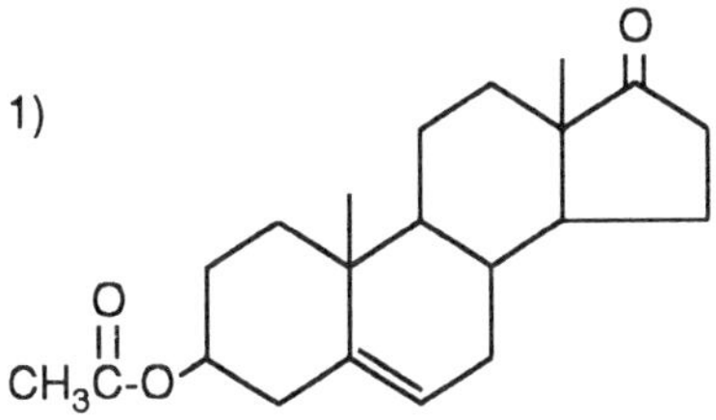

2)

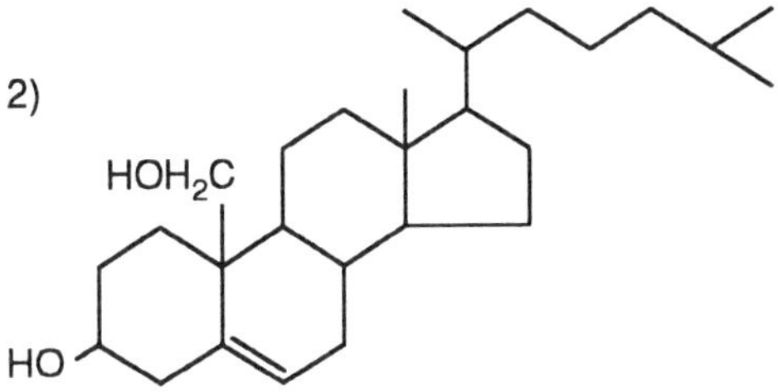

3)

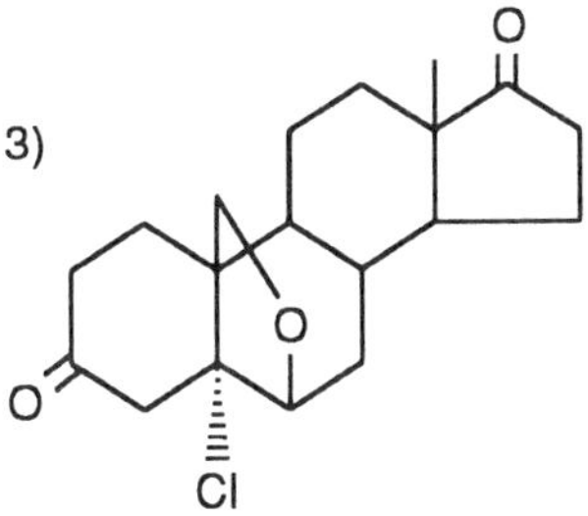

4)

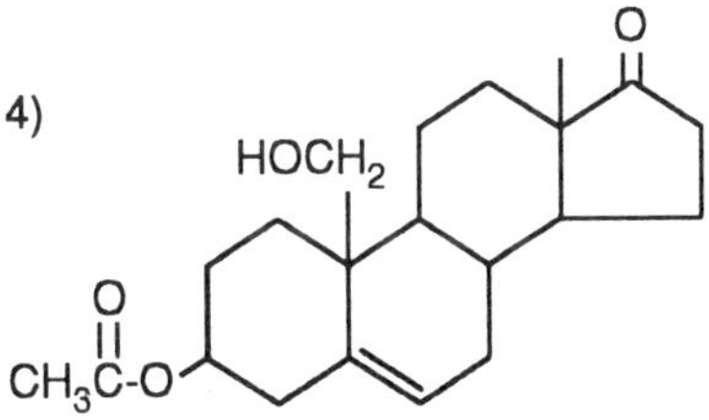

a) Predict where or not and to what extent the organism will grow on these substrates.

b) Predict the likely metabolic products from incubating this organism with these substrates.

SAQ 9.2

A medium containing bile salts such as lithocholic acid along with carbohydrates and peptone has been used to isolate gut organisms. One such isolate has been shown to completely degrade the lithocholic acid (structure given in Figure 9.1).

In an experiment in which a student was examining the tolerance of this organism to Pb^{2+} ions, it was observed that the organism grew quite well in 0.1 mmol l^{-1} Pb^{2+} but that the lithocholic acid was only partially catabolised yielding a range of products, including the following:

main products

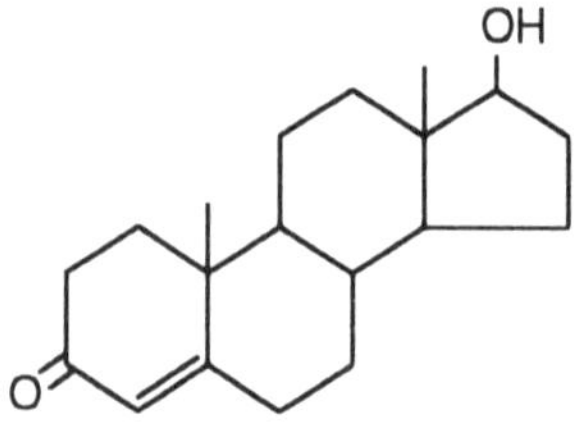

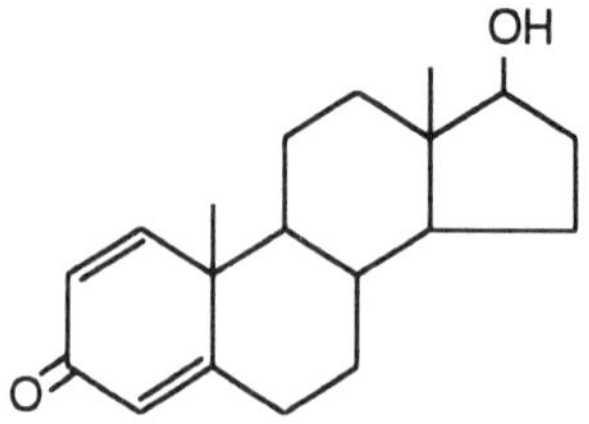

minor product

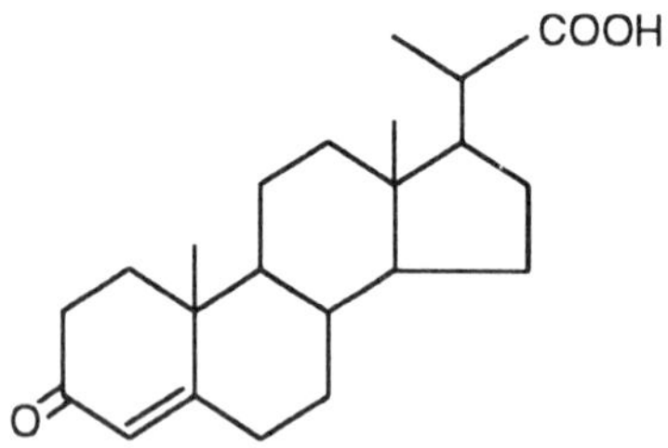

What is the likely explanation for this observation?

9.3.3 Use of mutants

A wide variety of mutants with modified sterol (steroid) metabolism has been produced using conventional mutagens such as N-methyl-N'-nitroso-guanidine and ultraviolet light. The organism most commonly, but not exclusively, used is *Mycobacterium fortuitum*. Wild type members of this species can catabolise a range of sterols. Mutants have been produced that are blocked at various stages of the catabolism, giving rise to a wide range of products. Some of the products are used by a variety of commercial organisations as intermediates in the manufacture of medically valuable steroids. We have reported a few of the many possible examples in Table 9.4. We would not expect you to recall the details of these. You should, however, realise that this is a very valuable commercial market and companies involved in this area hold extensive patent rights protecting these processes.

patent rights

The strains described in Table 9.4 are all of commercial value since they produce compounds which are either pharmacologically active or can be converted to pharmacologically important compounds. For example, the production of 1,4-androstadiene-3,17-dione from β-sitosterol provides material which can be readily converted to estrone, while 4-androstene-3,17-dione can be converted to testosterone.

Organism	Principle Product	User
Mycobacterium fortuitum NRRL B-8119	9α-hydroxyandrost-4-ene-4,17-dione	Upjohn
M. fortuitum NRRL B-11359, 11045	4-androstene-3,17-dione	Upjohn
M. fortuitum NRRL B-8119	9α-hydroxy-3-oxo-23,24-dinorchol-4-ene-22-ol	Upjohn
M. fortuitum FERM P-4809	3,9-dihydroxy-9,10-secoandrosta -3,5-triene-17-one	Mitsubishi
Mycobacterium parafortuitum FERM P-4926	9α-hydroxyandrost-4-ene-3,17-dione	Mitsubishi
M. parafortuitum MCI 0807	4-androstene-3,17-dione	Mitsubishi
M. parafortuitum MCI 0617	3-oxo-23,24-dinorchola-1,4-diene-22-ol	Mitsubishi
Mycobacterium sp NRRL B-3683	1,4-androstadiene-3,17-dione	Searle
Mycobacterium sp NRRL B-3805	4-androstene-3,17-dione	Searle
Rhodococcus corallinus FERM P-4812	3β-hydroxy-23,24-dinorchol-5-ene-22-oic acid	Mitsubishi
Arthrobacter simplex FERM P-4477	3-hydroxy-9-oxo-9, 10-sec pregna-1,3,5 triene-20-carboxylic acid	Mitsubishi
Corynebacterium equi	9-*a*-hydroxy-3-oxo-23,24-dinorchola-4,17-diene-22-oic acid	Mitsubishi

Table 9.4 Some examples of mutants used industrially for the removal of the side chain of sterols.

9.4 Specific steroid interconversions and reactions

In section 9.3, we discussed in general terms the use of microbial metabolism to selectively remove the side chain from sterols to produce steroids. This removal may also be accompanied by some modification to the ring structure. We did not, however, discuss in any detail any specific reactions. In this section we will focus on some specific reactions.

Π Before we begin this discussion, see how many types of enzyme catalysed reactions you can think of that may be used to carry out specific modification of sterols/steroids. (Do this without looking at Table 9.5).

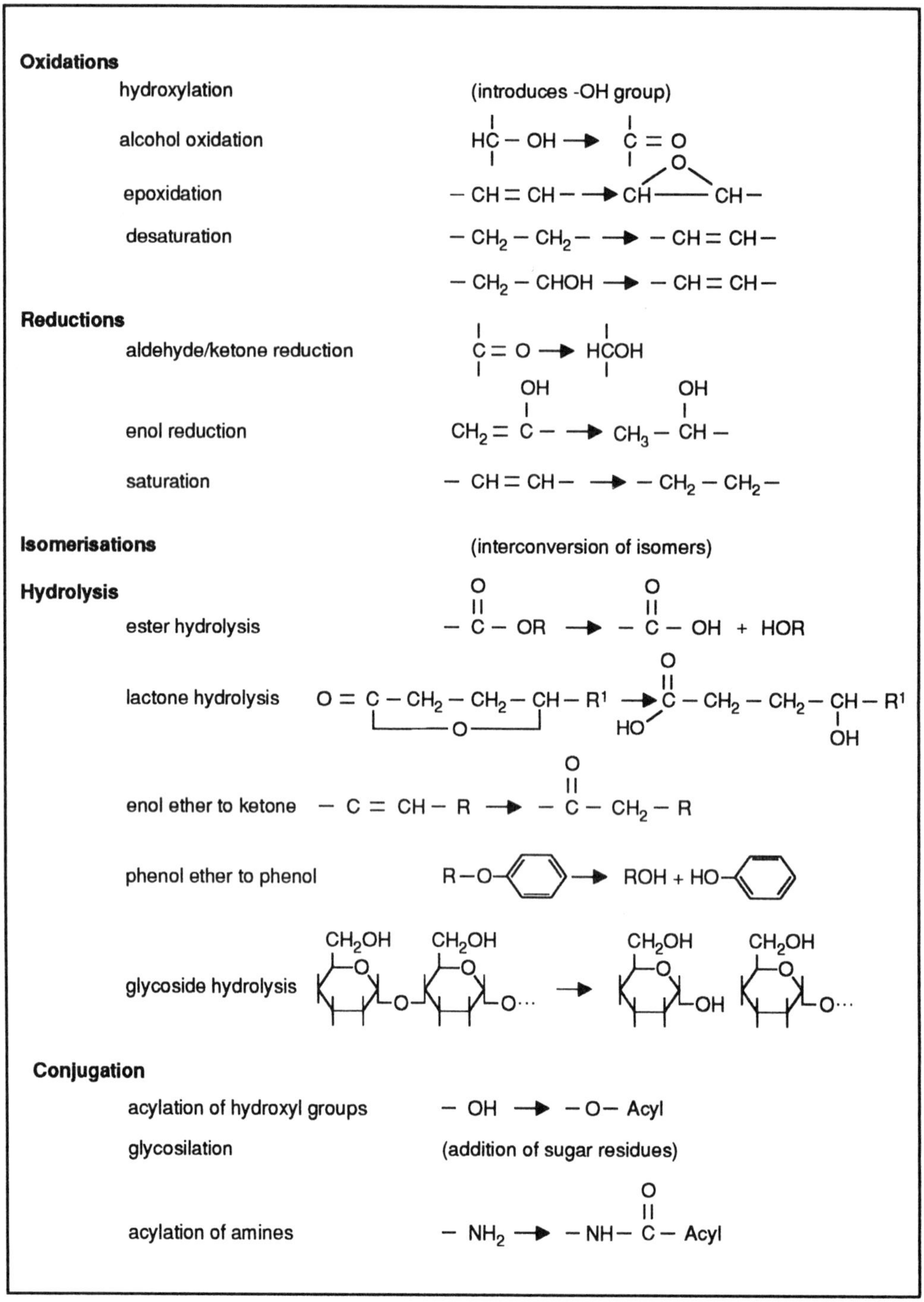

Table 9.5 Some reaction types commonly encountered in sterol/steroid interconversions.

In Table 9.5 we have listed a large number of reaction types. For many of these reaction types you may be able to think of examples from central metabolism. For example, the oxidation of alcohols to ketones is a very commonly encountered reaction. Thus:

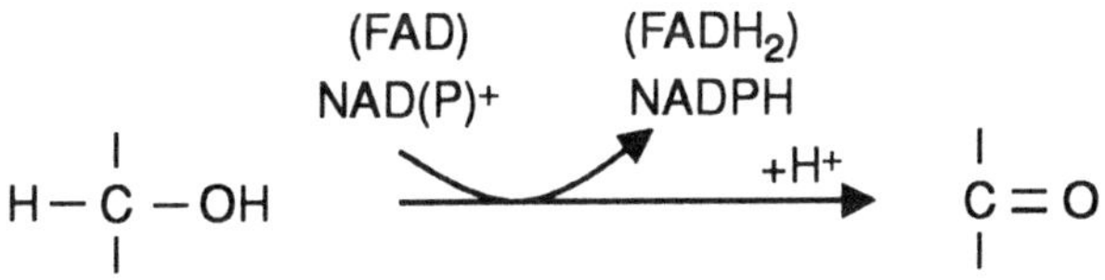

Similarly, the introduction of double bonds, isomerisation or hydrolysis are also frequently encountered reactions in central metabolism. Many of these reactions have their analogues in sterol/steroid interconversions. Below we will confine ourselves to a limited number of examples.

SAQ 9.3

Consider the following reaction sequences.

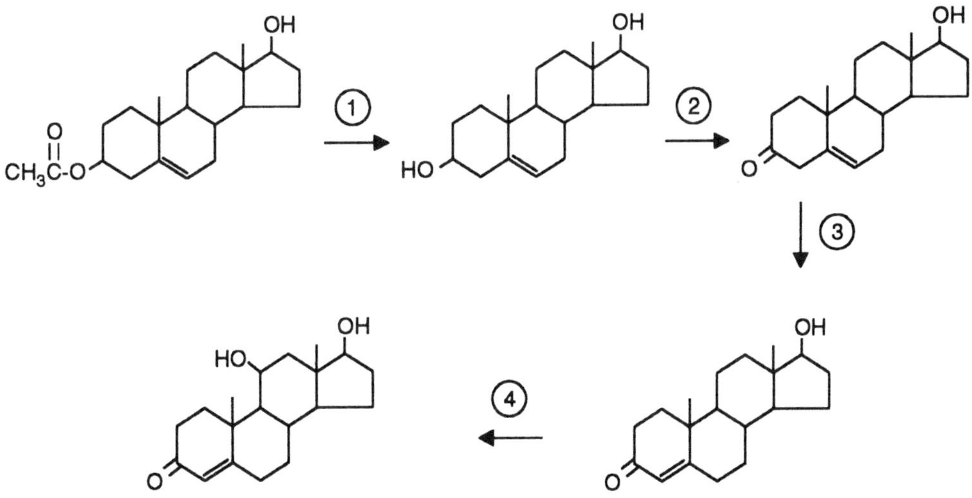

Using the terms given in Table 9.5, name the types of reactions shown as 1, 2, 3 and 4.

9.4.1 Hydroxylations

Many examples of microbial hydroxylation of sterols/steroids have been reported. These hydroxylations usually involve mixed function oxidases which utilise molecular oxygen and cytochrome P-450. The reaction can be represented by:

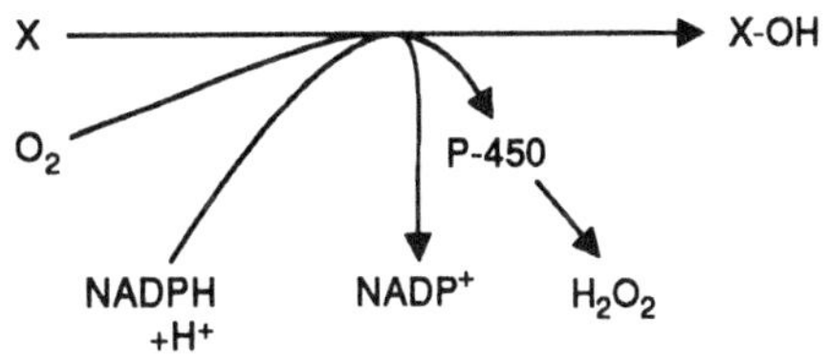

Π Which positions of the steroid rings could be hydroxylated?

Virtually all positions of the steroid nucleus may be hydroxylated. Of course, any particular enzyme is fairly specific. Any particular organism may carry out single or multiple hydroxylations. Where di-hydroxylations take place, the usual combinations are hydroxylations at positions 6 and 11 or 12 and 15. You should also note that not only are the positions of hydroxylation specific, the orientation of the added hydroxyl groups is specific as well. Thus hydroxylation at position 6 is often in the β configuration, while at position 11 the hydroxyl group is often added in the α configuration. Multiple hydroxylations usually involve more than one enzyme. Thus addition of hydroxyl groups in position 6β, involves a different enzyme from that which inserts hydroxyl groups in position 11α. Frequently substrates containing an 11α hydroxyl group are required to induce the enzyme which catalyses the introduction of a hydroxyl group at position 6β.

The key questions are: can hydroxylations be carried out *in vitro* at desirable positions on the steroid nucleus, and can this be done to achieve the desired configuration?

The answer is yes in both cases. A very large number of hydroxylating enzymes have been identified and it is theoretically possible to hydroxylate any of the 19 carbon atoms of the steroid nucleus. However, these enzymes will each only utilise a specific range of substrates. The sources of some hydroxylating enzymes and the reactions they catalyse are given in Table 9.6. We have been selective in compiling this table, giving only a single example for each hydroxylation. Again we would not expect you to remember all of the details given in Table 9.6; we have provided these data primarily as an illustration of the potential of using enzymes to bring about specific chemical transformations.

Reaction	Substrate	Product	Micro-organism
1α-Hydroxylation	Androst-4-ene-3,17-dione	1α-Hydroxyandrost-4-ene-3,17-dione	*Penicillium* sp.
1β-Hydroxylation	Androst-4-ene-3,17-dione	1β-Hydroxyandrost-4-ene-3,17-dione	*Xylaria* sp.
2α-Hydroxylation	Androstane-7,17-dione	2α-Hydroxyandrostane-7,17-dione	*Wojnowicia graminis*
2β-Hydroxylation	Androst-4-ene-3,17-dione	2β-Hydroxyandrost-4-ene-3,17-dione	*Penicillium* sp.
3α-Hydroxylation	Androstane-7,17-dione	3α-Hydroxyandrostane-7,17-dione	*Diaporthe celastrinia*
3β-Hydroxylation	17β-Hydroxyandrostan-11-one	3β,17β-Dihydroxy-androstan-11-one	*Wojnowicia graminis*
4β-Hydroxylation	17α-Methylestra-1,3,5 (10)-triene-3,17β-diol	17α-Methylestra-1,3,5 (10)-triene-3,4,17β-triol	*Aspergillus flavus*
5α-Hydroxylation	Nor-5α-Pregnane-2, 20-dione	5α-Hydroxy-nor-pregnane-2,20-dione	*Cokeromyces recurvatus*
5β-Hydroxylation	3β,14-Dihydroxy-5β, 14β-card-20(22)-enolide	3β,5,14-Trihydroxy-5β, 14β-card-20(22)-enolide	*Absidia orchidis*
6α-Hydroxylation	Androstane-1,17-dione	6α-Hydroxyandrostane-1,17-dione	*Calonectria decora*
6β-Hydroxylation	17β-Hydroxyestr-4-ene-3-one	6β,17β-Dihydroxyestr-4-ene-3-one	*Helminthosporium rasonoi*
7α-Hydroxylation	Androst-4-ene-3,17-dione	7α-Hydroxyandrost-4-ene-3,17-dione	*Mucor griseocyanus*
7β-Hydroxylation	Androst-4-ene-3,17-dione	7β-Hydroxyandrost-4-ene-3,17-dione	*Xylaria* sp.
8β-Hydroxylation	17,21-Dihydroxypregn-4-ene-3,20-dione	8β,17,21-Trihydroxy pregn-4-ene-3,20-dione	*Cercospora melonis*
9α-Hydroxylation	Androst-4-ene-3,17-dione	9α-Hydroxyandrost-4-ene-3,17-dione	*Norcardia corollina*
9β-Hydroxylation	9β,10α-Pregn-4-ene-3,20-dione	9β-Hydroxy-9β,10α-pregn-4-ene-3,20-dione	*Cephalothecium roseum*

Table 9.6 Examples of hydroxylation of steroid nuclei by micro-organisms. (Data derived from Neidleman, SL "Industrial Chemicals: Fermentation and Immobilised Cells" in "Biotechnology the Science and the Business", edited by Moses V and Cape RB, published by Harwood Academic, London 1991).

10β-Hydroxylation	17β-Hydroxyestr-4-ene-3-one	10β,17β-Dihydroxyestr-4-ene-3-one	*Botrytis paeoniae*
11α-Hydroxylation	Progesterone	11α-Hydroxyprogesterone	*Rhizopus*
11β-Hydroxylation	11-Deoxycortisone	Hydrocortisone	*Curvularia lunata*
12β-Hydroxylation	17β-Hydroxyestr-4-ene-3-one	12β,17β-Dihydroxyestr-4-ene-3-one	*Collectotrichum derridis*
14α-Hydroxylation	Androst-4-ene-3,17-dione	14α-Hydroxyandrost-4-ene-3,17-dione	*Dematiaceae* Strain M202
15α-Hydroxylation	11α-Hydroxy-5α-pregnane-3,20-dione	11α,15α-Dihydroxy-5α-pregnane-3,20-dione	*Calonectria decora*
15β-Hydroxylation	Androst-4-ene-3,17-diore	15β-Hydroxyandrost-4-ene-3,17-dione	*Xylaria* sp.
16α-Hydroxylation	Androst-4-ene-3,17-dione	16α-Hydroxyandrost-4-ene-3,17-dione	*Streptomyces roseochrogene* sp.
16β-Hydroxylation	17β-Hydroxyestr-4-ene-3-one	16β,17β-Dihydroxyestr-4-ene-3-one	*Mycosphaerella latebrosa*
17α-Hydroxylation	Pregn-4-ene-3,20-dione	17α-Hydroxypregn-4-ene-3,20-dione	*Cephalothecium roseum*
17β-Hydroxylation	Androstane-3,11-dione	17β-Hydroxyandrostane-3,11-dione	*Wojnowicia graminis*
18β-Hydroxylation	11β,21-Dihydroxy-pregn-4-ene-3,20-dione	11β,18,21-Trihydroxy-pregn-4-ene-3,20-dione	*Corynespora cassiicola*
19-Hydroxylation	6α-Hydroxyandrostane-3,17-dione	6α,19-Dihydroxyandrostane-3,17-dione	*Calonectria decora*
21-Hydroxylation	Nor-pregn-3-ene-2,20-dione	21-Hydroxy-nor-pregn-3-ene-2,20-dione	*Aspergillus niger*

Table 9.6 .... Continued

Π From your knowledge of the structure of pharmacologically active steroids such as the corticosteroids, which of the hydroxylations shown in Table 9.6 are likely to be of most commercial value? (You may find Figure 9.1 helpful).

commercially important hydroxylations

You probably concluded that hydroxylation at position 11 is of potential value. This is in fact true. However, hydroxylation at position 16 is also important. The three commercially most important hydroxylations are 11α, 11β and 16α hydroxylations.

11α-Hydroxylation of progesterone is used by Upjohn, whilst 11β-hydroxylation of various substrates is used by a number of companies including Gist Brocades, Pfizer and Schering AG. 16α Hydroxylation of 9-fluorohydrocortisone is used by Squibb.

11α-hydroxylation

The first microbial hydroxylation to be exploited was the 11α-hydroxylation. This reaction is catalysed by a wide variety of organisms but, after a considerable amount of screening, the preferred species is *Rhizopus nigricans* ATCC 6227b. This filamentous fungus is used to hydroxylate progesterone. Normally large bioreactors are inoculated with a large inoculum and the culture allowed to grow for about 12-20h with moderate aeration. The media used are usually rich in carbohydrates (usually 5% dextrose) and corn steep liquor (3%) although there are many variations. The pH is maintained at about pH 4.2-4.7 and the incubation is carried out at 28°C. In practice, the actual composition of the medium used depends on the current economics and availability of the feedstocks.

After the initial growth, progesterone (0.5-5 g l^{-1}) is added as a powder or as an acetone solution. If a powdered form is used, it is wetted with a small amount (0.01%) of Tween to facilitate its dissolution into the reaction mixture. A single addition of progesterone at a concentration of about 5g l^{-1} enables about 86% hydroxylation to take place within about 50h. The remaining progesterone remains unaltered.

Alternatively, hydroxylation may be carried out in a continuous process. In this the organism is first cultured in one vessel and then transferred to a production vessel. Progesterone is fed into this second vessel at a concentration of 0.5g l^{-1}. Over 50% of the progesterone is hydroxylated when the residence time in the production vessel is about 5h.

Π Bearing in mind that hydroxylation requires molecular oxygen and that *Rhizopus* is a filamentous fungus, what factors in the reaction vessel do you anticipate are critical to the successful hydroxylation of progesterone?

damage to mycelia

You should have anticipated that aeration and mixing are of critical importance. The mycelial nature of *Rhizopus* restricts the rate at which we can use impeller mixing. High impeller speeds will disperse air within the medium but will also cause breakage of the mycelium. Thus fairly low impeller speeds have to be used. This, of course, restricts the distribution of air (oxygen) within the system. This could, in principle, be circumvented by using high aeration rates but this, in turn, may cause foaming problems. If anti-foaming agents are used, these may interfere with the subsequent collection and purification of the product. In practice a compromise has to be achieved between impeller speeds and aeration rates which produces tolerable damage to mycelia without slowing the hydroxylation too greatly. Considerable attention has to be paid to optimising and controlling these factors.

Π Can you foresee any major problems in monitoring the health of the culture and the conversion of substrate to product, in the process described above?

Both of these factors are difficult. Obtaining representative samples of filamentous organisms from cultures is difficult and both substrate and product are relatively insoluble. Thus it is difficult to monitor these processes. Nevertheless, in practice satisfactory procedures have been developed to generate data that are sufficiently accurate to enable reasonable monitoring of cultures.

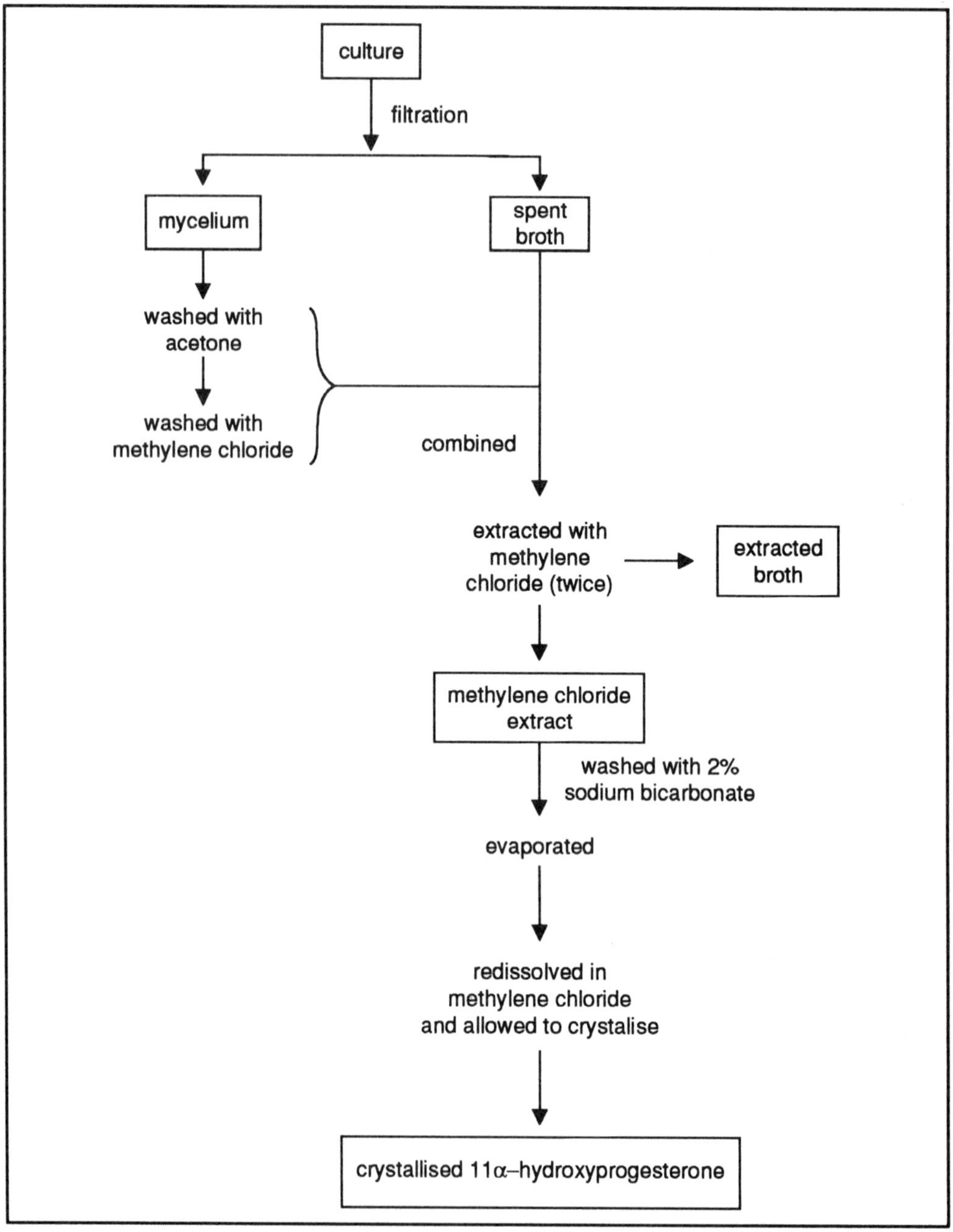

Figure 9.5 Outline of the recovery of 11α-hydroxyprogesterone from cultures of *Rhizopus nigricans*.

The recovery of the product is described in outline in Figure 9.5. Essentially the process involves separating the broth and mycelium by filtration, extracting the mycelium with acetone and methylene chloride. Combining these extracts with the broth and re-extracting with methylene chloride. The extract is washed with 2% sodium bicarbonate, evaporated and re-dissolved in methylene chloride. The product is allowed to crystallise from the methylene chloride.

Π Would you expect the product to be pure 11α-hydroxyprogesterone?

A single crystallisation is unlikely to lead to the isolation of pure crystals. In practice the product recovered in this process contains about 90% 11 α-hydroxyprogeterone with low levels of other products (especially 5 α-pregnane-3,20-dione and 6β, 11α-dihydroxyprogesterone). An example of a manufacturer who uses microbial 11α hydroxylation is Upjohn; progesterone is used as substrate.

11 β-hydroxylation

11β-Hydroxylation is achieved in a process analogous to that described above for 11α-hydroxylation. In this case, however, the organism of choice is *Curvularia lunata* NRRL 2830. The use of this organism is particularly attractive as it will carry out the hydroxylation using a wide variety of substrates, including steroids that have already been substantially modified. This has, however, a serious draw back: 11β-hydroxylations by *C. lunata* suffer from the production of a wide variety of other products, especially other hydroxylated steroids. Substrate purity, incubation times and recovery processes are, therefore, of crucial importance. Commercial 11β-hydroxylations are carried out by Pfizer, Gist-Brocades, Shering AG and Merck Darmstadt.

16α-hydroxylation

16α-Hydroxylation is used in the production of triamcinolone and fluocinolone. The organism of choice is *Streptomyces rosechromogenes*. In commercial use, this organism is used to catalyse the 16α-hydroxylation of 9α-fluorohydrocortisone. The product is subsequently dehydrogenated, as shown in Figure 9.6. This two stage process replaced a fourteen-step chemical-based process.

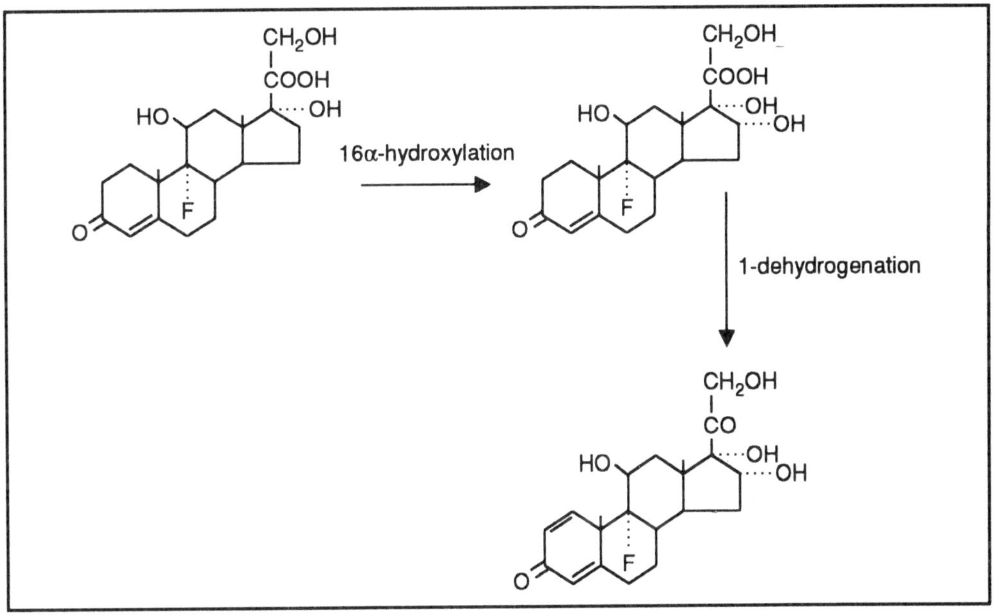

Figure 9.6 An example of 16α-hydroxylation catalysed by *Streptomyces roseochromogenes*. Commercial 16α-hydroxylations are carried out by Squibb.

We will leave the hydroxylation of the steroid nucleus at this stage. You should however appreciate that the examples we have described are but a very small part of the potential of enzymatic chemical transformations. Considerable efforts are being made to find easier (cheaper/more efficient) ways of using biocatalysis. This may involve developing new strains, exploring the possibilities of using immobilised cells or using extracted (cell-free) enzymes. Although microbial steroid hydroxylation has been achieved using cell-free extracts (see Table 9.7), there are many problems to be overcome before these systems become commercially applicable. Thus, although currently there is only limited commercial interest in such systems, developments in enzymology, especially the application of site directed mutagenesis to produce enzymes with desirable characteristics, may rekindle this area of development.

Enzyme source	Substrate	Site(s) of hydroxylation
Fungi		
Aspergillus niger	progesterone	11α
Aspergillus ochraceus	various	11α
Curvularia lunata	various	10β, 11β, 14α
Rhizopus nigrans	progesterone	6β, 11α, 17α
Prokaryotes		
Bacillus megaterium	progesterone	15α, 15β
Nocardia restrictus	progesterone	9α
Streptomyces roseochromogenes	progesterone	16α

Table 9.7 Examples of steroid hydroxylation using cell-free extracts. (Data cited by Martin CKA, Steroids in Biotechnology Vol 6a. Edited by Kieslich K 1984 Verlag Chemie, Weinheim).

So far we have focused on hydroxylation in the steroid nucleus.

Π From what you read earlier, where else on sterol molecules may hydroxylation take place?

You should have recalled that the breakdown of the side chain involves an initial hydroxylation (see Figure 9.2). Hydroxylation most frequently occurs on the terminal or side chain methyl groups.

SAQ 9.4

Select the most appropriate organism from the list below to catalyse each of the following conversions.

1)

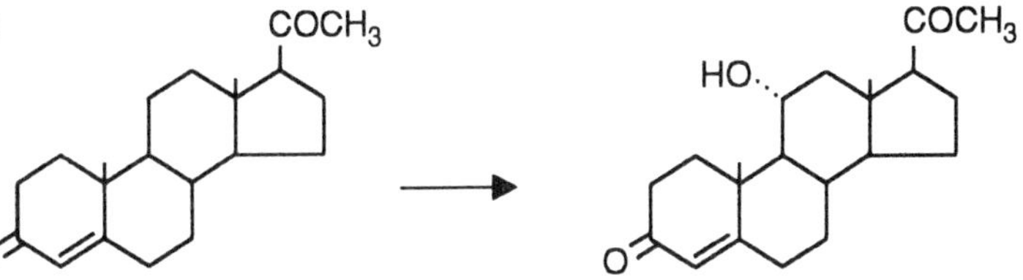

2)

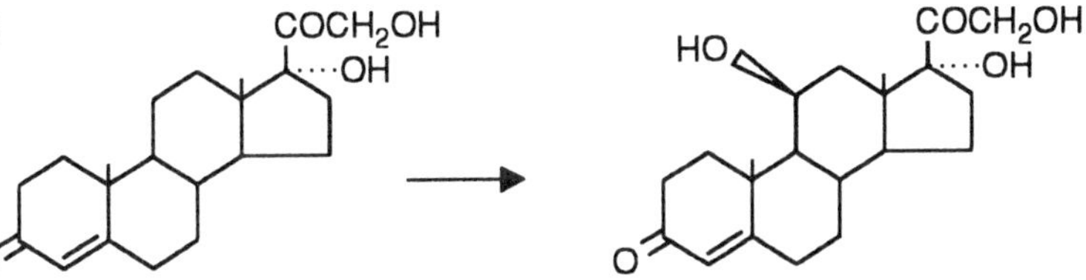

3)

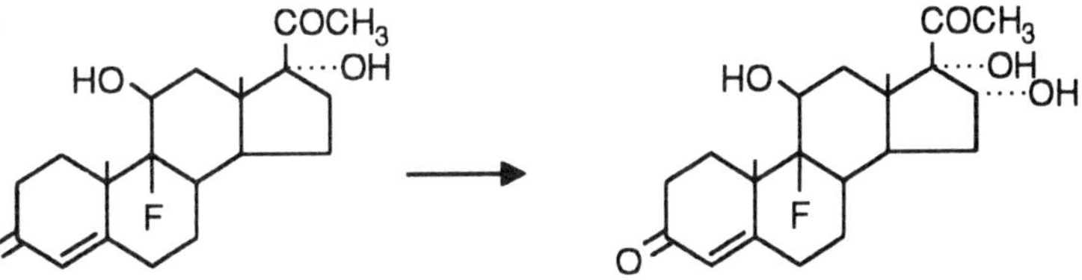

List of organisms:

Saccharomyces cerevisiae
Steptomyces roseochromogene
Rhizopus nigrans
Curvularia lunata
Corynebacterium sp

9.4.2 Alcohol - ketone interconversions

The interconversion of alcohols to ketones is a common biochemical reaction. The introduction of hydroxyl groups into the steroid nucleus and side chain creates a variety of secondary alcohols. Some of these, especially at positions 3, 7, 11 and 17 are frequently oxidised to ketones.

We have illustrated this type of reaction by showing the conversion of a 11α-hydroxy-steroid to a ketone.

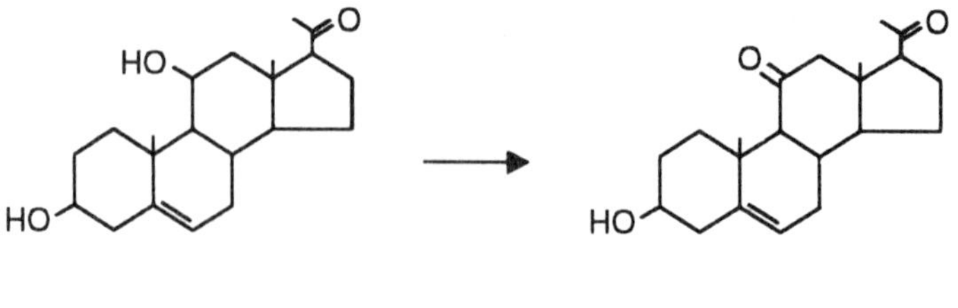

11α-hydroxy-steroid 11 keto-steroid

reduction of 17-keto steroids

The major commercial application of this type of reaction is the reduction of 17-keto steroids to the corresponding secondary alcohol. The manufacturers Schering AG use enzymes from *Saccharomyces* sp to reduce secosteroid (*rac*-3-methoxy-8, 14, secoestra-1,3,5(10),9(11)-tetraene-14,17-dione) and androst-4-ene, 1,17 dione.

9.4.3 Desaturation

Although there are many potential places around the steroid nucleus where desaturation may occur, only a few are of commercial value. The most important is the introduction of a second double bond in ring A already containing a double bond at position 4. Thus:

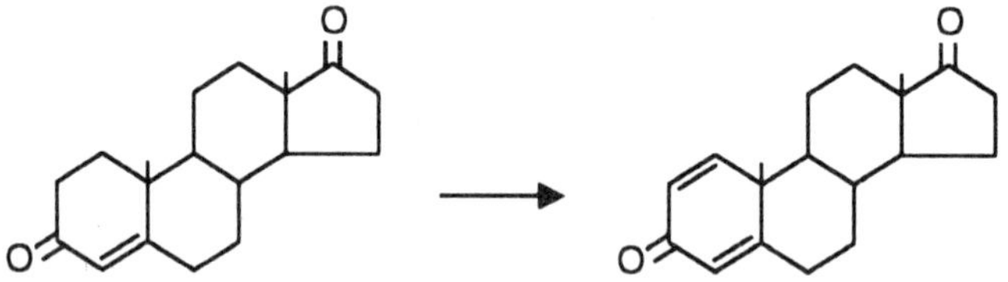

The desaturation takes place by the stereospecific removal of hydrogens from C-1 and C-2 (in fact it is the 1α and the 2β hydrogens that are removed).

ring aromatisation

Dehydration at position one often leads to ring A aromatisation. Thus:

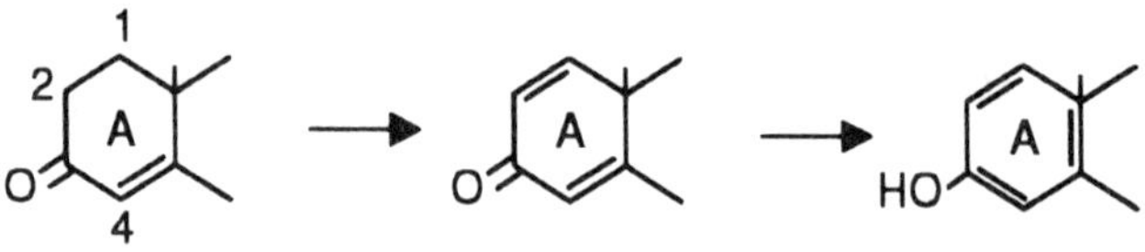

The use of microbial 1-dehydrogenations is essential to the manufacture of corticosteroids, as chemical dehydrogenation processes are commercially non-competitive.

organisms

A variety of enzymes are available from various organisms including *Arthrobacter simplex, Bacillus cyclooxydans, B. lentus, B. sphaericus, Mycobacterium globiforme* and *Septomyxa affinis. Arthrobacter simplex* is particularly useful as it will dehydrogenate a range of sterols/steroids. Other dehydrogenation processes show varying specificities. The Shering Corporation uses *Arthrobacter simplex* to 1-dehydrogenate hydrocortisone, whilst Upjohn employs *Septomyxa affinis* to 1-dehydrogenate dienodiol (11β, 21-dihydroxypregna-4,17(20)-diene-3-one).

9.4.4 Isomerisation

Although many isomerisations are possible, relatively few are of commercial value. Most involve the movement of double bonds, typically illustrated by the conversion of Δ^5-3 ketones to Δ^4-3-ketones.

movement of double bonds

The isomerase (EC 5.3.3.1) from *Pseudomonas testosteroni* has been studied in detail. This enzyme transfers a hydrogen from position 4 to the 6β-position. Although several isomerases have been detected, their presence is often seen as presenting problems as they frequently lead to product diversification.

9.5 Transformation of other terpenoids

monoterpenoids

In section 9.4 we described some of the transformations that are employed in the manufacture of valuable steroids. Sterols and steroids are, however, only a small fraction of the total range of terpenoids produced within the biosphere. Others include monoterpenoids based on the structures:

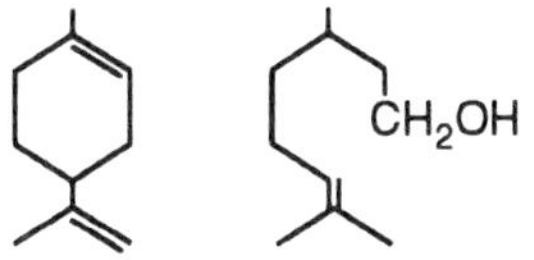

These include a wide variety of compounds used as flavours and fragences. Others, including citronellal, citral and limonene, are used as starting points for the production of more valuable terpenoids.

menthol

As an illustration of the use of microbial transformation of monoterpenoids we can cite the production of menthol from citronellal. Various organisms, including *Pseudomonas aeruginosa* and *Penicillium digitatum* have been used for this conversion. High yields (65% for *P. aeruginosa*; 93% for *P. digitatum*) have been claimed. We have illustrated these processes in Figure 9.7.

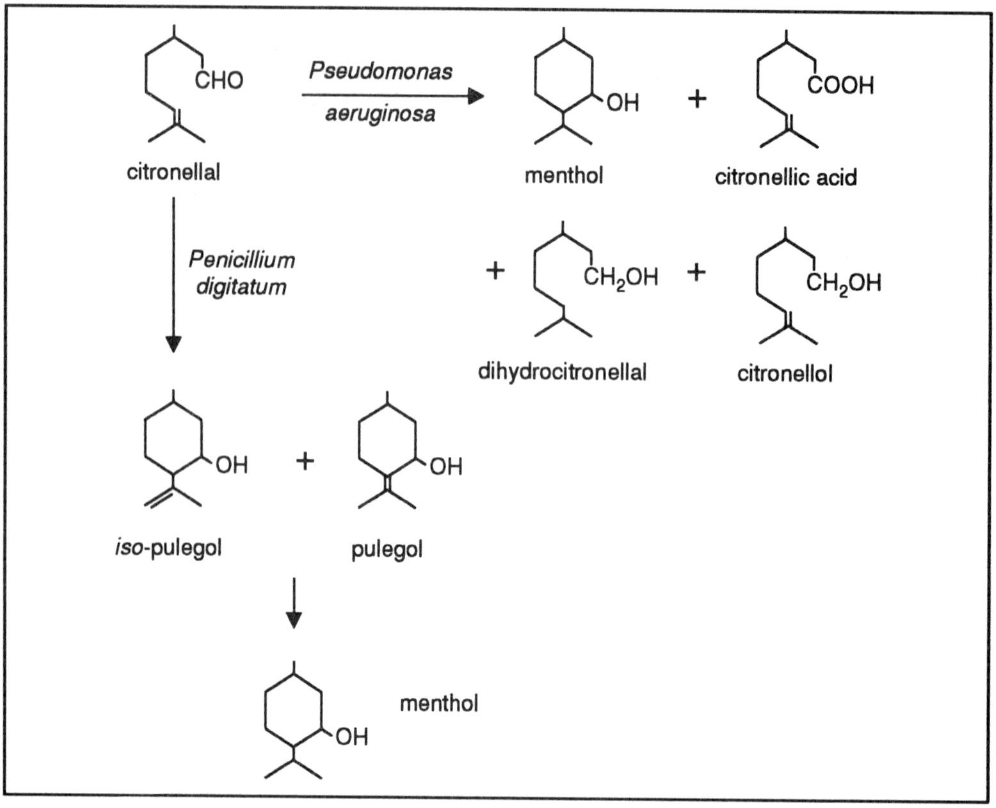

Figure 9.7 Examples of the microbial transformation of citronellal to menthol.

Cyclical monoterpenes such as limonene have also been used as substrate for the production of valuable products. A good example is the conversion of limonene to α-terpineol by *Cladosporium* sp. Thus,

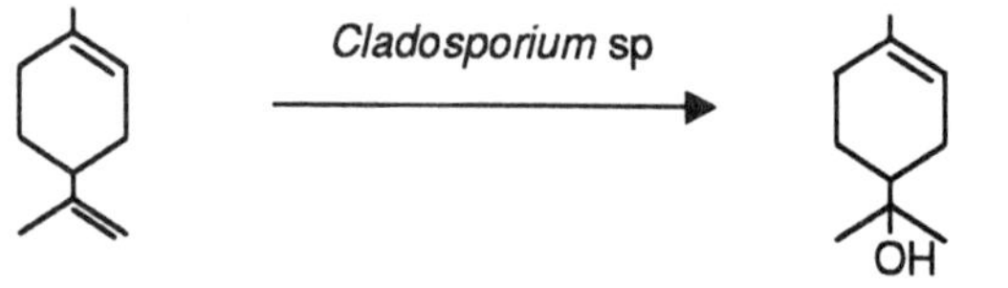

limonene α-terpineol

ionones

Related to the monoterpenes are the ionones based on the structure:

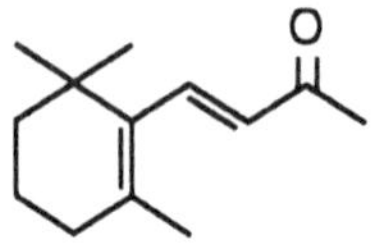

Using a variety of organisms, a range of chemical modifications may be carried out. We have summarised these in Figure 9.8.

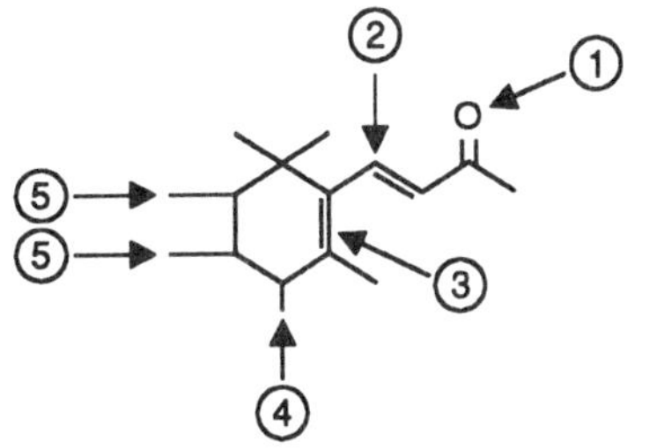

Figure 9.8 Typical enzyme-mediated modifications to ionones.

A wide variety of products are, therefore, possible. Many find commercial use as constituents of essential oils.

It is beyond the scope of this text to examine all of these processes in depth. Much literature is available, but we would recommend Volume 6a of Biotechnology, edited by Kieslich K (published by Verlag Chemie, 1984) as a good source if you wish to pursue this aspect further.

resolution of racemic mixtures

Before we leave the enzymatic modification of terpenoids, we should point out that enzymes are also employed to resolve racemic mixtures of terpenoids. The principles of this are similar to those employed in the resolution of racemic mixtures of amino acids (see Chapter 8).

Π See if you can suggest a strategy for resolving a racemic mixture of DL-menthol to produce L-menthol. Assume that the DL-menthol is produced as the succinate ester.

You probably came up with the suggestion that by using an esterase which selectively hydrolyses the succinate ester of L-menthol, you would be able to isolate L-menthol from the mixture. This is in essence the way the process is carried out commercially. We can represent this process by:

$$\text{DL-menthyl succinate} \xrightarrow[\text{esterase}]{\text{L-specific}} \text{L-menthol + D-menthyl succinate}$$

The process uses cells of *Rhodotorula minuta* entrapped in polyurethane. These cells selectively hydrolyse the L-ester. The remaining D-menthyl succinate is then hydrolysed and the D-menthol racemised via D-menthone and then recycled. We have represented this process in Figure 9.9.

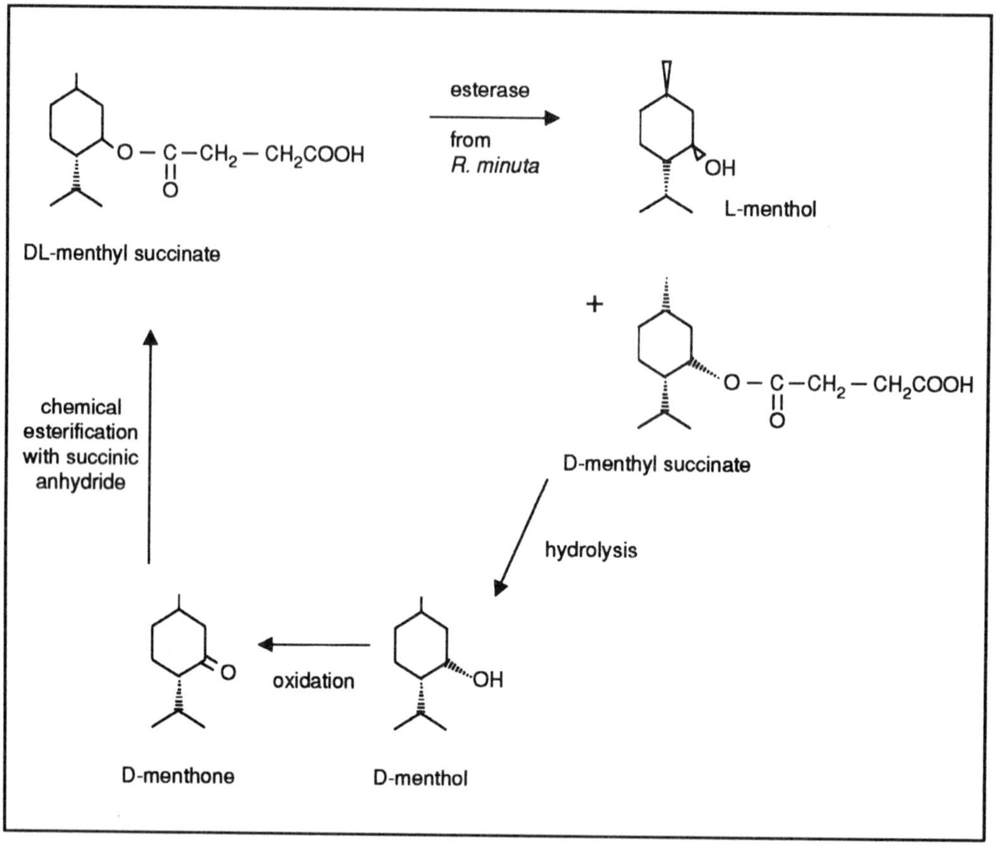

Figure 9.9 Resolution of DL-menthol using an esterase from *Rhodotorula minuta* var *texensis*.

The resolution of DL-menthol is important industrially. L-Menthol has a mint taste and gives a cooling sensation. It finds use in a number of important products including toothpaste and confectionary. D-Menthol does not have the same taste nor the same cooling properties. DL-menthol can be produced relatively simply using a variety of chemical routes.

SAQ 9.5

Using terms given in the list below, identify the nature of the following biotransformations.

1)

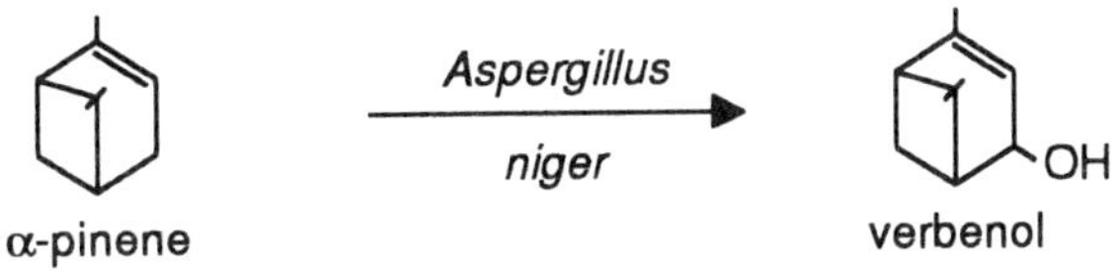

2)

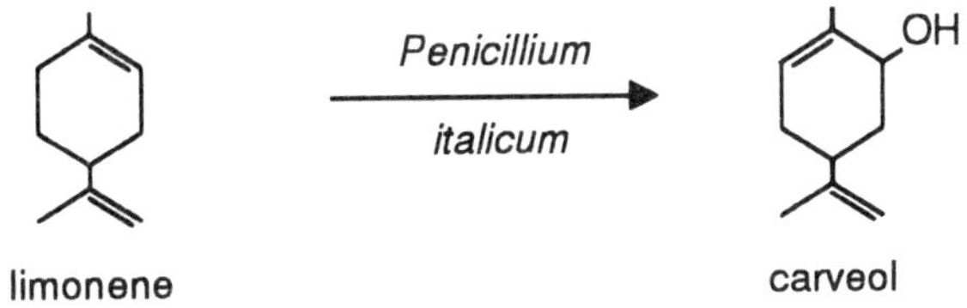

3)

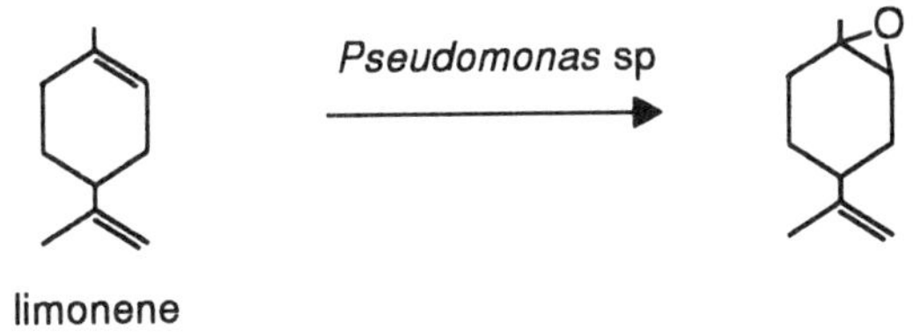

4)

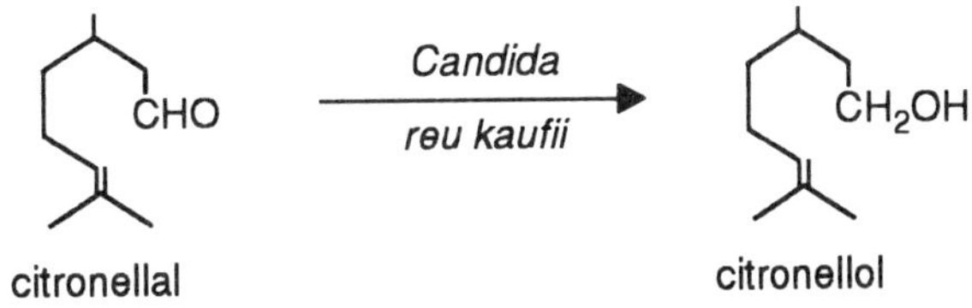

Term list

- reduction;
- hydrogenation;
- epoxidation;
- oxidation;
- hydroxylation.

SAQ 9.6

Candida cylindricae produces a lipase which will esterify L-menthol using 5-phenyl valeric acid.

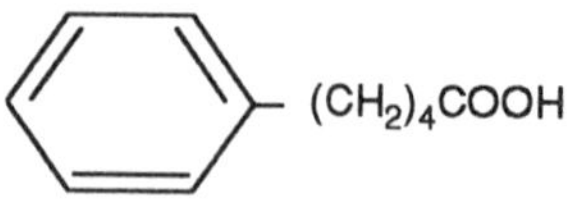

5-phenyl valeric acid

Explain how this enzyme may be used to resolve a racemic mixture of DL-menthol. Assume you want to prepare both D-menthol and L-menthol.

9.6 Chemical conversion of miscellaneous organic compounds

In the previous sections, we have given illustrative examples of biologically mediated chemical transformation of various terpenoids, particularly emphasising their application in chemically modifying sterols, steroids and monoterpenes. Analogous reactions are available for a whole range of organic molecules, be they of biological or chemical origin. In this section, we will briefly examine the application of biologically mediated chemical transformation to these compounds. Again we have been very selective, choosing a few examples simply for illustrative purposes. To give a comprehensive review of all of the possibilities would require a series of texts.

9.6.1 Biotransformation of alicyclic compounds

Π Consider that you have a racemic mixture of the following compound.

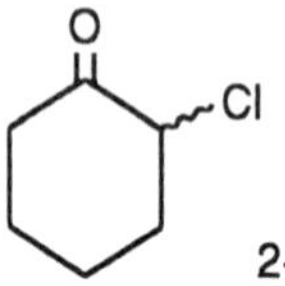

2-chlorohexanone

See if you can come up with an extensive range of possible applications of biological transformations that may facilitate separation and use of the racemates. (Think of the application of enzymatically mediated transformations which we describe earlier, this should help you to come up with several ideas).

The list of possibilities you could have come up with is quite extensive. You might, for example, have suggested that enzymatic processes may be used to resolve the racemic mixture.

reduction of ketone

This has, for example, been achieved using strains of *Saccharomyces cerevisiae*. These strains reduce the ketone to a secondary alcohol only if the chlorine on position 2 is in the R configuration. Thus:

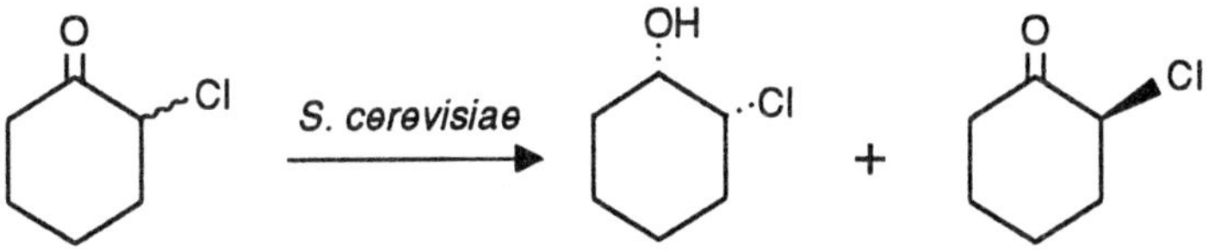

2 chlorohexanone — 1S hydroxy 2R chloro cyclohexane — 2,S-chloro cyclohexanone

With this single example we have in fact described two uses of enzymes in alicyclic chemistry, the reduction of ketone groups and the resolution of racemic mixtures.

Other possible transformations that are theoretically feasible are:

- the introduction of double bonds in the ring by using dehydrogenation reactions;
- specific hydroxylations.

The products of these reactions may be further modified by, for example, epoxidation or esterification.

Using this simple example, we have illustrated how the principles we established in discussing steroid transformation may be applied to a much wider range of organic materials.

Π In Figure 9.10 we have illustrated a few compounds. It would be instructive for you to take a separate sheet of paper for each compound, draw the compound at the top of the sheet and then draw out some products that may be produced from each of these using enzymatically mediated transformations. For each, you should consider the possibilities that arise from:

- hydroxylation reaction;
- dehydrogenation/hydrogenation reaction;
- alcohol/ketone interconversions;
- esterification;
- amidation;
- epoxidation.

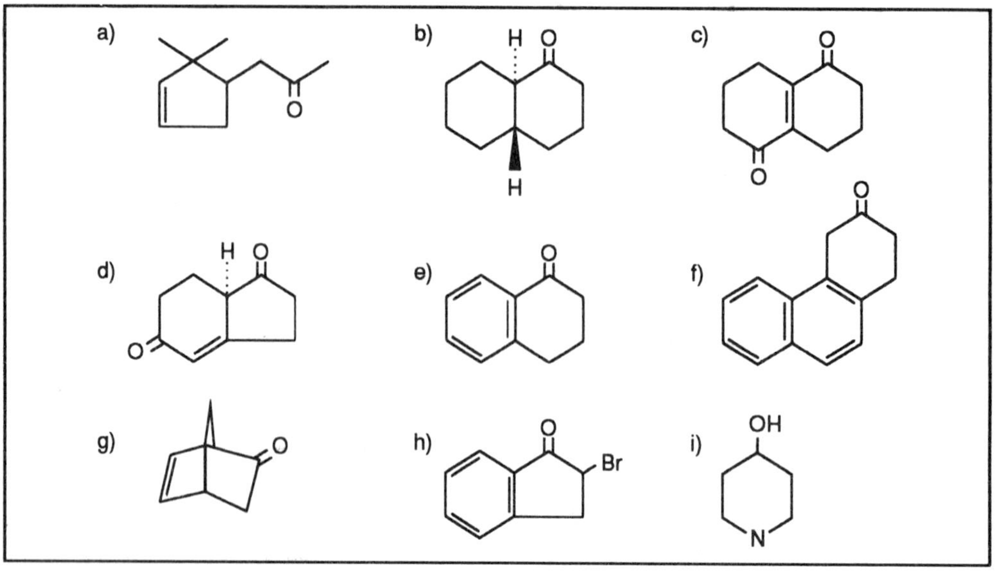

Figure 9.10 Compounds that are to be used for the in-text activity.

If you have carried out this in-text activity, you should be impressed by the enormous range of possible products that could be made. Whether or not these products are of commercial value, or the enzymatic conversion can be conducted in a commercially viable manner is another matter. Nevertheless, chemists have, over the past few decades, turned increasingly to microbial systems to mediate desirable chemical changes. An extensive review of the biotransformation of alicyclic and heteroalicyclic compounds is given by A. Kergomard in Biotechnology - Vol 6a edited by Kieslich K (published by Verlag Chemie, Weinheim 1984), for those who would like to examine this area in greater detail. Kergomard not only reviews the chemical modifications that may be carried out by biological systems, but also lists examples of organisms which will carry out these modifications. For example, he lists 21 species which will hydroxylate substituted phenylcyclohexanes. Such lists also indicate that the organisms that may be of value are drawn from diverse genera and include both prokaryotic and eukaryotic species.

wide range of chemical transformations and organisms

In the same text, P. R. Wallnöfer and G. Engelhardt have reviewed the application of biotransformation of aromatic and heterocyclic compounds, again describing a wide range of chemical transformation and organisms.

The metabolic processes underpinning the catabolism of aliphatic and aromatic compounds are described in the BIOTOL text "Energy Sources for Cell".

Before we leave the chemical transformation of these compounds we draw your attention to the fact that these transformations may be of value not only to produce desirable compounds, but also as a means of removing undesirable materials from the environment. This aspect is examined in the BIOTOL text "Biotechnological Innovations in Energy and Environmental Management".

9.7 Production and use of fatty acids and their derivatives

No discussion of the use of biotransformation in lipid chemistry would be complete without some mention of chemical transformation relating to fatty acids. Fatty acids are a major component of the lipid fraction of organisms. They are mainly found as components of triglycerides and phospholipids, although they may occur in smaller quantities as free fatty acids or as esters of other moieties. Fatty acids, either as free acids or as esters, are valuable commodities in the food and cosmetics industries. They may also serve as precursors of a variety of other compounds.

9.7.1 Production of desirable triglycerides

degree of saturation and chain length

Much of the efforts in the food industry centres on creating oils/fats with desirable characteristics, especially fats that will melt at about body temperature. These give the desired "melt in the mouth" feel. The melting point of triglycerides is governed by the degree of saturation and chain length of the fatty acids which esterifies the glycerol. We have illustrated the composition of cocoa butter and palm oil in Table 9.8. Palm oil, with a high unsaturated fatty acid content, has a low melting point (hence it is an oil). The fatty acids of cocoa butter are, on average, longer chained and more saturated and hence these triglyceride are solid at ambient temperature.

Triglyceride	Palm oil mid-fraction (% dry weight)	Cocoa butter (% dry weight)
StStSt	5	1
POP	58	16
POSt	13	41
StOSt	2	27
StLnSt	9	8
StOO	4	6
Others	2	1

Triglyceride	Fatty acids	Triglyceride	Fatty acids
StStSt	stearate stearate stearate	POP	palmitate oleate palmitate
POSt	palmitate oleate stearate	StOSt	stearate oleate stearate
StLnSt	stearate linoleic stearate	StOO	stearate oleate oleate

Table 9.8 Triglyceride composition of palm oil and cocoa butter.

Cocoa butter has the desired "melt in the mouth" property and is of high commercial value in comparison with palm oil. On the other hand, palm oil is more abundant than cocoa butter. The question is, can we convert palm oil to a product which has the desired properties of cocoa butter? The answer is yes, by using lipases.

Selection of lipases

The enzymes used for modification of oils and fats are extracelluar microbial lipases. They are excreted by micro-organisms into the growth medium to catalyse the degradation of lipids, and can be produced on a large scale by fermentation.

lipase action

Lipases catalyse the hydrolysis of oils and fats to give diglycerides, monoglycerides, glycerol and free fatty acid. The reaction is reversible, and consequently microbial lipases also catalyse the formation of glycerides from glycerol and free fatty acid. Because of the reversibility of the lipase reaction, hydrolysis and re-synthesis of glycerides occur when the enzymes are incubated with oils. These cause an exchange of fatty acid groups between triglyceride molecules giving interesterified products (Figure 9.11).

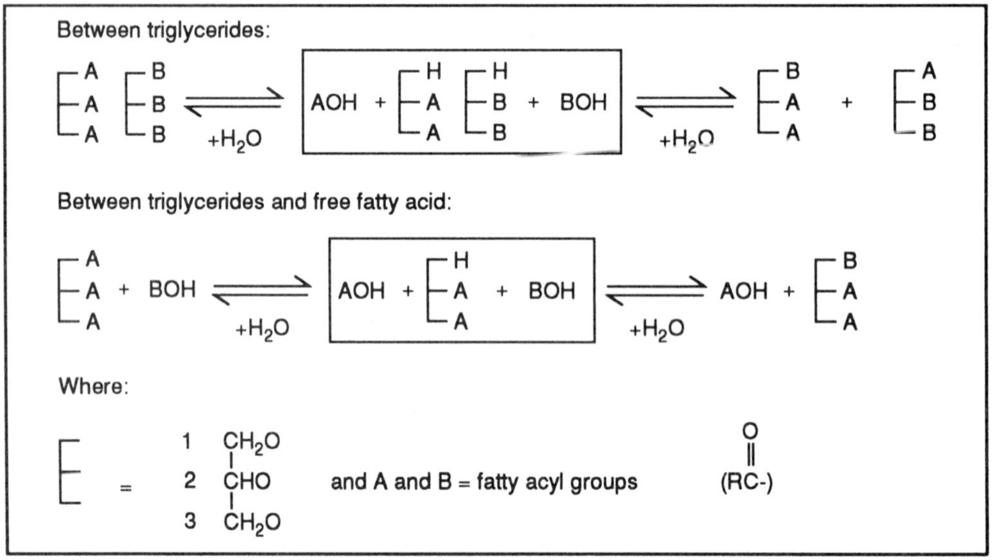

Figure 9.11 Mechanisms of lipase-catalysed interesterifications.

lipases active in hydrophobic environments

Microbial lipases are catalytically active in a predominantly organic environment containing very small amounts of water. Under these conditions the hydrolytic action of the enzymes is restricted by the limited availability of water, and high yields of interesterified triglycerides can be obtained. Mixtures of triglycerides and free fatty acids can be used as reactants for lipase catalysed interesterifications. In these cases, free fatty acid exchanges with the fatty acyl groups of the triglycerides to form new triglycerides enriched in the added fatty acid.

The substrate specificities of lipases are crucial to their application as catalysts for modification of triglycerides. The enzymes can show specificity with respect to both the fatty acid and glycerol parts of triglycerides. However, most extracelluar microbial lipases are not highly specific with respect to the fatty acid groups found in the oils and fats used as raw materials for the edible fats industry, although reaction rates can vary with the chain length, and extent and position of unsaturation of the fatty acid group.

With regards to the glycerol part of triglycerides, the specificities of lipases are of technical significance. Some microbial lipases are not specific and catalyse reactions at all three positions of glycerol. When lipases of this type are used as catalysts for interesterification of triglyceride mixtures, products containing a random distribution of fatty acid groups are obtained. These products have similar composition to those obtained by chemical interesterification.

regiospecificity

stereospecificity

A second group of lipases catalyse reactions specifically at both the outer (1- and 3-) positions of glycerol. These enzymes are said to show regiospecificity. This specificity can be exploited to produce triglycerides which are difficult to obtain by conventional chemical procedures. Regiospecific lipases catalysing reactions selectively at either the 1- or 3-positions of triglycerides would have very useful commercial applications. Microbial lipases show stereospecificity (specificity for stereoisomers) in reactions with many types of esters, but unfortunately stereospecificity in reactions with triglycerides has not been detected to date.

lipase stability

In addition to having the required specificity, lipases employed as catalysts for modification of triglycerides must be stable and active under the reaction conditions used. Lipases are usually attached to supports (ie they are immobilised). Catalyst activity and stability depend, therefore, not only on the lipase, but also the support used for its immobilisation. Interesterification reactions are generally run at temperatures up to 70°C with low water availability. Fortunately many immobilised lipases are active and resistant to heat inactivation under conditions of low water availability, but they can be susceptible to inactivation by minor components in oils and fats. If possible, lipases resistant to this type of poisoning should be selected for commercial operations.

Interesterification catalysts

reaction mixture

The reaction systems used for modification of triglycerides usually consist of a lipase catalyst and a small amount of water dispersed in a bulk organic phase containing the reactants and, if required, a water immiscible solvent. The small amount of water in the reaction system partitions between the catalyst and the bulk organic phase.

Lipases catalyse reactions at interfaces, and to obtain a high rate of interesterification the reaction systems should have a large area of interface between the water immiscible reactant phase and the more hydrophilic phase which contains the lipase. This can be achieved by supporting the lipase on the surface of macroporous particles.

enzyme adsorption onto macroporous particles

Highly active catalysts have been produced by adsorption of lipases onto macroporous acrylate beads, polypropylene particles and phenol-formaldehyde weak anion exchange resins. Protein is bound, presumably essentially as a monolayer, within the pores of the particles. The large surface area of the particles ($10m^2\ g^{-1}$) means that substantial amounts of protein can be adsorbed, and the pores are of sufficient size to allow easy access of reactants to this adsorbed protein.

Π When choosing a support for an immobilised enzyme, what other factors (apart from activity and access to the substrate) do you think need to be considered? (Think about the cost of producing the immobilised system and how it will be used).

As well as being active, the immobilised enzyme also needs to be stable (active for a long period) and the support must promote this. The support must also have appropriate mechanical characteristics: it should not disintegrate if used in a stirred tank reactor; it should produce even flow (without channelling) in a packed bed reactor. The cost of the support is also important.

The interesterification processes

Mixtures of triglycerides, triglycerides plus free fatty acids or triglycerides plus fatty acid alkyl esters are used as reactants in fat modification processes. These mixtures are exposed to lipases supported on macroporous particles in the presence of a small amount of water. Liquid substrates (oils) can be reacted without use of a solvent, but with solid reactants (fats) it is necessary to add a solvent to ensure that the reactants and products are completely dissolved in the organic phase. Various water immiscible solvents can be used, but hexane is preferred for commercial operation because this solvent is already used industrially for the processing of oils and fats.

The fat modification processes can be operated either in batches using stirred tank reactors or continuously with packed bed reactors.

Π Which process do you think would be preferred: stirred tank reactors operated batch-wise, or packed bed reactors operated continuously? (Before reading on consider the likely overall yields of each type of operation and the cost of operation. Then make your decision).

productivity

Continuous reactors are likely to give the greatest overall specific productivity (quantity of product formed in a given time from a given quantity of enzyme), and therefore could be most cost-effective. In addition, in batch systems, the longer residence time involved can result in side reactions, leading to a decrease in the yield of triglycerides (the triglycerides are degraded). Continuous systems are also easier to monitor and to regulate by automation.

In a typical reaction, a feedstream consisting of refined palm oil and stearic acid dissolved in petroleum either, is almost saturated with water (water content 0.06%), and then pumped through a bed of regiospecific lipase from *Mucor miehei* supported on diatomaceous earth. High catalyst activity as measured by an increase in the stearoyl content of the triglycerides can be obtained throughout 300 hours of continuous operation. Analysis of the triglyceride products shows that stearoyl groups are incorporated exclusively into the 1- and 3-positions, mostly in exchange for palmitoyl groups. This stearoyl incorporation results in the formation of 1(3)-palmitoyl-3(1)-stearoyl-2-oleoylglycerol (POSt) and,1,3-distearoyl-2-oleoylglycerol (StOSt). POSt and StOSt are the major triglycerides of cocoa butter, a valuable confectionary fat.

Thus when palm oil is incubated in this way its composition shifts and becomes more like cocoa butter (see Table 9.9 and compare with Table 9.8).

Triglyceride	Palm oil (% dry weight)	Post lipase treatment (% dry weight)
StStSt	5	3
POP	58	16
POSt	13	39
StOSt	2	28.5
StLnSt	9	8
StOO	4	4
Others	2	1.5

Table 9.9 The triglyceride content of palm oil pre- and post-incubation with lipase and stearic acid as described in the text.

use of interesterification products

Other reactants have been used for the production of cocoa butter equivalents by the enzyme technique. For example products enriched in StOSt and POSt can be produced by reaction of olive oil, high oleate safflower and sunflower oils, sal fats and shea oleine with stearic and palmitic acids or their esters. It is also possible to use lipases specific at 1,3 sites to produce triglyceride mixtures having useful functional properties in products such as margarines, low-calorie spreads and bakery fats. An example is the formation of triglycerides containing two long chain saturated fatty acid groups and one medium or short chain fatty acid group. These fats are effective hardeners for margarines and other spreads.

high value and low value products

It has been shown that lipase-catalysed reactions can be used for the large-scale production of modified triglycerides. At present the technology is being targeted to the production of comparatively high-value products such as confectionary fats. Wider application of the reactions to lower-value, higher-tonnage products will be dependent on the development of cheaper processes using more-productive and/or cheaper catalysts.

better supports for enzyme immobilisation

Fortunately there are indications that immobilised lipase catalysts will become more efficient and cheaper in the future. In the past, because of low fermentation yields, lipases have been expensive in comparison with the other main groups of extracelluar microbial enzymes such as proteases. Application of gene transfer technology to lipases could make them available at lower cost in the future. Considerable attention is also being given to the development of more effective supports for enzyme immobilisation. A range of organic and inorganic materials are being investigated as potential enzyme supports, and parameters which affect the activity expressed by immobilised enzymes are being studied.

9.7.2 Production of fatty acids and related compounds

natural sources

The fatty acids commonly encountered in biological systems are straight chained alkanoic or alkenoic acids, containing an even number of carbon atoms (usually C_{14}-C_{22}). In general, these fatty acids can be produced readily by extraction of the lipids from natural sources and saponifying the neutral triglycerides. This is satisfactory providing a mixture of fatty acids is acceptable. Purification of specific fatty acids from the saponification mixture increases the costs considerably.

Considerable interest arose during the 1970's and 1980's in the use of micro-organisms to produce useful fatty acids and related compounds from hydrocarbons derived from the petroleum industry. During this period, a large number of patents were granted in Europe, USA and Japan protecting processes leading to the production of alkanols, alkyl oxides, ketones, alkanoic acids, alkane dioic acids and surfactants from hydrocarbons. Many of these processes involved the use of bacteria and yeasts associated with hydrocarbon catabolism.

bacteria and yeasts

Let us see if we can first establish some principles.

Π If you were attempting to produce the alkanol, C_{16}-OH, what would be a suitable substrate?

We hope you would suggest a C_{16} alkane.

Π If you were attempting to produce a C_{10} alkanoic acid, what would be a suitable substrate?

Again, we hope you would suggest C_{10} alkane.

The point we are trying to establish is that to some extent we control the nature of the product by selecting a particular substrate.

Π Do you think we need to select a different organism for each product we would like to produce?

The answer is, generally, no. Most of the enzymes involved are specific in terms of the reaction they catalyse, but will work with a range of substrates. Some *Candida* sp for example contain an enzyme system which will convert *n*-alkanes into alkane dioic acids:

$$CH_3(CH_2)_nCH_3 \longrightarrow HOOC(CH_2)_n\,COOH$$

choice of substrate

Thus if provided with C_{12} the product will be C_{12} dioic acid, whilst if provided with C_{18} the product will be C_{18} dioic acid. If supplied with a mixture of alkanes, then a mixture of dioic acids will be produced. There are, however, some restrictions on this and not all alkanes will be oxidised equally. Thus the *Candida* sp generally use C_{12}-C_{18} alkanes, whilst *Corynebacteria* will preferentially utilise C_{12}-C_{22} alkanes.

We should not, however, mislead you into believing that the product is always predictable. Strains of *Candida topicalis* (FERM P3291) will preferentially produce C_{10} dioic acid even if a mixture of C_{11}-C_{18} alkanes are used.

In Table 9.10 we have listed some examples of substrates, products and organisms that have been cited in the patent literature for the production of various alkanols, alkanoic acids, ketones and dioic acids. These are meant to act as illustrations, we would not expect you to remember them all.

Π Which type of organism, prokaryotic or eukaryotic, seem to dominate the list of organisms which appear to be useful for the chemical modification of alkanes?

Most of the organisms listed in Table 9.10 are prokaryotes. The exceptions are strains of *Candida* and *Torulopsis*.

Products	Substrates (alkanes unless otherwise specified)	Organisms
Alkanols		
C_1-OH	C_1	*Methylococcus*
	C_2H_4-C_4H_8	*Methylosinus trichosporium*
C_2-OH	C_2H_4-C_4H_8	*Methylosinus trichosporium*
C_6-C_{16}-OH	C_6-C_{16}	*Methylosinus trichosporium*
C_3-C_6-OH	C_3-C_6	*Methylosinus* sp
		Methylococcus sp
		Methylobacter sp
C_8-C_{16}-OH	C_8-C_{16}	*Acinetobacter* sp
C_{11}-C_{20}-OH	C_{11}-C_{20}	*Arthrobacter* sp
		Brevibacterium sp
		Corynebacterium sp
		Micrococcus sp
		Nocardia sp
Alkyloxides		
C_2-oxide	C_3H_6	*Methylocystis* sp
C_3-oxide	C_4H_8	*Methylomonas* sp
C_4-oxide	C_4H_6	*Methylobacter* sp
C_2-oxide	C_2H_4-C_4H_8	*Methylamonas* sp
C_8H_{16}-oxide	C_8H_{14}	*Pseudomonas oleovorans*
Ketones		
C_3-C_6 ones	C_3-C_6	Wide range of methanotrophic bacteria
C_6-C_2 ones	C_{16}	*Arthrobacter* sp
C_6-C_3 ones	C_6H_{14}	*Arthrobacter* sp
C_6-C_4 ones	C_4H_{14}	*Arthrobacter* sp
C_6-C_{20} ones	C_6-C_{20}	*Nocardia* sp
Alkonoic acids		
C_{10}-C_{20}-oic	C_{10}-C_{20}	Wide range of organisms especially *Candida* sp, *Torulopsis* sp, *Arthrobacter* sp, *Corynebacterium* sp
Note that, by using alkane derivatives such as alkylchloride, corresponding derivatives of the alkanoic acids may be produced		
Alkane dioic acids		
C_x-dioic acid	C_x	(x can be between 10-22) Wide range of organisms including *Candida* sp, *Torulopsis* sp, *Corynebacterium* sp.

Table 9.10 Examples of microbial products produced from alkanes and related compounds. All the examples cited are the subject of patents.

Prostaglandins

Prostaglandins are important derivatives of unsaturated fatty acids. Below we have drawn the structure of prostaglandin E_2 as an illustration of the general structure of the prostaglandins.

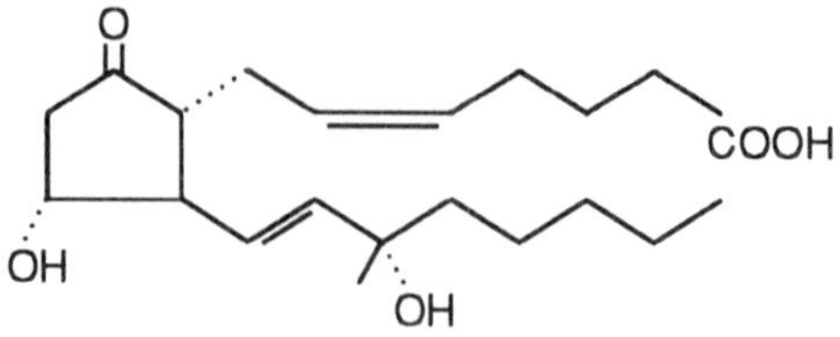

prostaglandin (PGE_2)

There are however a very large number of prostaglandins and these show differing pharmacological properties. They have therefore different applications in health care. These compounds are, however, only produced in very small quantities by natural systems. Therefore a variety of strategies have been developed to produce these compounds in larger quantities *in vitro*. This usually involves a mixture of chemical and biological procedures. For example, biological catalysis is used to:

biological catalysis

- synthesise chiral reactants for subsequent chemical modification;
- resolve racemic mixtures produced by chemical modification;
- undertake stereo specific reactions.

Since you are now familiar with the types of chemical modifications that may be mediated by biological systems, you should be able to answer the following SAQ regarding the modification of prostaglandins.

SAQ 9.7

Consider the structure of prostaglandin PGE_2

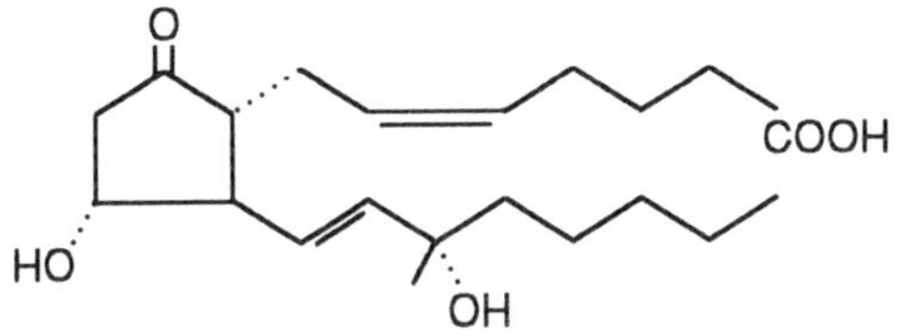

Prostaglandins belonging to the PGF_α group have the following ring structure:

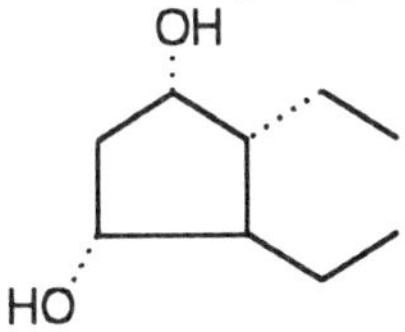

whilst prostaglandins belonging to the PGF_β group have the ring structure:

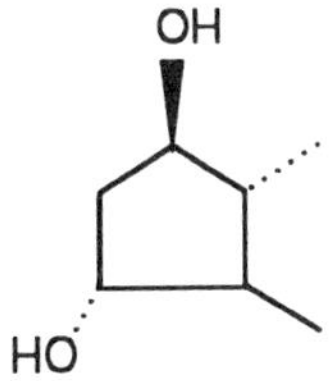

Assume you have a plentiful supply of PGE_2, suggest two strategies for producing samples of PGF_α and PGF_β prostaglandins, using PGE_2.

9.8 Selection of production systems for the biotransformation of lipids

Generally speaking we consider that most micro-organisms live and grow in aqueous environments, and that the cytoplasm within cells in which enzymes function is also aqueous. On the other hand, most lipids are only sparingly soluble in aqueous media. Cholesterol, for example, has a solubility of less than 2 mg l^{-1} (equivalent to a concentration of less than 5 μmol l^{-1}). Even at much lower concentrations (25-40 nmol l^{-1}) it tends to aggregate into micelles. There is, therefore, a general problem of how to supply lipid substrates at sufficient concentration to produce reaction kinetics that are appropriate for industrial purposes.

solubility problem

Π See if you can write down at least two ways that this problem may be overcome.

The two most commonly used strategies are:

selection of solvent system

- reduction of the aggregation of lipids in aqueous solvents by using non-ionic surface active agents, such as Tween. To achieve an even, finely dispersed distribution of the substrate, this is usually dissolved in an organic solvent (acetone, ethanol, dimethyl sulphoxide) prior to being mixed with the culture broth containing Tween;

- use of the enzyme in a non-aqueous solvent or in a finely dispersed aqueous emulsion within an organic solvent containing the substrate. The industrial application of enzymes using non-aqueous systems has been discussed in detail elsewhere in the BIOTOL text "Technological Applications of Biocatalysts", so we will not elaborate on this aspect here.

immobilised systems

In addition to selection of solvent system, we also have a choice of using freely suspended enzymes/organisms or immobilised systems. Increasingly, attention is being paid to immobilised systems because of their potential advantages (cost, purity and ease of recovery of products). Again, this aspect of industrial enzymology is discussed in detail in the BIOTOL book "Technological Applications in Biocatalysts", so we will not discuss this aspect further.

Summary and objectives

In this chapter, we have examined the use of cells and enzymes to chemically transform lipids. We have had to be selective and have predominantly focused attention on the transformation of sterols and steroids. We first explained why these compounds were commercially important and why they only occur in low concentrations in natural systems. We pointed out that a very large number of reaction types are possible, but those which have found greatest use include stereospecific hydroxylations, alcohol/ketone interconversion, hydrolysis, conjugation and isomerisation.

We included some discussion of other terpenoids and other organic molecules, including aliphatic and alicyclic materials. We also considered the use of enzymes in producing triglycerides with desirable characteristics.

Now that you have completed this chapter you should be able to:

- explain why steroids are of value and why they only occur in low concentrations in nature;
- describe the major molecular differences between sterols and steroids;
- list the options for producing steroids commercially and explain the advantage and disadvantages of these options;
- explain, using examples, the options available for selectively removing the side chains of sterols, including using modified sterols, selective enzyme inhibition and mutants;
- list the important reaction types used in sterol/steroid interconversions;
- identify suitable organisms for conducting particular chemical transformations for a number of substrates, but especially for steroids and other terpenes;
- identify a wide range of reaction types;
- explain how enzymes may be used to produce lipids, especially triglycerides, with desirable characteristics;
- list some examples of how hydrocarbon-utilising organisms may be used to introduce functional groups into hydrocarbons.

Responses to SAQs

Responses to Chapter 2 SAQs

2.1 The reasons are:

1) True: Growing and dividing cells need to use substrate to provide energy and materials for growth, maintenance and product formation. In immobilised (non-growing) systems the energy and materials are only required for cell maintenance and product formation.

2) True: Separation of the biocatalyst (enzyme or cells) from the product stream is not required and relatively few other substances are present.

3) True: The concentration of immobilised biocatalyst in a reactor can be much greater than for systems using soluble of free cells.

4) True: Higher substrate concentrations can be used, which often enables higher product concentrations to be achieved, thus reducing volume of effluent for a given product yield.

2.2 The biotransformation should be carried out using whole cells because:

1) The multi-component enzyme system is likely to have much lower activity in enzyme preparations.

2) The reaction has a cofactor requirement (see Table 2.1) which is relatively easy to satisfy when using whole cells.

2.3

1) Batch. Fermentation run times are relatively short and the same bioreactor can be used for making several different products.

2) Continuous. Long fermentation run times, with many generations, increase the chance of mutation or loss of plasmid DNA.

3) Continuous. Difficult to maintain sterile conditions over very long periods. Contaminants may grow faster (out compete) process organisms and take over the vessel.

4) Batch. Fermentation run times are relatively short.

5) Batch. All waste products (metabolites) accumulate in batch culture; in continuous culture much is lost through the outflow.

6) Batch. Batch cultures have a 'stationary phase' during which little or no growth occurs.

7) Continuous. Unlike batch it is not possible to identify all of the materials involved in the fermentation run.

8) Continuous. When in steady state the culture does not change with time and is, therefore, relatively easy to operate and control.

2.4 Costs of downstream processing for bioprocesses are increased by 1) low concentrations of products, 2) numerous impurities at low concentration and 3) intracellular materials (if cell disruption is necessary). However, the high specificity of biocatalysts is a benefit to downstream processing since products closely related to the desired product are less likely to be present. Waste products of bioprocesses are likely to be less environmentally damaging, which also reduces downstream processing costs.

Costs of downstream processing for purely chemical synthesis would be increased by 1) low specificity of reactions (giving rise to chemical contaminants closely related to the desired product) and 2) the toxic/corrosive nature of the chemicals.

2.5

Process A:	fed-batch mode free cells vacuum fermentation
Process B:	batch or fed-batch modes immobilised cells solvent extraction
Process C:	continuous mode free enzyme ultrafiltration with enzyme recycling

Responses to Chapter 3 SAQs

3.1 Reaction equation:

$a'\,CH_x\,O_y + b'O_2 + c'\,H_l\,N_n \longrightarrow C\,H_\alpha\,O_\beta\,N_\delta + d'\,H_2O + e'\,CO_2$

Balances for the four elements:

C: $a' = 1 + e'$

H: $a'x + c'l = \alpha + 2d'$

O: $a'y + 2b' = \beta + d' + 2e'$

N: $c'n = \delta$

Six moles of glucose are used for each mole of biomass produced, thus

$6C\,H_2O + b'O_2 + c'H_4N^+ \longrightarrow C\,H_{1.666}\,O_{0.27}\,N_{0.20} + d'\,H_2O + e'\,CO_2$

ie unknown coefficient are b', c', d' and e'

C: $6 = 1 + e'$	$e' = 5$
N: $c'1 = 0.20$	$c' = 0.2$
H: $6(2) + 0.2(4) = 1.666 + 2\,d'$	$d' = 5.567$
O: $6(1) + 2\,b' = 0.27 + 5.567 + 2.5$	$b' = 4.918$

The reaction equation then becomes:

$6\,CH_2O + 4.918\,O_2 + 0.2\,H_4N \longrightarrow C\,H_{1.666}\,O_{0.27}\,N_{0.20} + 5.567\,H_2O + 5CO_2$

3.2

1) Decrease. Decrease in degree of reductance of substrate increases the demand for NADH. See E - 3.7.

2) Increase. Increased efficiency of oxidative phosphorylation increases the P/O quotient. See E - 3.3.

3) Increase. A lowered energy demand for biomass synthesis increases Y_{ATP}^{max}. See E - 3.4.

3.3

1) Type 1.

2) Types 3 and 4.

3) Type 3.

4) Type 1.

3.4 Productivities at low and high dilution rates are 2.40 and 1.92 kg product m^{-3} h^{-1}. The process should therefore be operated at the low dilution rate.

Since biomass productivity is $D\bar{x}$, product productivity (P) can be calculated as follows:

$$P = \frac{Y_{p/s}}{Y_{x/s}} \cdot \bar{x} \cdot D = \text{kg product } m^{-3}\, h^{-1}$$

3.5 m = 0.013 C-mol substrate/C-mol biomass h^{-1}

$Y_{x/s}^{max}$ = 0.48 C-mol biomass/C-mol substrate

These values were obtained as follows.

D (h^{-1})	$\frac{1}{D}$ (h)	$Y_{x/s}$	$\frac{1}{Y_{x/s}}$
0.9	1.11	0.475	2.10
0.4	2.50	0.470	2.13
0.2	5.00	0.465	2.15
0.1	10.00	0.455	2.20
0.05	20.00	0.426	2.35

We will plot $\frac{1}{Y_{x/s}}$ against $\frac{1}{D}$ since at steady state $\frac{1}{D} = \frac{1}{\mu}$

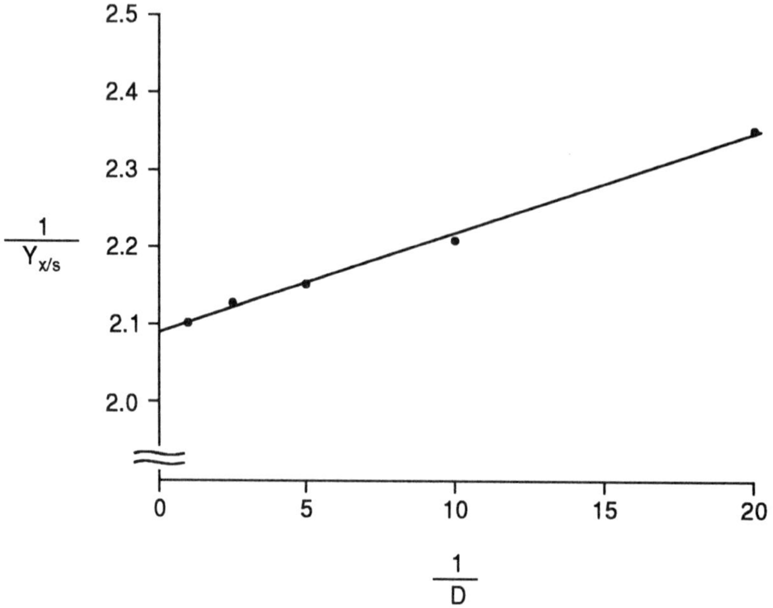

Slope = 0.013 C-mol substrate/C-mol biomass h^{-1} = m

Intercept = 2.09 C-mol substrate/C-mol biomass = $\dfrac{1}{Y^{max}_{x/s}}$

So, $Y^{max}_{x/s}$ = 0.48 C-mol biomass/C-mol substrate.

Y_{x/O_2} is determined from $Y_{x/s}$ using (E - 3.15):

$Y_{x/s} = 0.475$

$\gamma_o = -2 \times 2 = -4$

$\gamma_s = 4 + 2 - 2 = +4$

$\gamma_x = 4 + 1.666 - (3 \times 0.2) - (2 \times 0.27) = 4.526$

$$Y_{x/O_2} = \frac{4}{4.526} \times \frac{0.475 \times \dfrac{4.526}{4}}{1 - 0.475 \times \dfrac{4.526}{4}}$$

= 0.80 C-mol biomass/mol O_2

3.6 P/O quotient = 2.4

This is obtained using: $Y^{max}_{O} = Y^{max}_{ATP} \cdot P/O$

Firstly, $Y^{max}_{O_2}$ is determined graphically:

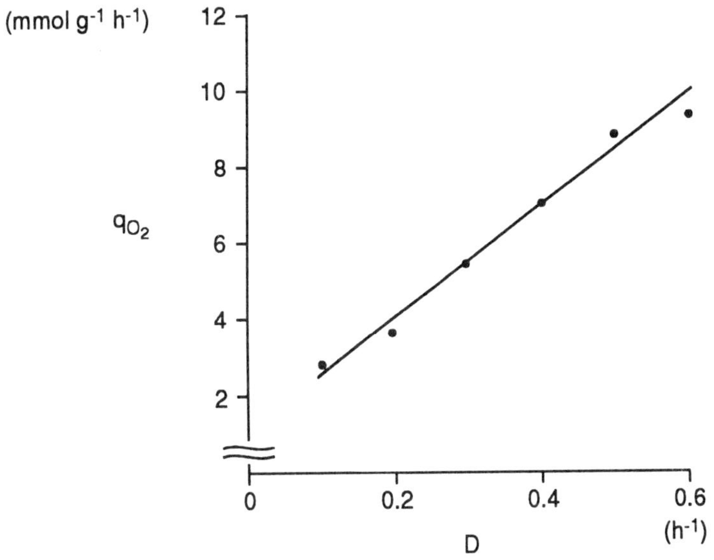

$Y^{max}_{O_2}$ = reciprocal of slope, ie $\dfrac{1}{14.87}$ = 0.067 g mmol^{-1}

= 67 g mol^{-1}

Y^{max}_{O} = 33.5 g mol^{-1}

Y^{max}_{ATP} = 13.9 g mol^{-1}

Therefore: $33.5 = 13.9 \, . \, P/O$ (E - 3.16)

$$P/O = \frac{33.5}{13.9} = 2.4$$

3.7 Statements a), b) and c) are applicable to class 2 metabolites. Since $Y_{O_2}^{max}$ is a measure of growth efficiency, statement d) is the converse of statement b) and is therefore not applicable.

3.8 1)

			Degrees of reductance
more reduced than glucose	↑↓	Ethanol	+ 5.00
		Glycerol	+ 4.67
		Sorbitol	+ 4.33
		Glucose	+ 4.00
		Xylose	+ 4.00
more oxidised than glucose	↑↓	Gluconate	+ 3.67
		Succinate	+ 3.50

Refer to section 3.1.2 if you were unable to calculate degrees of reductance.

2) The specific rate of exopolysaccharide production (q_p) is inversely related to growth efficiency ($Y_{O_2}^{max}$).

3) From Table 3.1: $Y_{x/s} = 80$ g dry wt mol^{-1}

$\mu = 0.2 \text{ h}^{-1}$ ($\mu = D$)

$$\text{So, } Y_{x/s} = \frac{80}{180} = 0.44 \text{ g g}^{-1}$$

We can see from the units of q_s that $Y_s \, . \, \mu = q_s$.

So, $q_s = 0.44 \, . \, 0.2 = 0.039 \text{ g g}^{-1} \text{ h}^{-1}$.

3.9 Class 1 or 2, depending on the substrate used. We can see from Table 2, for example, that succinoglycan biosynthesis leads to a net production of ATP (Class 2) with ethanol as substrate, but the biosynthesis is energy requiring (Class 1) with glucose as substrate.

3.10 1) The organism, as we have already seen, has a relatively high P/O quotient (high growth efficiency). It would therefore seem to have only a limited capacity for energy dissipation. During citric acid production, energy dissipation is desirable to ensure continued operation of glycolysis leading to citric acid formation.

2) A high growth efficiency (high $Y_{O_2}^{max}$) is desirable because sophorolipid production has a high demand for ATP.

Responses to Chapter 4 SAQS

4.1 Items (2) and (3) are the main factors encouraging development of SCP as alternatives to plant proteins. This is a method of converting waste organic materials to produce a valuable product. The speed with which micro-organisms can do this surpasses any form of agriculture.

Items (4) and (5) can be true of both plants and micro-organisms and are thus not a relative advantage to either.

Item (1) is not always true. Some micro-organisms are easier to digest than plants, whereas others (such as algae) are more difficult to digest than many plant foods.

4.2

1) Incorporating a proportion of SCP into manufactured foods can disguise unpleasant flavours or textures. Food technologists have a wide array of flavourings at their disposal, which can be used to produce particular flavours. If this cannot be done it might be possible to use the SCP as feed.

2) Processing the SCP to break up the cells (by milling or some other means), then incorporating it into prepared foods may overcome this problem. Otherwise it might be possible to use the SCP for feed.

3) Blending the SCP with foods containing high levels of the deficient amino acids, or adding the deficient amino acid to the SCP can overcome this problem. Genetic engineering could be used to induce the organism to manufacture deficient amino acids. Otherwise it might be possible to use the SCP for feed.

4) Toxicity of SCP is usually due to high RNA content. Processing the SCP can reduce the RNA content. Otherwise it should be possible to use the SCP for feed.

4.3 3) is the correct response. Algal cultures grown on molasses in open lagoons would become overgrown with bacteria, making the product useless as SCP. Responses 1 and 2 do not apply, as on molasses the organism grows as a heterotroph and requires neither sunlight or CO_2. Response 4 is incorrect as the yield would not be affected solely by the types of system used. In lagoons, yields of algae would in fact probably be less than in bioreactors, but the reason for this is 3) - contaminants would use much of the substrate, leaving less available for algal growth.

4.4

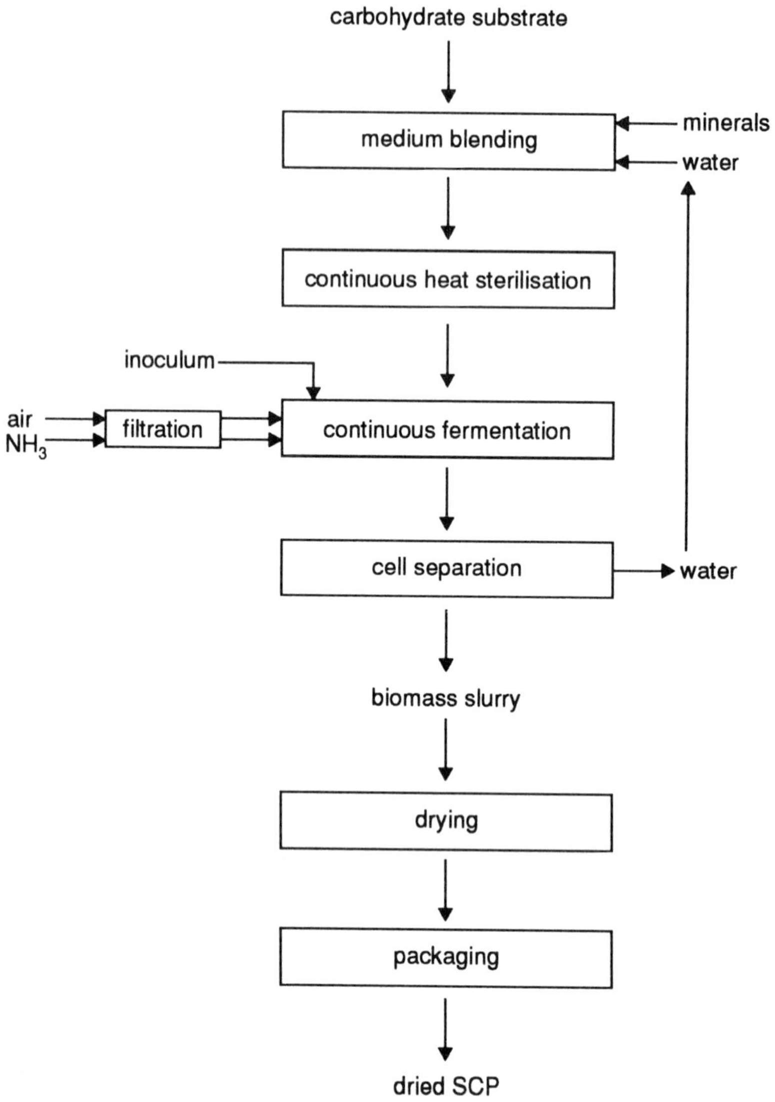

4.5 Your diagram may not look exactly like ours - but you can use ours to see if you have the stages in the process in the correct order and if you have omitted an essential step.

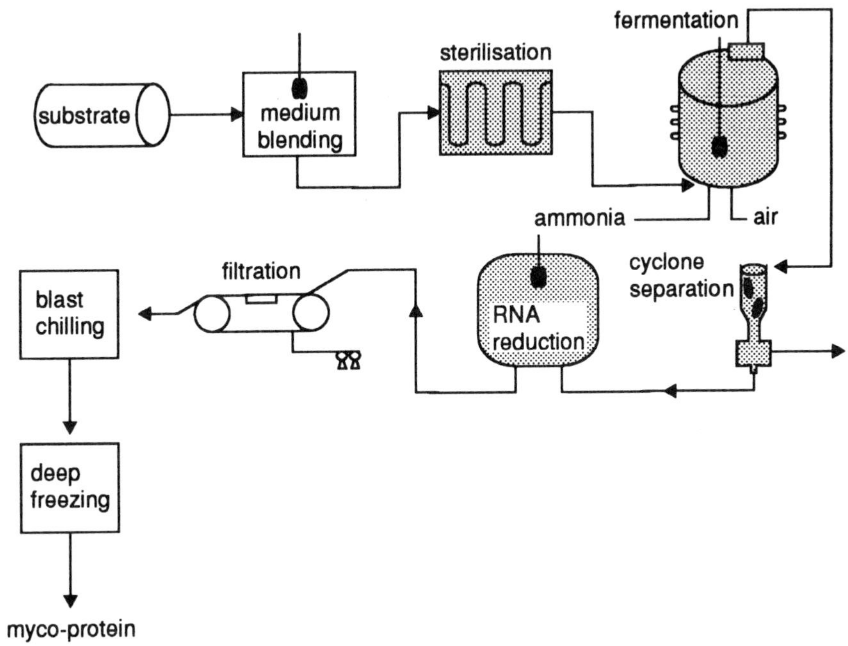

4.6 Your drawing should look similar to ours. Make sure you have all of the inputs in the right place.

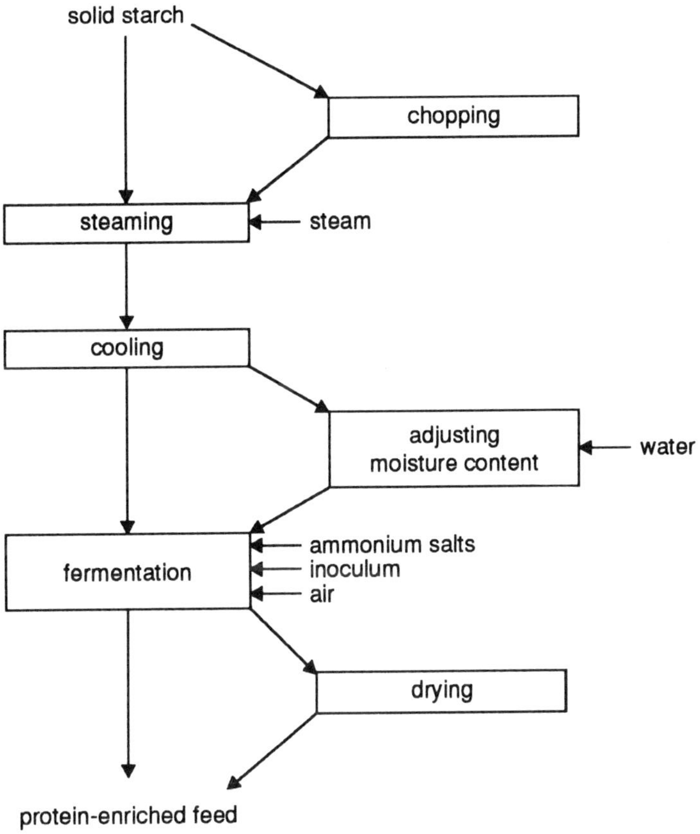

4.7 a) Minimum OTR. The oxygen requirement for the metabolism of carbohydrates we gave as 0.7 kg^{-1} of substrate during our discussion of carbohydrates. Did you remember? The oxygen requirement for biomass production on *n*-alkanes compared to carbohydrate is in the ratio 2.2/0.7 = 3.14. In other words, the oxygen requirement for biomass production on *n*-alkanes exceeds that for production on carbohydrate by a factor of 3.14. For the system described, the minimum OTR for biomass from carbohydrate is 1.89 kg O_2 m^{-3} h^{-1}. The minimum OTR for a similar system based on *n*-alkanes would be 1.89 x 3.14 = 5.94 kg O_2 m^{-3} h^{-1}.

b) Heat evolution rate. The heat evolution for biomass production on *n*-alkanes compared to carbohydrate is in the ratio of 27,100/12,300 = 2.2. In other words, the heat evolution (or cooling requirement if the operating temperatures are the same) for biomass production on *n*-alkanes exceeds that for production on carbohydrate by a factor of 2.2. For the system described, the heat evolution for production from carbohydrate is 33,210 k Joules m^{-3} h^{-1}. The heat evolution for a similar system based on *n*-alkanes would be 33,210 x 2.2 = 73, 062 k Joules m^{-3} h^{-1}.

4.8 i) Waste paper ... *Trichoderma viride* ... Solid substrate fermentation.

Only this fungus amongst those listed is capable of using the cellulose of which paper is composed. Solid substrate fermentation would be the easiest and cheapest production system.

ii) Exhaust gas emissions ... *Chlorella regularis* ... open lagoons.

Only *Chlorella* spp. can use the CO_2 in such exhaust gas emissions. Open (sunlit) lagoons would be necessary.

iii) Molasses ... All the organism listed ... bioreactors

All the organisms listed would grow on the sucrose in molasses. *Candida utilis* or *Kluyveromyces fragilis* would be the best organisms to use, as they are food-grade yeasts with high growth rates.

iv) Effluent ... *Aspergillus niger* ... bioreactors or open lagoons.

Only *Aspergillus niger* is strongly amylolytic and capable of using the starch in the effluent. Open lagoon systems operating at low pH, if effective, would be a cheaper method of production than aseptic bioreactors.

4.9 1) Output

The productivity is 30 x 0.3 = 9 kg m^{-3} h^{-1}. Thus the output is 9 x 36 = 324 kg biomass h^{-1}.

2) Minimum OTR (we can determine this in at least two different ways).

a) The yield is 0.5 kg biomass per kg methanol. From Figure 4.10 at yield = 0.5, oxygen requirement is about 0.05 moles/g cells. As 1 mole O_2 = 32 g, 0.05 moles = 1.6 g. Therefore the oxygen requirement is 1.6 g O_2 per g biomass, or 1.6 kg O_2 per kg biomass. With productivity 9 kg m^{-3} h^{-1}, the minimum OTR is 1.6 x 9 = 14.4 kg O_2 m^{-3} h^{-1}.

b) The yield = 0.5, and productivity = 9. From Figure 4.11 at yield = 0.5 and estimating productivity = 9, the minimum oxygen-transfer rate is about 0.5 moles oxygen l^{-1} h^{-1}. This represents 32 x 0.5 = 16 g O_2 l^{-1} h^{-1} or 16 kg O_2 m^{-3} h^{-1}.

3) Heat Evolution Rate (again we have at least two ways to calculate this).

a) From Figure 4.12, with yield 0.5, the heat produced is about 27 k Joules per gram biomass (or 27,000 k Joules per kg biomass). With productivity 9 kg m^{-3} h^{-1}, the heat evolution rate is 27,000 x 9 = 243,000 k Joules m^{-3} h^{-1}.

b) From Figure 4.13, with productivity 9 and yield 0.5, the heat produced is about 251 k Joules l^{-1} h^{-1}, or 251,000 k Joules m^{-3} h^{-1}.

4) Methanol Concentration

To support a biomass concentration of 30 kg m^{-3} at yield 0.5, the methanol concentration in the incoming medium would need to be 30/0.5 = 60 kg m^{-3}. Thus at 90% utilisation, the concentration in the incoming medium would need to be 60 x 100/90 = 66.7 kg m^{-3}.

4.10

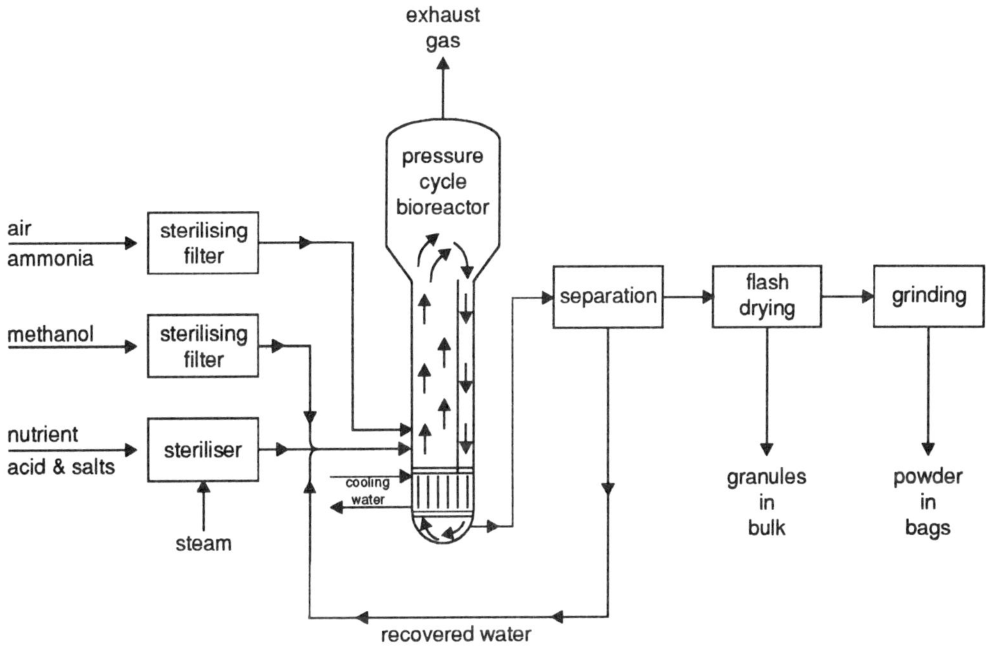

Your flow diagram might not look exactly like the one shown: for instance you will probably have represented the fermentation step by 'fermentation', rather than drawing the bioreactor. Nevertheless you should have all the unit processes and inputs shown above in your own diagram.

4.11 1) Productivity = biomass concentration x D.

2) Percentage utilisation = 100 - [(concentration in outgoing medium/ concentration in incoming medium) x 100].

3) Minimum OTR = productivity x 1.6 (see SAQ 4.9).

D (h^{-1})	Productivity (kg biomass m^{-3} h^{-1})	Minimum OTR (kg O_2 required m^{-3} h^{-1})	Methanol Utilisation (%)
0.1	3.43	5.49	98.0
0.2	6.62	10.59	94.8
0.3	9.33	14.93	88.9
0.4	10.52	16.83	75.3
0.5	1.25	2.0	7.2

Dilution rate of 0.5 h^{-1} produces the fastest growth rate (since μ = D), but gives poor productivity and substrate utilisation. This is a feature of chemostat cultures, when μ approaches μ_{max}.

Dilution rate 0.4 h^{-1} gives highest productivity, but 25% of the substrate remains unused. Dilution rate 0.3 h^{-1} gives better substrate utilisation but with reduced productivity. Both these dilution rates require minimum OTRs greater than the bioreactor can supply. This means that such biomass concentrations could not be produced in practice.

Dilution rate 0.2 h^{-1} produces efficient substrate conversion and has an OTR that the system can just about provide. Dilution rate 0.1 h^{-1} gives better substrate conversion but lower productivity. The chosen dilution rate would be just below 0.2 h^{-1}. In practice dilution rates of between 0.1 h^{-1} and 0.2 h^{-1} are used to operate the production system.

4.12 1) Output

With 1500 m^3 at D = 0.2 h^{-1}, volume of culture produced is 300 m^3 h^{-1} (ie 1,500 x 0.2 m^3 h^{-1})

At 36 kg biomass m^3, the output is 10,800 kg h^{-1} or 259,200 kg day $^{-1}$ (259 tonnes day^{-1}) or 94,608 tonnes yr^{-1}.

2) Biomass production of 94,608 tonnes would require 94,608/0.5 = 189,216 tonnes methanol.

4.13 The most likely order is as outlined below.

1) Establish the size of the market.

2) Isolate organisms that can use the substrate

3) Measure temperature optimum.

4) & 5) Measure affinity for substrate/Measure protein content.

6) Perform feeding trials in animals.

7) Apply for sales licence.

8) Marketing

9) Full-scale production.

The reasons for this sequence are:

1) The size of the market must be established first of all. There is no point in developing a product you cannot sell.

2) Once a substrate is chosen you must select organisms that can grow on it.

3) The temperature optima of isolated organisms would be measured early on in development, usually in shake-flask culture.

4) & 5) The affinity for the substrate would be measured in small-scale continuous culture. The protein content of organisms would probably be measured at about the same time, or perhaps earlier with 3).

6) Feeding trials would only be performed on selected organisms, perhaps on just one organism that appears suitable for production. There is no point performing such expensive trials until the characteristics of the organism in culture and chemical profiles (i.e. the operations in 3), 4) and 5)) have been carried out.

7) A sales licence can only be applied for when extensive feeding and toxicity trials have been carried out.

8) There is no point marketing a product until you have a sales licence. However you need to market the product at pilot-scale, to establish the market before you begin full-scale production.

9) Full-scale production is the last step.

4.14

1) The most significant production cost in SCP production is the cost of raw materials, ranging from 55-77% of the total production cost. See Table 4.9.

2) The substrate contributes least to production cost when it is a waste. See Table 4.9 - Sulphite waste liquor is the waste and contributes only 17% to total production cost.

3) For liquid substrates the most significant equipment cost is fermentation, ranging from 43 to 51% of the total equipment costs. See Table 4.10.

4) The most significant running cost is fermentation, ranging from 53 to 77% of the total running costs. See Table 4.11.

5) The most significant running cost of the fermentation is aeration, ranging from 71 to 92% of fermentation running costs. See Table 4.12.

6) The most significant cost of fermentation equipment is usually the bioreactor. However, for fungal SCP processes from sulphite waste liquor, the cooling system is the most significant cost. See Table 4.13.

7) The most significant cost of harvesting equipment is drying. See Table 4.14.

8) The most significant running cost of harvesting and drying is drying. See Table 4.15.

4.15 The cost of soya protein is \$0.29 kg^{-1} (from Table 4.7).

The cost of yeasts from *n*-alkanes is \$0.42 kg^{-1} (from Table 4.7). By comparison to yeasts from *n*-alkanes, bacteria from methanol cost 0.98 in proportion (Table 4.8), and so the cost of bacterial SCP must cost 0.42 x 0.98 = \$0.41 kg^{-1}. At 60% protein, the cost of bacterial protein is 0.41 x 100/60 = \$0.68 kg^{-1}

Methanol as a substrate contributes 47% of the total production cost of bacterial SCP (and therefore of bacterial protein) (Table 4.9). Therefore of the \$0.68 kg^{-1} that bacterial protein costs 0.68 x 0.47 = \$0.32 kg^{-1} is accounted for by the cost of methanol, and 0.68 x 0.53 = \$0.36 kg^{-1} is accounted for by the remaining production costs.

If methanol costs are discounted altogether, the remaining production cost of bacterial protein is still \$0.36 kg^{-1}, ie more than the cost of soya protein. At these levels the particular SCP process in question is clearly not competitive.

4.16 Yeast from *n*-alkanes has a production cost of \$0.42 kg^{-1} (Table 4.7).

1) Aeration comprises 88.9% of the running costs of fermentation (Table 4.12). Running costs of fermentation comprise 67.9% of the Total running costs (Table 4.11). Running costs comprise 23.8% of the total production cost (Table 4.9).

   Aeration thus contributes 0.42 x 88.9/100 x 67.9/100 x 23.8/100 = \$0.060 per kg biomass

2) Cooling comprises 7.9% of running cost of fermentation (Table 4.12). Cooling thus contributes 0.42 x 7.9/100 x 67.9/100 x 23.8/100=\$0.005 per kg biomass.

3) Ammonia comprises 11.1% of total production costs (Table 4.9). Ammonia thus contributes 0.42 x 11.1/100 = \$0.046 per kg biomass.

4) Labour comprises 8.4% of total production costs (Table 4.9). Labour thus contributes 0.42 x 8.4/100 = \$0.035 per kg biomass.

5) Drying comprises 77.7% of the running costs of harvesting and drying (Table 4.15). Harvesting and drying costs comprise 5.9% and 21% respectively (=26.9%) of total running costs (Table 4.11). Running costs represent 23.8% of total production costs (Table 4.9).

   Drying thus contributes 0.42 x 77.7/100 x 26.9/100 x 23.8/100 = \$0.020 per kg biomass.

   Aeration contributes most to the production cost of the yeast, at \$0.060 per kg. Cooling comprises least, at \$0.005 per kg.

4.17 To produce 1 kg *Candida* sp. would require 1/0.48 = 2.08 kg substrate. At \$0.2 kg^{-1} this represents a cost of \$0.417 per kg biomass. With the additional fixed cost of \$0.205, the total fixed cost of producing *Candida* sp. would be 0.417 + 0.205 = \$0.622 per kg biomass.

To produce 1kg *Fusarium* sp. would require 1/0.45 = 2.22 kg substrate. At \$0.2 kg^{-1} this represents a cost of \$0.444 per kg biomass. With the additional fixed cost of \$0.205, the total fixed cost of producing *Fusarium* sp. would be 0.444 + 0.205 = \$0.649 per kg biomass.

1) To produce *Candida* sp. as feed, the cheapest option would be non-aseptic fermentation, centrifugation and drying. These add only \$0.005 and 0.02 to the cost, for centrifugation and drying. The cost of feed from *Candida* sp. is thus 0.622 + 0.005 + 0.02 = \$0.647 per kg biomass or = 0.647 x 100/60 = \$1.078 per kg protein.

2) To produce *Candida* sp. as food additive the above process could be operated, except that fermentation would have to be aseptic (with sterilisation costs). The cost would be 0.647 + 0.04 = \$0.687 per kg biomass or \$1.145 per kg protein.

3) To produce *Fusarium* sp. as feed, the cheapest option would be non-aseptic fermentation, filtration and drying. The cost is thus 0.649 + 0.001 + 0.02 = \$0.670 per kg biomass.

   or = 0.670 x 100/45 = \$1.489 per kg protein.

4) To produce *Fusarium* sp. as high-protein food additive, the above system could be used, with aseptic fermentation and grinding the product after drying. The cost is thus 0.670 + 0.04 + 0.01 = \$0.720 per kg biomass.

   or 0.72 x 100/45 = \$1.600 per kg protein.

5) To produce *Fusarium* sp. for meat substitutes the aseptic system must be operated with recovery by filtration followed by deep-freezing. The cost is thus 0.649 + 0.04 + 0.001 + 0.04 = \$ 0.730 per kg biomass.

   With competing feeds at \$0.80 per kg protein, and *Candida* sp. and *Fusarium* sp. at \$1.078 and \$1.489 per kg protein respectively, neither SCP process is profitable for producing feed.

   With high-protein food additives at \$1.55 per kg protein the process producing *Candida* sp. at \$1.45 per kg protein would make a small profit of \$0.10 per kg protein (or 0.1 x 60/100 = \$0.06 per kg biomass). The process producing *Fusarium* sp. at \$1.600 perkg protein would not be profitable.

With meat substitutes at \$1.05 per kg, and *Fusarium* sp. at \$0.730 per kg biomass, this product would make a profit of 1.050 - 0.730 = \$0.32 per kg. This would be the most profitable product.

Responses to Chapter 5 SAQs

5.1

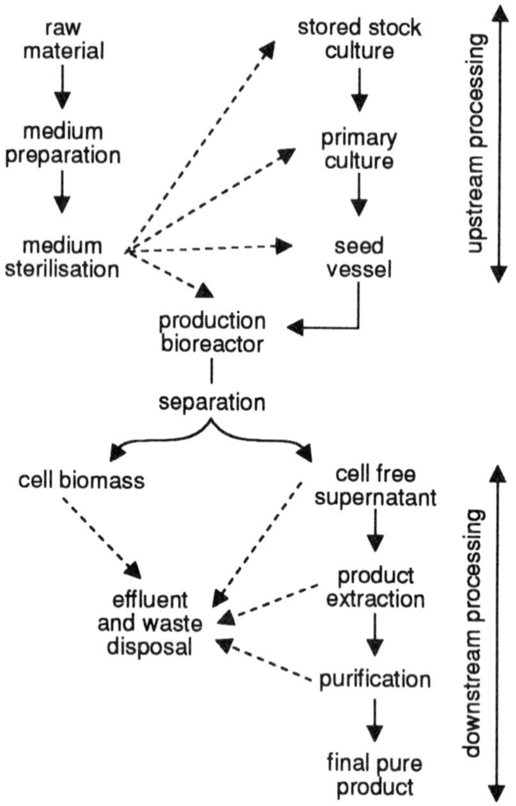

5.2

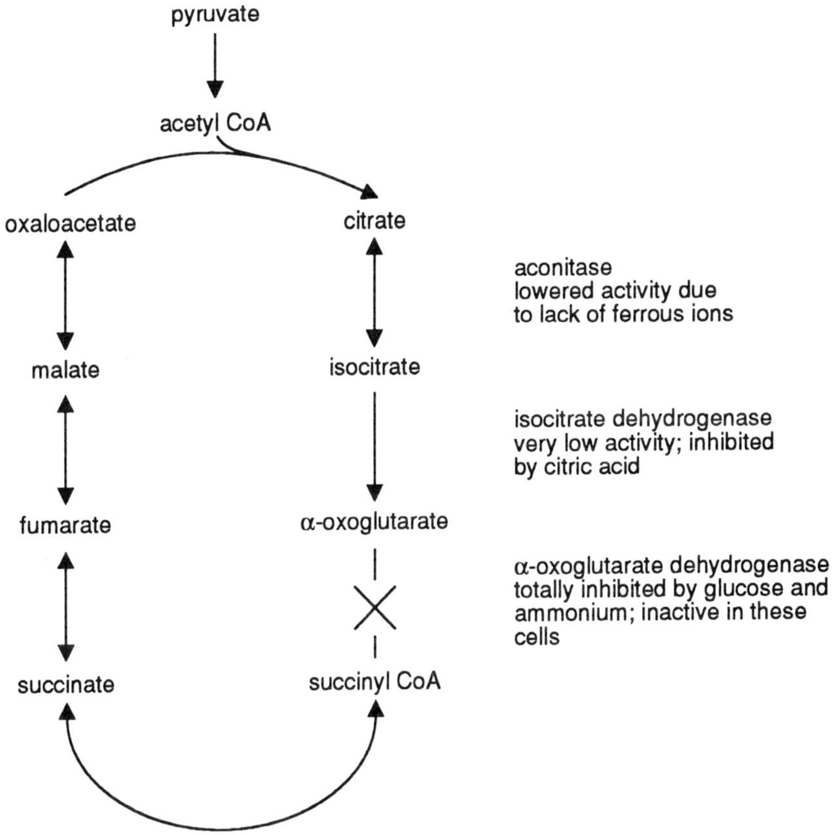

5.3 The missing words in the paragraph are:

a) ammonium.

b) manganese.

c) ATP.

d) constitutive.

e) citrate.

f) isocitrate.

5.4

	balanced growth		citric acid accumulation	
Enzyme	**Inhibits**	**Stimulate**	**Inhibits**	**Stimulates**
Phosphofructo-kinase	ATP	ADP	Citrate	-
Pyruvate kinase	glucose-6-phospate ATP	-	-	-
Pyruvate carboxylase	-	-	-	-
Pyruvate decarboxylase	GTP	AMP	-	-
Citrate synthase	ATP	-	-	-
Isocitrate dehydrogenase	Citrate	ADP AMP	Citrate	-

5.5 The medium must contain:

- a suitable carbon source in high concentration;
- a nitrogen source, preferably ammonia;
- low levels or the absence of trace metals, particularly manganese and ferrous ions.

Aeration is mandatory and must be constant.

It is essential to produce and maintain highly acidic conditions with a pH of 2.0 or below to minimise potential problems caused by contamination.

5.6 Statement 1, the submerged culture process.

Statement 2, the Koji process.

Statement 3, submerged culture process.

Statement 4, the submerged culture process.

Statement 5, the Koji process.

Statement 6, the surface culture process.

Statement 7, the submerged culture process.

5.7

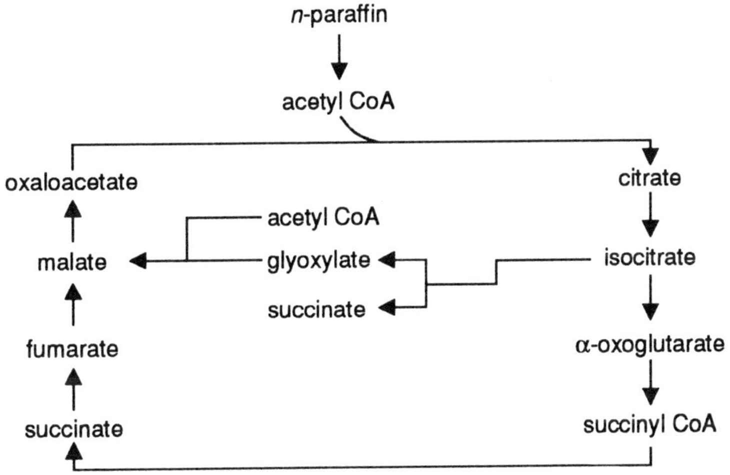

The overall reaction for this cycle is:

2 acetyl CoA → succinate + 2 CoASH

Using this as our starting point and then taking another set of reactions around the pathway starting from oxaloacetate and acetyl CoA we have:

2 acetyl CoA → succinate + 2 CoASH

succinate → oxaloacetate

oxaloacetate + acetyl CoA → citrate + CoASH

citrate → isocitrate

isocitrate → succinate + glyoxylate

glyoxylate + acetyl CoA → malate + Co ASH

succinate → malate

---

4 acetyl CoA → 2 malate + 4 CoASH

5.8 There are in fact several reasons, the main ones being:

1) The presence of a condensing enzyme which combines pyruvate and acetyl CoA to yield citramalic acid has been confirmed.
2) Conversion of itaconic acid via citraconic acid and itatartaric acid has been demonstrated.
3) Copper ions increase the yield of itaconic acid.

5.9

1) True.
2) True.
3) False. The reverse is true, because sodium gluconate is far more soluble than calcium gluconate high glucose concentrations can be used to produce higher production yields of sodium gluconate.

4) False. D-gluconolactone is produced directly from glucose via glucose oxidase. 6-phosphogluconolactone is an intermediate in the hexose monophosphate pathway.

5) True.

6) False. Glucose oxidase is specific for β-D-glucose.

5.10

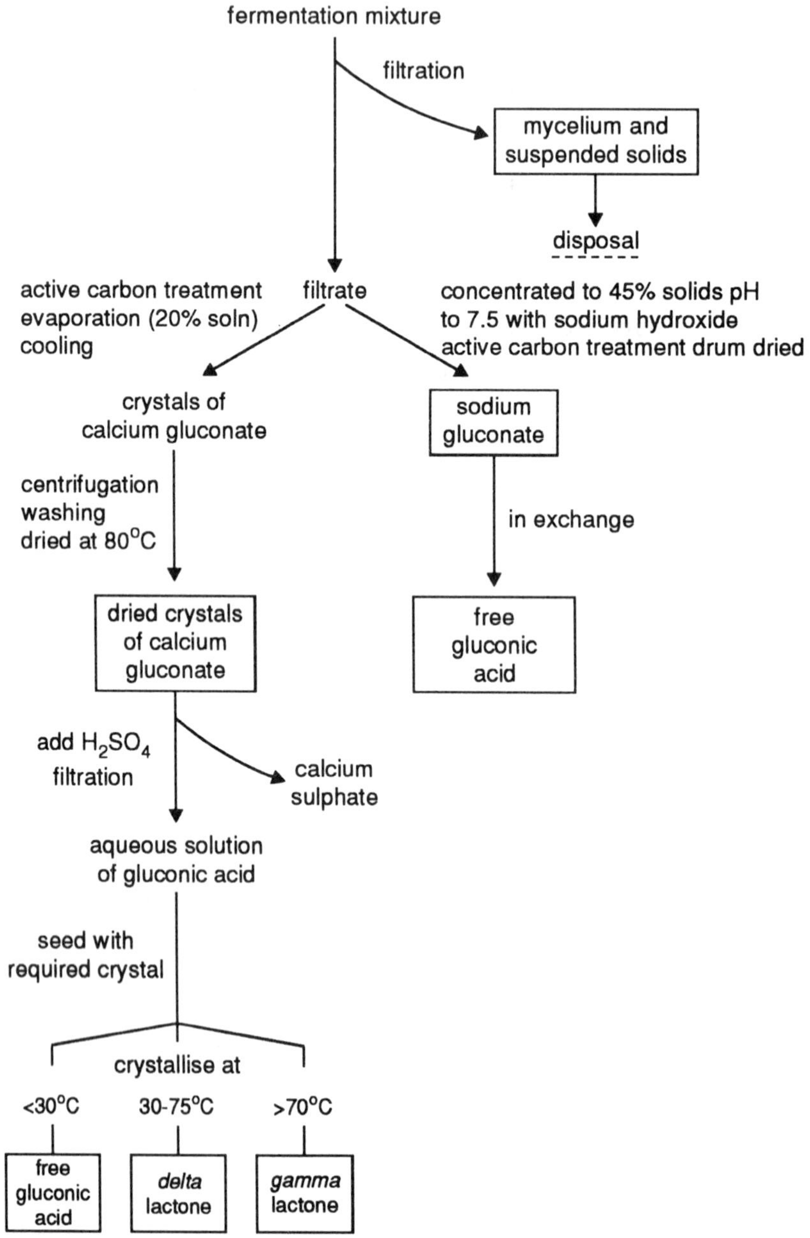

Responses to Chapter 6 SAQs

6.1 Both penicillin G and amoxycillin conform to the general structure.

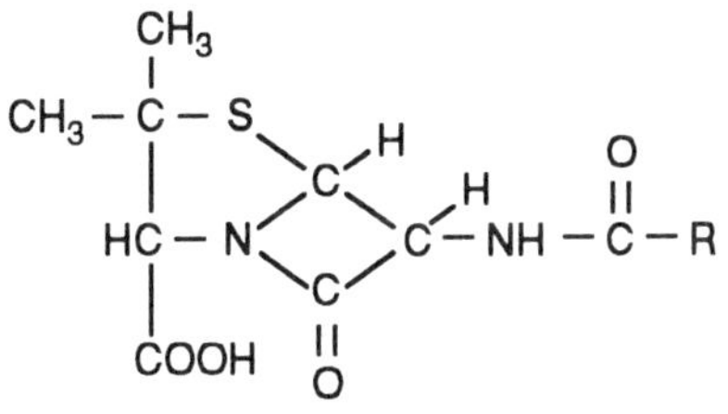

In the case of penicillin G, R =

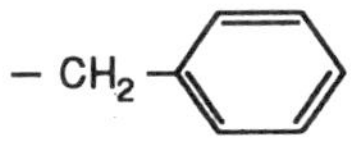

whilst with amoxycillin, R =

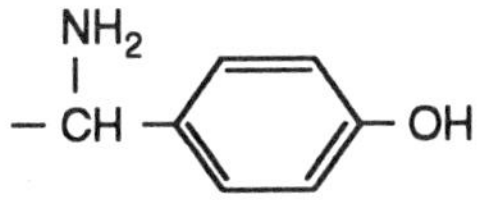

When written in this form, it should be apparent that penicillins can be diversified by substituting different groups in place of R. The penicillin class of antibiotics does, in fact, contain members that have different R groups. We will examine these later in the chapter and describe the strategies used to generate penicillins with different substitutions.

6.2

1) Penicillin G production can be encouraged by adding β-phenylacetic acid to the medium. This was described in the text and will not be elaborated upon here.

2) The obvious addition to produce penicillin X is pOHβ-phenylacetic acid.

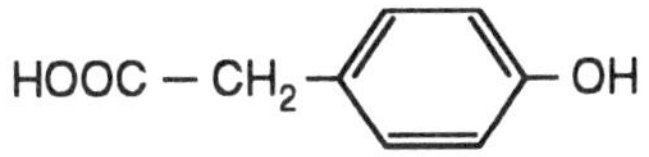

This could be derived from tyrosine via the amine derivative, in a process similar to the production of β-phenylacetic acid in CSL.

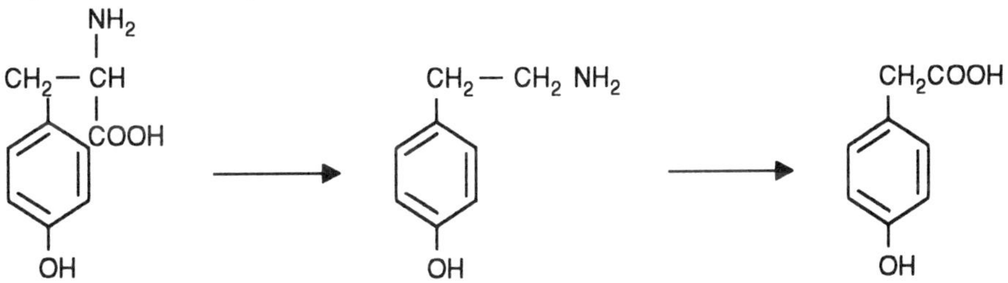

3) We would expect you to suggest the addition of:

$HOOC - CH_2 - CH = CH - CH_2 - CH_3$

to produce penicillin F.

4) Similarly, we would anticipate that the addition of $HOOC-CH_2-(CH_2)_5-CH_3$ would lead to the production of penicillin K.

5) Penicillin V was mentioned in the text. The addition of phenoxyacetic acid

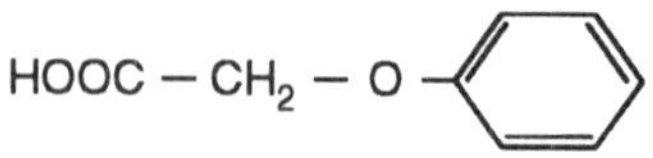

encourages the production of this penicillin.

6) The compound we would anticipate you would suggest to produce penicillin N is

D-configuration

$$HOOC-CH_2-(CH_2)_3-\underset{\displaystyle COOH}{\underset{|}{CH}}-NH_2$$

This is Dα-aminoadipic acid. You will learn a little later in the chapter that the normal biosynthetic pathway of pencillin production involves the incorporation of L-α aminoadipic acid. The product is called isopenicillin N.

Its side chain has the structure:

L-configuration

$$-\underset{\displaystyle O}{\underset{||}{C}}-(CH_2)_4-\underset{\displaystyle COOH}{\underset{|}{CH}}-NH_2$$

Conversion to penicillin N is achieved by an epimerase that converts the L configuration amino group into the D configuration.

Thus, in the case of penicillin N, we may not have to add a special precursor providing an active epimerase was present.

6.3 With glucose as substrate, substrate consumption is fast causing rapid growth and a depression of the pH. Penicillin production only begins towards the end of the growth phase. At this stage the pH is below the optimum (pH 6.5-7.0) for penicillin production. Thus penicillin production only proceeds for a short period and at a slow rate. Yields are, therefore, low. Note also that the biomass is unstable and, in the absence of external carbohydrate, begins to lyse.

With X as a substrate, substrate consumption is quite slow. Growth is slower on this substrate and even when maximum biomass levels are reached, there is still substantial amounts of substrate left. Thus the cells are able to utilise this to maintain themselves and also to support penicillin production. Because of this there is no violent swing in pH towards the acid end during the growth phase, nor to the alkaline end because of cell lysis. In this case the pH is maintained within the optimum range for penicillin production. Penicillin production is therefore extended, resulting in greater yields. Carbohydrates which give this type of result include lactose.

6.4 No, the enzyme IPN-acyltransferase is not highly specific. We learnt in section 6.3 that a range of penicillins could be produced by incubating penicillin-producing cells in the presence of analogues of acetic acid. Inclusion of β-phenoxyacetic acid in the media leads to the production of penicillin V, while inclusion of β-phenylacetic acid leads to the production of penicillin G. The implication is that IPN-acyltransferase will add a range of acyl groups onto the amino penicillinic acid moiety. If you re-examine Table 6.2 you will see that the transferase appears to be capable of using a wide range of acyl groups.

6.5 This was a fairly open ended question. The advantages we hoped you would identify were:

- the biocatalytic process is carried out in water, whereas the chemical deacylation techniques require halogenated solvents. This has environmental consequences. Obviously using water is cheaper, safer and less polluting than using halogenated organic solvents;
- the biocatalytic process yields acyl residues (for example phenylacetic acid) which may be reused as precursors for β-lactam production;
- the biocatalytic process is cheaper if the enzyme can be used many times;
- the biocatalytic process is specific and yields few by-products;
- the desired product is easy to separate from the biocatalytic reaction mixture. With the chemical catalytic process several other components (PCl_5; $(CH_3)_3$ SiCl; ROH) may also be present.

6.6 1)

a) Yes, the 6-O-methyldeoxystreptamine would be incorporated in place of 2 deoxystreptamine.

Its structure would be:

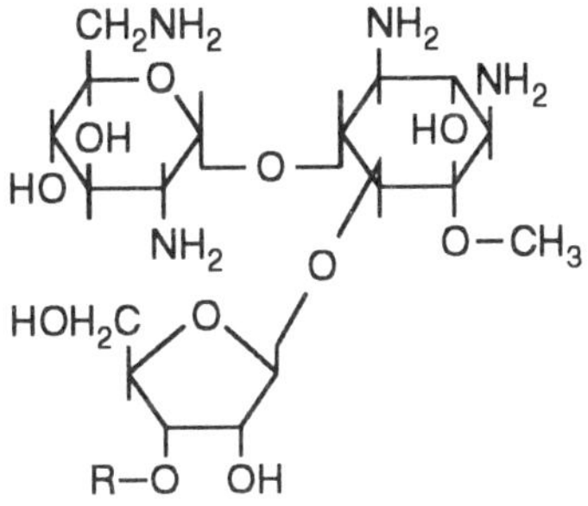

b) Yes, the 3-N-methyl-deoxystreptamine would be incorporated to form

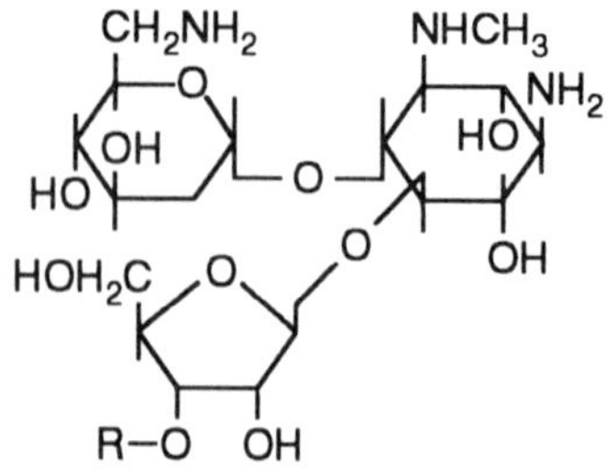

c) No, the methyl group at position 5 would block addition of the 5-0-methylstreptomycin into the neomycin moiety.

2)

The assumptions we have made in coming to our conclusions in 1) are that:

- the compounds are taken up by the cells;
- the enzymes responsible for adding the streptamine derivatives to the rest of the neomycin will work with a range of substrates and can work providing the hydroxyl groups in position 4 and 5 are available;
- the cells do not further metabolise the added analogues prior to their incorporation into the neomycin product.

Responses to Chapter 7 SAQs

7.1

1) False. D-xylose is a pentose sugar which are very rarely found in microbial exopolysaccharides.

2) False. Pyruvate ketals contribute to the anionic nature of exopolysaccharides.

3) True.

4) False. However, it is true that only relatively few yeasts and filamentous fungi produce exopolysaccharides.

5) True. Uronic acids contribute to the anionic nature of exopolysaccharides.

7.2

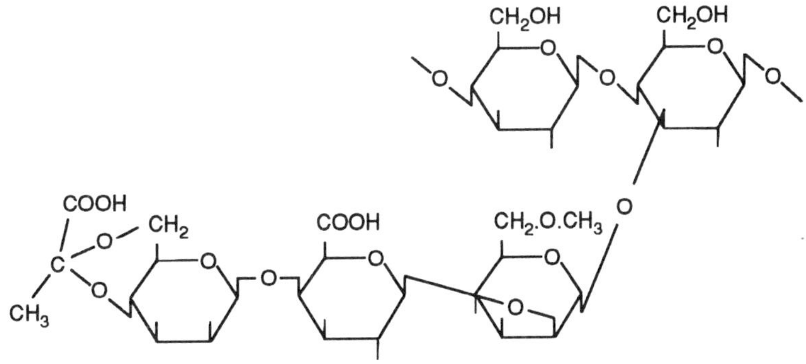

7.3

1) Glucose: source of carbon and energy for growth and exopolysaccharide production.

Succinic acid: source of carbon and energy; improves metabolic balance between carbon flux from glucose and oxidation throgh TCA cycle.

Ammonium chloride: source of nitrogen (required for synthesis of proteins, nucleic acids and co-enzymes).

Yeast extract: source of growth factors (amino acids, vitamins, nucleotides).

Potassium: the principal inorganic cation; cofactor for some enzymes.

Phosphate: constituent of nucleic acids, phospholipids and co-enzymes.

Magnesium sulphate: magnesium is an important cellular cation; inorganic cofactor for many enzymatic reactions, including those involving ATP; functions in binding enzymes to substrate.

Iron: constituent of cytochromes and other heme or nonheme proteins; cofactor for a number of enzymes.

Calcium: important cellular cation; cofactor for some enzymes; important in enzyme stability.

Trace elements: inorganic constituents of special enzymes.

2) Biomass concentration = 10 x 2.5 = 25 g l^{-1}.

3) Biomass concentration = 25 g l^{-1} (ammonia is the limiting nutrient).

7.4

1) Xanthan is synthesised during exponential growth of the culture, whereas succinoglycan is synthesised after growth has ceased (biomass constant).

2) Since succinoglycan synthesis commences after the growth phase there would be no carbon available for succinoglycan formation.

3) From Figure 3.6: since growth and xanthan synthesis ceases simultaneously, we can deduce that the limiting nutrient is glucose. The rapid decrease in glucose concentration during the growth phase also suggests this.

4) Glucose limits succinoglycan production (ammonia limits growth). It is necessary to limit succinoglycan production to avoid an excessively viscous fermentation liquor which would affect mixing and subsequent downstream processing to recover exopolysaccharide.

5)

$$Y_{ammonia} = \text{biomass produced/ammonia consumed} = \frac{2.5 - 0.8 \text{g l}^{-1}}{140 \text{ m mole l}^{-1}}$$

$$= 0.0121 \text{ g m mole}^{-1}$$

$$= 12.1 \text{ g mole}^{-1}$$

$$= 205.7 \text{ g g}^{-1}$$

7.5

An appropriate order is:

Addition of cations
Heat treatment
Addition of solvent
Distillation
Centrifugation
Vacuum drying
Milling
Cool dry storage
Propylene oxide treatment
Sterile nitrogen gas treatment
Aseptic packaging

7.6

1) The missing words are:

a) pseudoplastic;

b) viscoelasticity;

c) ordered, melting;

d) above.

2) Alginates with a high content of L-guluronic acid have a relatively high capacity for cross-linking, involving the binding of divalent cations, and thus form brittle gels. D-mannuronic acid does not bind divalent cations, so the gels are more flexible (less cross linking).

7.7

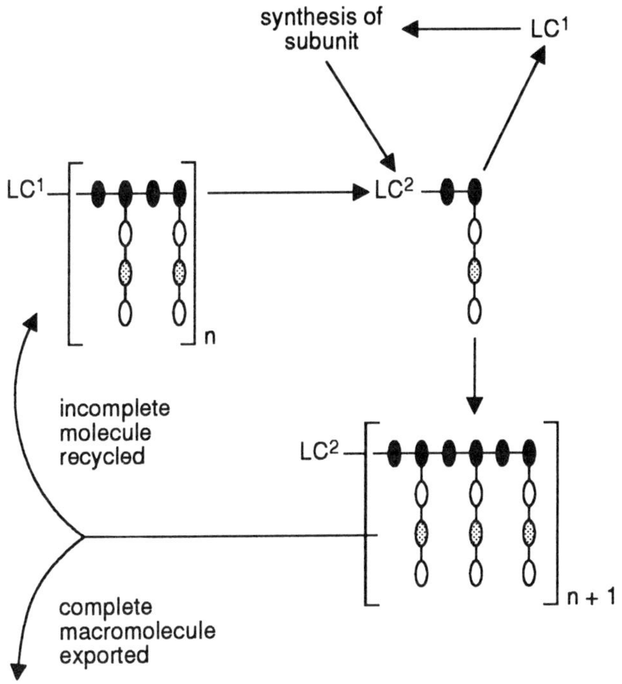

7.8 The missing words, in order, are: intracellularly, bisphosphates, energy, lipid, subunits, enzyme, homopolysaccharides.

7.9 Deficiency in *gum*D will mean that synthesis is not started and there will be no label incorporated either from the UDP[^{14}C] glucose or the GDP[^{14}C] mannose. A deficiency in *gum*M will give labelled lipid carrier if glucose is used but not if mannose is the labelled sugar nucleotide. A mutant with a non-functional *gum*K gene will incorporate label into the lipid and soluble fractions from UDP[^{14}C] glucose and GDP[^{14}C] mannose, producing the polytrimeric polysaccharide lacking the terminal mannose.

7.10 The genes responsible for encoding the proteins specifically involved in exopolysaccharide synthesis are clustered in one large operon. The genes encoding the proteins for sugar nucleotide phosphates, which are not necessarily specifically used for exopolysaccharide synthesis, also tend to be clustered.

7.11

A, sensor

B, effector protein

C, degrading enzyme

D, activator protein

E, activated complex

F, catabolite activator protein

G, pathogenicity regulating genes

H, exopolysaccharide gene cassette

I, pathogenicity genes

7.12

Exopolysaccharide	Application	Important Physical Properties
welan	enhanced oil	high viscosity over broad temperature range; stable to 149°C.
emulsan	cleaning agent	stabilises hydrocarbon/water emulsions at low concentrations; low toxicity.
dextrans	plasma substitute	equivalent viscosity and osmotic properties to blood plasma; non-irritant and non-toxic.
curdlan	food additive	retains shape in cooked food and gels at relatively low temperatures (55-80°C)
polyethylene glycol/dextran water mixtures	purification of biological materials	aqueous two-phase with low interfacial tension
alginate	immobilising enzymes	gel formation under mild conditions
gellan	use in glazes, jellies and icings	heat stable and gives a very clear thermostable gel which sets at lower concentrations and more rapidly than most other polysaccharides; superior flavour release characteristics.

Responses to Chapter 8 SAQs

8.1

1) Cysteine, cystine, methionine.

2) The eight essential amino acids are listed immediately after Table 8.2.

3) L-Proline and L-tryptophan (see Table 8.2).

4) Glycine (see Table 8.2).

5) Cysteine.

6) L-Methionine (see Table 8.1).

7) L-Cysteine (see Table 8.1).

8.2

Possible advantages of using enantiomerically pure compounds as drugs include:

1) avoids detrimental side effects

2) enables more specific drug action

3) reduces amount of drug administered

4) avoids work involved in metabolism of inactive enantiomer.

8.3

1) Enzymes in the pathway to L-phenylalanine are subject to feedback inhibition by products (amino acids) arising from pathway intermediates.

2) Auxotrophic mutant lack one or more enzymes involved in the synthesis of amino acids (such as tyrosine). This prevents accumulation of the amino acid and thus avoids feedback inhibition of enzymatic steps in the L-phenylalanine pathway.

3) Regulatory mutants are not subject to feedback inhibition, even by L-phenylalanine itself.

4) A Try^- mutant would not be subject to feedback inhibition by overproduction of tryptophan. Also, the mutation may allow more chorismate to proceed to prephenate via E3 (see Figure 8.4) and thus through to L-phenylalanine.

8.4

1) In fed-batch mode residual substrate concentration may be maintained at very low levels. This would reduce substrate costs, may remove catabolic repressive effects and may avoid possible toxic effects of the substrate. The fed-batch mode of operation may also avoid oxygen depletion of the culture during rapid growth.

2) Monoseptic fermentations are easier to characterise and to repeat, compared to fermentation involving more than one micro-organism.

3) Addition of antibiotics to the fermentation broth may be used to: avoid problems associated with growth revertants (eg auxotrophic back mutation); ensure that genetic material (eg plasmid DNA) is maintained within the process micro-organism.

4) Use of fresh starting material for each fermentation run may avoid problems of back mutation and loss of genetic material.

5) Addition of chelating agents to the fermentation medium may help to inhibit phage multiplication by prevention of phage adsorption to the cell wall.

8.5

1) Solvent extraction.

2) Crystallisation, ion exchange, electrodialysis.

3) Crystallisation.

4) Decoloration.

5) Evaporation.

8.6

1) From the data 6 g l^{-1} glucose is used = 0.0333 mol l^{-1}.

2) During this period 1.4 g l^{-1} of phenylalanine is produced $= \frac{1.4}{165} = 0.0085$ mol l^{-1}.

Number of moles of glucose needed to provide the carbon for this phenylalanine is 0.0085 x 2 mol l^{-1} = 0.0170 mol l^{-1}.

During this period 0.3 g l^{-1} acetate are produced = 0.005 mol l^{-1}.

Number of moles of glucose needed to provide the carbon for this acetate is $\frac{0.005}{2} =$ 0.0025 mol l^{-1}.

Thus, to produce the acetate and phenylalanine,
0.0170 + 0.0025 mol l^{-1} glucose are carried
= 0.0195 mol l^{-1}.

But 0.0333 mol l^{-1} of glucose is consumed.

This means that $\frac{0.0333 - 0.0195}{0.333}$ x 100% = (41%) of the consumed glucose is not used to produce phenylalanine and acetic acid but is used for maintenance (note: the cells are in stationary phase and thus there is no cell growth over this period).

8.7

Total sales	= 500,000 kg x \$50	= \$25 million a year
Net sales	= \$25 million - 20%	= \$20 million
Gross profit	= \$20 million - (500,000 kg x 24.5\$)	
	= \$7.75 million	

Taking into account administration and R and D costs.
Profit = \$7.75 - (12.5% of £20 million)
= \$5.25 million

Profit for the first year = \$5.25 million - (10% of \$40 million)
= \$1.25 million

Net earnings in first year (after taxes) = 50% of \$1.25 million
= \$0.625 million
(A in Equation 8.4)

Net earnings per annum in the period to come
= 50% of \$5.25 million = \$2.625 million
(B in Equation 8.4)

Working capital = 25% of \$20 million = \$5 million
(C in Equation 8.4)

Total fixed capital = \$40 million (D in Equation 8.4).

Using Equation 8.4, for the period of investment (15 years) we have:

$$\text{ROI} = \frac{[0.625 + 14\,(2.625)]/15}{5 + 40} \times 100 = 3.4\%$$

8.8

1) phenylpyruvic acid: L-amino acid aminotransferase

2) N-acetyl-DL-phenylalanine: acylase

3) acetamidocinnamic acid: acylase
L-amino acid aminotransferase

4) phenylpuruvic acid: L-amino acid dehydrogenase

5) *trans*-cinnamic acid: phenylalanine-ammonia-lyase

6) phenyllactic acid: L-hydroxy acid dehydrogenase
L-amino acid dehydrogenase

8.9

1) D- or L-phenylalanine induces the enzyme phenylalanine aminoaminotransferase, which is required for L-phenylalanine synthesis.

2) An alkaline pH (~ pH 11) is desirable in order to: achieve high conversion rates; increase solubility of L-phenylalanine; inhibit enzymes catalysing degradation of L-phenylalanine and formation of byproducts; reduce inhibition of the reaction by the keto form of phenylpyruvic acid.

3) Concentrations of 4% phenylpyruvic acid leads to end product inhibition by the L-phenylalanine produced, resulting in lower final yields.

4) Cell immobilisation is relatively costly because of the cost of the immobilisation procedure and of supplying cofactor.

8.10 Cost price of L-phenylalanine by direct fermentation is 28.50$ kg^{-1}.

From Table 8.8:

1) Cost price of L-phenylalanine by precursor feeding is 35.05 $ kg^{-1} and substrate costs are 18.05$ kg^{-1}.

   It follows that for precursor feeding to be competitive, substrate costs would have to be reduced by: 35.05 - 28.50 = 6.55$ kg^{-1}. So, the percentage reduction in substrate costs would be: $\frac{6.55}{18.05} \times 100 = 36\%$.

2) A similar approach for bioconversion, based on the date given in Table 8.8, shows that the percentage reduction in substrate costs required is 11%.

   (30.22 - 28.5$ kg-1 = 1.72$ kg-1 reduction is required.

   Substrates cost 15.66$ kg-1. Thus % reductase is $\frac{1.72}{15.66} \times 100 = 11\%$).

Responses to Chapter 9 SAQs

9.1 a)

1) The organism would probably grow quite well. The structure shown is simply analogous to the acetyl ester of cholesterol except that the side chain has been removed. The organism would, presumeably, degrade the sterol nucleus by its normal route.

2) The organism would probably only show slight growth. It would be able to degrade the side chain in the normal way, but the hydroxyl group in position 19 blocks the metabolism of the sterol nucleus. Thus only a small portion of the molecule can be catabolised.

3) The modification to positions 19, 5 and 6 prevents metabolism of the ring structure, thus the organism would not be able to utilise this substrate (no growth). There is no side chain to catabolise.

4) The organism would not be able to grow on this substrate as the hydroxyl in position 19 blocks metabolism of the ring structure (see 2). There is no side chain to catabolise.

b) The likely metabolic products from compound 1 are CO_2 and H_2O, as this organism can completely metabolise this substrate. Incubation with compound 2 is likely to lead to removal of the side chain, so a major product might be:

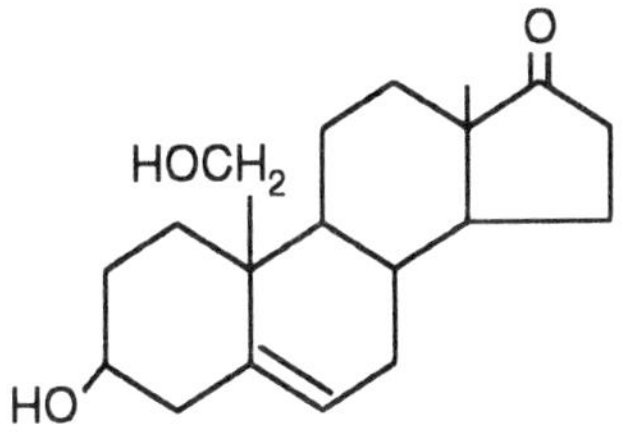

However, there may be some further, minor modifications to this basic structure.

Compounds 3 and 4 are likely to remain unaltered.

9.2 The isolate appears to produce the enzymes for the complete catabolism of lithocholic acid. However, in the presence of Pb^{2+} ions, some of these catabolic enzymes are inhibited, leading to the accumulation of partial breakdown products. It appears that enzymes involved in catabolism of the ring structure are more susceptable to inhibition by Pb^{2+} ions than are the enzymes involved in side chain catabolism.

9.3 1) hydrolysis (ester hydrolysis)

2) oxidation (alcohol oxidation)

3) isomerisation - the double bond has been moved from position 5-6 to position 4-5. We would normally denote such an isomerisation by $\Delta^5 \longrightarrow \Delta^4$.

4) hydroxylation - a hydroxyl group is added to position 11.

9.4

1) You should have identified this reaction as an 11α-hydroxylation. The most commonly used organism to carry out 11α-hydrolylations is *Rhizopus nigrans*. The substrate in this case is progesterone.

2) This is an 11β-hydroxylation, usually carried out using *Curvularia lunata*. The substrate in this case is known as substrate S.

3) This is a 16α-hydroxylation, usually carried out using *Streptomyces roseochromogenes*.

9.5

1) This is a hydroxylation. It may also be regarded as an oxidation.

2) This is also a hydroxylation.

3) This is an epoxidation.

4) This is a reduction in which the aldehyde group is reduced to an alcohol.

9.6

The way to achieve resolution is to use the lipase to selectively esterify the L-menthol. Thus:

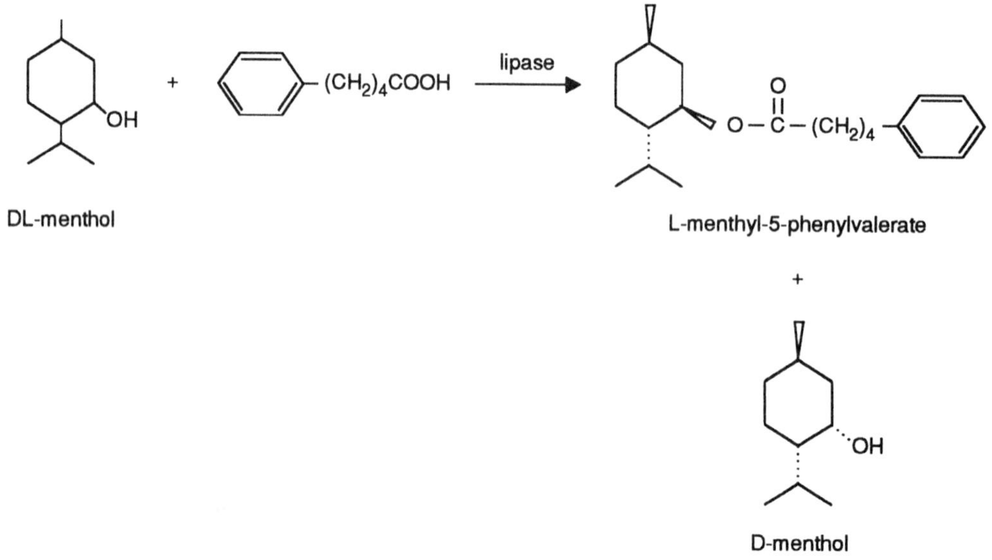

We can, therefore, isolate D-menthol from the mixture.

The L-menthol could be recovered by hydrolysing the L-menthyl-5-phenylvalerate. You might anticipate using an esterase to carry out this hydrolysis such as the enzyme from *Rhodotorula minuta* described in the text.

9.7

This is a fairly open-ended question and there are many options available. Here we will give some illustrative examples.

First, the chemical difference between the ring structures of PGE and PGF prostaglandins is that in PGE there is a keto group at position 9, whilst in PGF this position is taken up by a hydroxyl (secondary alcohol) group.

The conversion of a ketone to a secondary alcohol is, biochemically, fairly straightforward involving a hydrogenation. Hydrogenases tend to be stereospecific. Thus, incubating portions of PGE_2 with appropriate hydrogenating enzymes would yield samples of either PGFα or PGFβ.

An alternative approach would be to chemically reduce the keto group at position 9, forming a racemic mixture of PGFα and PGFβ. These could then be resolved enzymatically by, for example, esterfying the 9-hydroxy and using specific esterases to selectively de-esterify PGFα or PGFβ.

Index

A